Plastering Skills

F. VAN DEN BRANDEN   THOMAS L. HARTSELL

With the collaboration of Old House Journal

 AMERICAN TECHNICAL PUBLISHERS, INC.
HOMEWOOD, ILLINOIS 60430

Preface

The second edition of *Plastering Skill and Practice* is designed to fulfill three objectives. First, and most importantly, it provides a thorough and practical groundwork for the apprentice at all stages of his training. Second, the expanded and updated text (accompanied by a great many new illustrations and charts) can serve as a refresher and reference for the practicing journeyman. He will be particularly interested in the treatment of plain and enriched cornice work, ornamental work, and the development of arches, domes, lunettes, etc. Third, the new text briefly covers several closely related building trades as these affect the plasterer's operations. This information has been included especially for the tradesman who may have ambitions to broaden his scope with a view to advancing to foreman, superintendent, or contractor.

While these aims are extensive, the practicality of the material presented is well established. It has been gathered and used by the authors over many years, and each item of information and each method discussed has been further checked by the Detroit Apprentice School and tested in actual applications. This does not imply that alternate methods may not be valid, because there are various ways of performing most of the operations of the plastering trade. The important point here is that confidence in methods that have been proved successful is essential, especially to the beginner.

In preparing this new edition the authors have been mindful of many local variations and preferences and have taken these into consideration. The result is a book of wider than regional practices.

Underlying all differences in techniques and materials used in various localities, there is the universal requirement of safety. This requirement is stressed throughout the book. Safety is a matter of growing concern because today's plasterer is no longer simply a hand craftsman; in much of his work he is also applying his materials by machines of various kinds. While it is true that this mechanization has greatly revived and enlarged the plastering trade, it has also increased the hazards that accompany the operation of power equipment.

Plastering is both a craft and an art, combining many skills. It is known to be older than the ancient Sumerian temples built centuries before the Egyptian pyramids; yet it is as new as this morning's weather report. In its long history there have been flourishing times and times of relative stagnation. In recent years, following a period when plastering lagged behind some of the other building trades, there has been a decided revival brought about by several developments.

Materials for plastering have been greatly improved, and new materials are continually being developed that not only produce stronger, lighter, more fire-resistant and attractive coatings, but also allow plastering over bases that only a few years ago were considered impractical or impossible to plaster. Variations of texture, color, exposed aggregates, and simulated stone and wood offer almost unlimited choices for architectural motifs, fireproofing and sound control. This book covers uses of all these materials, as well as latex and epoxy base binders, chemical accelerators and retarders, and related additives.

The use of lightweight aggregates is destined to increase in view of the additional resiliency, fire resistance, sound deadening, and weight reduction they offer compared with conventional mortars made with sand. These topics are extensively covered in the text.

Since it is clearly impossible to publish a book on such subjects that will not be somewhat obsolete before the ink dries on the pages, the authors have added Appendix B, listing manufacturers from whom further current information can be obtained on request.

Also, for quick and handy reference and review, Appendix A, entitled *Math for Plasterers*, has been added. This section does not attempt to replace regular texts on mathematics, but is confined to those subjects and formulas most frequently and urgently needed in the plastering trade. While omitting many subjects found in a mathematics book, this section contains some valuable shortcuts for computing volumes as these relate to areas to be covered and quantities of materials needed. Of particular interest to the plasterer are diagrams and procedures for developing approximate and exact ellipses, ovals, arches, and like geometrical figures needed in the construction of templates, molds and full size wall and ceiling layouts.

Thanks should be made to Mr. Gilbert A. Wolf, Director, National Plastering Industry's Joint Apprenticeship Trust Fund, for continued assistance and advice, and for supplying the authors with crucial photographs. Mr. Charles Breen, Local 96, Apprenticeship School, Washington, D. C. also provided valuable assistance.

THE PUBLISHERS

Contents

A man who works with his hands is a Laborer.

A man who works with his hands
and his head is a Craftsman.

A man who works with his hands,
his head, and his heart is an Artist.

Plastering as a Trade

Plastering is the art or skill of applying various materials over surfaces producing both exterior and interior walls, ceilings and other surfaces in the construction or remodeling of buildings, homes and other structures.

The plasterer today uses many new materials, and these involve the use of equipment, techniques and skills unheard of a few years ago. The problem of keeping abreast of new developments is a never ending one, and the plasterer, to be successful, must be constantly alert to these changes.

History of the Trade

The trade or craft is one of the most ancient building handicrafts. Historical evidence shows that primitive man plastered mud over a framework of sticks and reeds to enclose a protective structure to keep out the elements.

The Pharoahs of Egypt used plaster surfaces in their palaces. The pyramids, which the Pharoahs built as monuments to themselves, contain plasterwork which may be seen to this day. It is known that this plasterwork, and the decoration upon it, was applied earlier than 4,000 years ago. These plaster surfaces still exist in a hard and durable state today.

Research has also indicated that the principal tools of the plasterer of ancient Egypt were practically identical to those we use today.

The finest plasterwork accomplished by the Egyptians was made of a plaster produced from calcined

1

gypsum (gypsum made powdery by heat action) just like the plaster of Paris of the present time.

The method of applying plaster was also very similar to the methods used today. The Egyptians plastered on reeds—a method which resembles in every way our method of plastering on lath. Hair was introduced to strengthen the plaster even at this early date.

A study of ancient Greek architecture reveals that plaster and stuccowork (plaster was primarily interior, while stuccowork meant exterior) were used by the Greeks at least 500 years before the birth of Christ. It is from the Greek, incidentally, that we get the word "plaster," for in the ancient Greek language that word meant "to daub on."

The sanitary value of using plaster was apparent to those early users. The density of the material, plus its smooth surface, provided both protection and a surface ideal for decorative treatment. Later lime and sand were combined as a mortar to cover both the reed lath and masonry walls and ceilings. The antiseptic value of lime was used by ancient people in preventing the spread of vermin and disease.

Plaster was recognized long ago as a protection against fire. Its value as a fire retardant was demonstrated in the many fires that ravaged London during the Thirteenth Century. The king, at that time, issued an order that all buildings were to have plastered walls. Houses that did not meet meet this specification within a stated period were to be torn down.

During this period and through the Sixteenth Century, the plasterer's skill was developed to a height unequaled in history. The Renaissance period (14th, 15th, 16th centuries) of art and architectural development began in Italy and spread to other countries. This was the transition period from the medieval world or dark ages of art to the modern revival of the ancient classical orders and designs.

During this period there developed first the Baroque style, a gaudily ornate mass of carved figures and foliage covering both ceilings and walls. This was followed by the Rococo style, an overpowering display of ornamentation which reached its greatest development in France during the 18th century. Plaster ceilings during this period contained life-sized figures carved in almost full relief, interwoven with scrolls and foliage completely covering the total area.

The style was so overpowering it soon faded and the work became more subdued. The style pendulum swung to the light and delicate ornament or repeating patterns. This ornamentation was revived from the ancient Greek, Persian, Egyptian and Indian works which decorated the temples and mosques. English plasterers developed this type of work to its perfection in the English manor house ceilings during the 18th and 19th centuries.

Fig. 1-1 illustrates a ceiling in the Flemish style (1890 - 1910 period) which combines many types of light, natural ornamentation (leaves, flow-ers, etc.) with plain moldings run in place. This is a style in between the heavy Rococo style and the light geometrical style which developed in England.

During all this period of history from almost the first use of plastering to the middle of the 19th century,

Fig. 1-1. Flemish ceiling from the 1890-1910 period.

the plasterer used lime and sand for the basic plain work of covering walls and ceilings. This mortar took about two weeks to set (harden) under favorable conditions.

The gypsum plaster used set faster, but it was too costly for ordinary plain work. It was used only in the ornamental work and for various imitation marble finishes called scagliola, a skill developed in Italy in the 15th century. Fig. 1-2 illustrates the final finishing of scagliola.

Gypsum and its product, plaster

Fig. 1-2. Rubbing scagliola with stone and water during first polishing. (John David Snyder, Detroit.)

of Paris, have gradually replaced lime as the binding agent for sand in plastering mortar since the beginning of the 20th century. Its rate of set could be controlled, allowing the plasterer to build up layers or coats of plaster in a matter of hours rather than the days and weeks required with lime mortar. Speed became an important factor in the continued growth and development of the crafts.

A number of other factors helped to change the centuries-old style of plastering. Some of these were the invention of Portland Cement by Joseph Aspdin, a bricklayer in Leeds, England in 1824; Keanes Cement, a slow-setting but extremely hard plaster by R. W. Keane, of England in 1838; metal lath in mesh form developed in England in 1841; and plaster board or gypsum lath first produced in England in 1890 which later, in the early years of the 20th century, developed into the modern lath used today.

The development of new lathing and plastering materials increased the demand for plasterers during the first half of the 20th century. Then the age of mechanical power overtook the plasterer and he began to lose ground to substitutes. However, the trade picked up the challenge, and today the plasterer has at his command unlimited types of machines to mix and apply the numerous materials he has to work with. Fig. 1-3 illustrates the spraying of a scratch coat of plaster to gun lath.

In the natural evolution of the trade certain specializations developed. For thousands of years the plasterer performed all phases of the work involved in producing the complete job: developing finished walls, ceilings, and in many cases floors. Later, when Portland cement became available in many areas, the plasterer also became involved in the finishing of concrete road pavement, sidewalks and similar work.

As the volume of work grew, a natural subdivision of the work developed. Certain plasterers became *lathers* and soon did nothing but apply wood lath, the popular plaster base at that time. Later, other plasterers stayed with the Portland cement part of the trade that was involved in laying concrete floors, sidewalks and roads. These men were then called *cement masons*. In this way three different trades developed from one because the various areas of skill became too complex for one man successfully to do all of them.

Within the present plastering trade itself there are four divisions which have naturally evolved. These are the *plasterer*, or applier of the material on the job-site; the *architectural sculptor*, an artist who produces original models and elaborate ornamentation in clay and plaster; the *modeler*, a highly skilled craftsman who develops all types of models and ornaments in clay and plaster;

5

Fig. 1-3. Spraying scratch coat of plaster to gun lath. (K-Lath Corp.)

and the *caster*, a skilled worker who reproduces in volume (using various types of molds) the ornamentation developed by the other two men. Fig. 1-4 shows casters pouring plaster into a mold.

The plasterer, in his phase of the trade, has developed into two classes of craftsman: (1) the *plain hand*, or person who by choice or training does only plain plastering or the applica-tion of plaster, cement and other materials to walls and ceilings; and (2) the *cornice hand*, or plasterer who specializes in the ornamental part of the trade, but who is qualified to do all phases of the work. Some men prefer to work in the residential field while others tend to stay in the commercial or large structure area.

Today, a group of men is developing who now specialize in the applica-

Fig. 1-4. Casters pouring plaster into a rubber mold.

tion of the various exposed aggregate finishes plus many other materials combining various plastics. Fig. 1-5 shows an exposed aggregate job. The plasterer, to keep up-to-date, must be aware of all these developments and should learn the specialized techniques involved in using them.

The use of machinery of all types is now changing this trade from a

Fig. 1-5. Exposed aggregate exterior finish. (Ceram-Traz Corp.)

purely manipulative (handwork) trade, to one that requires the ability to operate, repair and clean numerous machines, each one designed to do specific jobs. Along with the use of machinery, there has developed a large body of knowledge on new materials designed to be applied mechanically, and in many cases applied to surfaces that were previously thought impossible to plaster. Examples are steel decking insulation and structural steel fireproofing. Fig. 1-6 shows a plasterer spraying fire resistant plaster to structural steel.

In doing his work the plasterer must coordinate what he is to do with the *lather*, who prepares the surfaces to receive the various materials. The surfaces must usually be

Fig. 1-6. Plasterer spraying fire shield on structural steel. (National Gypsum Co.)

either level or plumb and set back enough to permit the plasterer to apply the required coats of plaster. The lathing must also be checked to insure that it is properly applied with the right channels, studs, and lath used. Fig. 1-7 shows a lather applying gypsum lath.

After the job has been completed, if any trouble develops later, it is the plasterer who is blamed, even though the actual fault may lie with another trade. This same problem also involves the *plasterer tender*, the person who mixes the materials for the plasterer, brings these materials to the work site, erects scaffolds and cleans up after the work is done. The plasterer is responsible for whatever problems may develop from the work

Fig. 1-7. Lather screwing gypsum lath to steel partition studs using power screwdriver. (National Gypsum Co.)

performed by the plasterer tender, or *hod carrier* as he is sometimes called. The plasterer must supervise the work of proportioning and mixing materials, erection of scaffolds, etc., plus making sure all the required materials are available to do the work. ·

Plastering is divided into two main areas of work: *exterior* and *interior*. The basic difference in these two areas is in the types of materials

used. For exterior work the materials must be water resistant and structurally strong. The common materials used to bind together the aggregates are Portland cement, lime and plastics.

For interior work almost any type of material can be used because the work will be protected from the elements. The most common binding materials are gypsum, lime, and

Portland cement. Gypsum is not water resistant and should not be used where high moisture conditions exist.

Plaster is applied in coats or layers to the desired thickness; the total thickness may vary from ⅛" for veneer plastering to 2" thick for solid plaster partitions. Various types of plastering materials are also used to provide sound control, insulation, fireproofing and ornamentation.

One of the prime advantages plaster has over most of the other interior and exterior finishes is its low cost; another is its versatility. It can be formed into monolithic surfaces of any shape or size. There are almost unlimited supplies of the basic raw materials used in plastering. Gypsum, limestone, clay, sand, vermiculite and perlite are available worldwide with deposits located close to all markets.

Structure of the Trade

The plastering trade is structured into various levels of craftsman: apprentice, journeyman, foreman, superintendent.

Apprenticeship or learning usually lasts from three to four years. From man's earliest history apprentices in the various crafts were indentured (a contract binding one person to work for another for a given length of time to learn a trade) to a master craftsman (a contractor) for a number of years to learn the trade.

In many cases the apprentice's father had to pay the master a fee to get him to teach his son the trade. From the medieval days down through most of the 19th century, the apprentice would live with the master and would get room and board plus some clothing. However, he was a virtual slave to the master, subject to his every wish. Both old and new certificates of apprentice indentures are reproduced on pages 12 and 13.

Note that today the apprentices are protected by Federal, State and the Local J.A.C. (Joint Apprenticeship Committee) in regard to hours of work, wages, conditions of employment, and there is no control over the apprentices outside of the working hours. Also, apprentices are now selected from applicants who meet the standards of the local J.A.C. In most cases the apprentice is indentured to the J.A.C. and they assign him to a contractor. If the contractor runs out of work the J.A.C. will place him with another contractor. This permits the J.A.C. to control the training and handle the federal and veteran's paperwork.

This Indenture *witnesseth, That*

Jacob Peterson, with the consent of his mother, *Mary Ann Griffith*, hath put himself, and by these presents, and for other good causes, doth voluntarily and by his own free will and accord, put himself apprentice to *Hiram Miller* to learn the art, trade, and mystery of a *House Carpenter* and, after the manner of an apprentice, to serve the said *Hiram Miller* from the day of the date hereof, for and during, and to the full end and term of *Three Years Ten Months and Fifteen Days* next ensuing. During all which time the said apprentice doth covenant and promise, that he will serve his master faithfully, keep his secrets, and obey his lawful commands—that he will do him no damage himself, nor see it done by others, without giving him notice thereof—that he will not waste his goods, nor lend them unlawfully—that he will not play at cards, dice, or any other unlawful game, whereby his master may be injured—that he will neither buy nor sell, with his own goods nor the goods of others, without license from his master—and that he will not absent himself day nor night from his master's service, without his leave—nor haunt ale-houses, taverns, nor play-houses; but in all things behave as a faithful apprentice ought to do, during the said term. And the said master, on his part, doth covenant and promise, that he will use the utmost of his endeavors to teach or cause to be taught or instructed, the said apprentice in the art, trade or mystery of a *House Carpenter* and that he will procure and provide for him sufficient meat, drink & lodging fitting for an apprentice during the said term—and that he will give him *Forty dollars per Year payable Quarterly in lieu of Clothing and two Quarters Night Schooling*—And for the true performance of all and singular the covenants and agreements aforesaid, the said parties bind themselves each unto the other, firmly, by these presents.

In witness whereof, the said parties have interchangeably set their hands and seals hereunto. Dated the *Twentieth* day of *July*, Anno Domini *1844*.

THE ABOVE IS AN EXACT TRANSCRIPT OF
ORIGINAL INDENTURE WRITTEN IN THE
YEAR 1844

☐ P. L. 346 ☐ P. L. 16 ☐ Other   ☐ Vet Applying   C No. ....
☐ P. L. 550 ☐ P. L. 894   ☐ Vet Not Applying   Serial No. ....

This agreement is sponsored by:
Detroit Plasterers' .... Joint Apprenticeship Committee
REPRESENTING THE EMPLOYER GROUP .Detroit.Contracting.Plasterers'.Assoc....................
REPRESENTING THE UNION GROUP .... Local.#67,.O..P..&.C..M..I..A...............................

APPRENTICESHIP AGREEMENT
Between Employer's Agent and Apprentice

THIS AGREEMENT, entered into this ...... day of ...., 19...., between .Detroit. Plasterers'...........
                                                                       (Name of Employer's Agent)

Joint Apprentice Committee
.................., hereinafter referred to as the EMPLOYER'S AGENT, and ....

...., born ...., hereinafter referred to as APPRENTICE (and if

minor) name of PARENT or GUARDIAN ...., hereinafter referred to as PARENT or GUARDIAN.

Witnesseth that the EMPLOYER'S AGENT, the APPRENTICE, and his PARENT (or GUARDIAN) desire to enter into an agreement of apprenticeship in conformity with the standards of the above-named Joint Apprenticeship Committee, and therefore, in consideration of the premises and of the mutual covenants herein contained, do hereby mutually covenant and agree as follows:

That the EMPLOYER'S AGENT shall use its best influence to find employment for the APPRENTICE with an EMPLOYER who

agrees to teach the APPRENTICE the trade of .Plastering.5-29.100..
in conformity with the terms and conditions outlined in the standards established by said Joint Apprenticeship Committee, which standards are made a part hereof.

That the APPRENTICE shall perform diligently and faithfully the work of said trade during the period of apprenticeship, complying with the terms and conditions contained in the said standards.

That the PARENT (or GUARDIAN) guarantees that the apprentice will duly perform all obligations undertaken herein.

That the apprenticeship term begins on the ...... day of ...., 19......, under guidance of said employer's agent, in said trade as stipulated in the said schedules; and shall be for 6,000......hours or ...3..........years (less ....hours or ....years granted for previous experience).

That any disagreement or difference in relation to the terms and conditions of employment and training under this agreement shall be submitted to the above-named Joint Apprenticeship Committee for adjustment, and the decision of the Joint Committee shall be final. However, either party may consult with the Bureau of Apprenticeship regarding any decision on labor standards and with the Michigan State Board of Control for Vocational Education regarding any decision on related and supplemental instruction.

Upon completion of the term of apprenticeship, or if this agreement is terminated for any other reason, the Bureau of Apprenticeship and the Michigan State Board of Control for Vocational Education shall be so notified by the Joint Apprenticeship Committee.

In witness whereof the parties hereunto set their hands and seals:

                                   Detroit Plasterers' Joint
....   Apprentice Committee
       (Apprentice)   By .... (SEAL)
                                  (Employer's Agent)
....(SEAL)
       (Address)   (Officer)
                                   1435 Livernois, Detroit, Michigan 48238
....
       (Parent or Guardian)   (Address of Employer's Agent)

Approved by the ..Detroit.Plasterers'...............................Joint Apprenticeship Committee

By .... on ...., 19.....
     (Chairman or Secretary of Committee)

Registered with the
Bureau of Apprenticeship and Training, U. S. Department of Labor
as incorporating the basic standards recommended by the
Federal Committee on Apprenticeship

By .... on ...., 19.....

Registered as an apprentice student with the
Michigan State Board of Control for Vocational Education

By .... on ...., 19.....

The *journeyman* or experienced craftsman is one who has completed an apprenticeship in the trade. He is now a free agent and can work for any contractor he pleases. He may travel from place to place, going where the work is to be found.

The *foreman* is a journeyman who has been placed in the job of supervising a group of men. He is given this job because of his ability as a craftsman and his knowledge of how to supervise other craftsmen.

The *superintendent* is usually a foreman who has been promoted to this important position. He is in charge of all the work in the field for his contractor and supervises the work of the foreman. Some large contractors have a *general superintendent* to oversee the *job superintendents*.

In the construction industry the *foreman* and *superintendent* keep their union membership. Many of the smaller contractors are permitted to retain their union cards in some unions. Plasterers, for one, permit this in many areas.

Large contractors will employ an *estimator* who works in their offices to estimate the cost of the jobs the contractor wants to bid on. Working from blueprints and specifications, he takes off (measures all the walls and ceilings) the yardage on a given job. He must be skilled in mathematics, blueprint reading, trade practices and the cost of labor and materials.

The last person in this team of workers, supervisors and estimator is the *contractor*. He must be knowledgeable in all phases of the business. He must know all the regulations governing the construction industry plus how to provide the money for payrolls and materials. Upon his overall ability to run the business successfully depends the livelihood of all his employees.

Rules for the Contractor

It is common in the trade for a man to work his way up the ladder of responsibility from apprentice to contractor staying at each level until he is proficient and conditions, plus drive, enable him to move ahead. The majority of the contractors in the various skilled trades have worked their way up this ladder. They are craftsmen as well as businessmen.

When a plasterer has worked his way up the ladder of experience and aspires to become a contractor he must prepare himself for operating under a whole new set of rules and conditions. As a contractor he will be risking his own money on the gamble that his knowledge and experience in estimating work, handling men and keeping up with the paperwork that is required today will pay a profit.

With these factors understood, the new contractor stands ready to bid against his fellow contractors on the work that is available. When there is a good supply of work ready to be bid and the new contractor has a crew of men that he knows can produce the work required, he will have a good chance of success. By starting small and bidding only on work that is fully understood, there is a chance to make money and progress to bigger and better jobs. Contracting is a rewarding endeavor for those who are suited for it by ability, temperament and aggressiveness to stick it out through good and bad times.

Therefore, to be successful the new contractor should consider the following rules, conditions, laws and facts which he will have to know and work with.

1. Develop a good knowledge of estimating yardage, material and labor costs to do a given job. Remember that no two jobs are ever alike. Conditions of weather, availability of men, changing costs for materials and labor, plus the differences in architects and general contractors can together or individually change the cost of doing a given job.

2. The contractor must know the laws and regulations he must follow, use and invoke at times to protect himself and his work force. These include the lien laws (used to protect against loss due to failure of the owner or general contractor to pay his bills), local permits and inspections as called for, the collection and payment of federal, state and local taxes on the men employed, sales taxes, use taxes, property and inventory taxes on office and storage yard.

Also there are a number of different types of insurances that must be carried. These include compensation, unemployment, accident, property damage, fire, theft, car and truck insurances, business loss, etc. Each one of these items costs quite a bit of money and must be included in the cost of doing business. Too many beginning contractors fail to consider these costs in their bidding of a job and end up having to take it out of the profit (if any) on the job—which is really taking these costs out of their own wages.

3. A contractor, to be successful, must be fully aware of the working rules of the unions he will be dealing with in doing various types of jobs. A residential job may involve only the lathing, plastering and plasterer tender unions. On a commercial job he will add the carpenter (to build large scaffolds), truck drivers, operating engineer (to run large air compressors and similar equipment in certain areas), painters (painting and sand blasting), electrical (temporary lighting and equipment hookup),

15

plumber and pipefitter (drainage and pipe connections on some equipment in certain areas)—all of these different unions must be dealt with and understood.

4. As the contractor advances in work he will find that each architect, general contractor and owner that he works for will have different conditions and rules governing the jobs they design and build. It is therefore very important to study and understand these variations so that the cost of working under these conditions can be added into the bid for the job. The plans and specifications must be gone over with a fine-tooth comb to pick out each condition, rule and method set down in both items.

Anything that is not clear should be questioned in writing to the architect before the bid is sent in. No verbal agreement should ever be accepted or given. Everything should be in writing and signed by the architect or the general contractor. Written and signed work orders should be issued for any changes, additions or deletions in the work.

It is the plastering contractor's responsibility to check the prints, specifications and the job site for omissions in these items because it is the accepted practice in the industry to do and complete the job in a workmanship manner. This means that even though the architect did not specifically mention or draw an item or condition on the prints or specifications and this item or condition is required to complete the job in the generally accepted trade practice the contractor must furnish it or do it to complete the job and be paid.

When the job is ready for the lather and plasterer it must be carefully examined to check for conditions which are not according to the prints or specifications. Some of these conditions are: walls or ceilings not in condition to receive the required lathing and plastering called for, unsafe job site conditions, other trades work or equipment incorrectly placed or interfering with the lathing and plastering work to be done.

Once the plastering contractor starts the job and tries to overcome these wrong conditions on his own—without calling them to the attention of the architect and general contractor in writing and receiving a written reply covering what is to be done—he will be responsible for the condition of his work regardless of what may occur later.

All changes in work to be done on the job site should be given in writing as work orders by the architect and general contractor. Temporary lighting, power, cleanup and removal of rubbish, use of scaffolds, use of hoist for lifting material and equipment to various floors, availability of work area in reasonable time to get the work done, storage of material at the job site and on various floors, use of water and many other items must all be agreed on by job conferences between the architect, general contractor and the plastering contractor. The other trades should be consulted on the cost of their services for anything that they will be asked to provide to the plastering contractor.

5. The plastering contractor must also have available for his men the required equipment which will enable them to do the job in the best and shortest time possible. Poor equipment or trying to do the job with insufficient equipment is a good way to lose money. Scaffolding is a good example of this rule; if the job is short of scaffolding then there will be a lot of unnecessary moving of scaffolds, resulting in high labor costs.

Not having the lathing and plastering materials in the required amounts and in the right place at the right time will increase the labor cost due to high-priced skilled men waiting for material to arrive or to be moved.

6. Labor relations and cost accounting tied in with job financing all combine to make or break a contractor. The contractor must be aware of his labor and overall operating costs under varying conditions, he must be able to keep accurate cost accounts or have an accountant do it for him. Labor costs can vary considerably depending on how many men are available on the labor market and what the contractor's reputation is as a person to work for. As an overall statement, poor cost accounting is one of the most common causes of contractor failure.

7. The beginning contractor must also remember that on most commercial jobs, the general contractor retains ten percent of the total contract cost as a protection for him to make sure the plastering contractor will repair any defects or other problems for a period of one year. This means that if the contractor does a number of jobs in a year each one will have this ten percent withheld for one year. This will tie up a lot of his working capital and he may have to borrow additional money to stay in business. The cost of this borrowed money must be included in the cost of the job to enable the contractor to stay in business.

Joint Apprenticeship Committee and Apprenticeship Standards

The plastering industry has, in cooperation with the U.S. Department of Labor, Bureau of Apprenticeship and Training, set up National Standards of Apprenticeship. These standards define what the term *apprentice* in the trade shall mean. The standards set forth age limits, educational requirements, length of apprenticeship, ratio of apprentices to journeymen, hours of work and wages.

The Joint Apprenticeship Committee, commonly known as the J.A.C., is composed of equal representation from labor and management, with consultants from the Bureau of Apprenticeship and the State or local Board of Education attending to act as advisors without a vote.

The J.A.C. has the delegated power to set the local standards consistent with the basic requirements established by the National Committee. The apprentice, when he signs the indenture agreement, agrees to live up to all its provisions and in turn is protected by its rules and regulations.

The J.A.C. also establishes the curriculum for the related classroom work plus supervising the on-the-job training the apprentice receives. When an apprentice completes his training, the J.A.C. notifies the Bureau of Apprenticeship and this

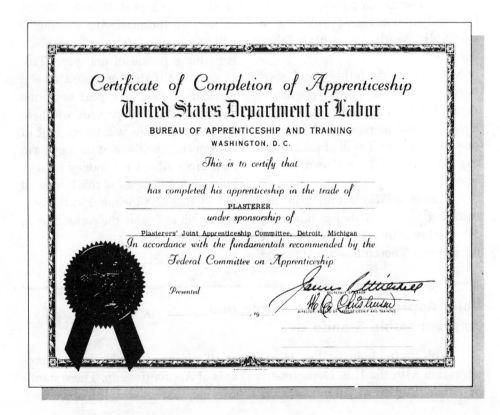

Certificate of Completion of Apprenticeship

United States Department of Labor

BUREAU OF APPRENTICESHIP AND TRAINING
WASHINGTON, D. C.

This is to certify that

_____

has completed his apprenticeship in the trade of

PLASTERER

under sponsorship of

Plasterers' Joint Apprenticeship Committee, Detroit, Michigan

In accordance with the fundamentals recommended by the

Federal Committee on Apprenticeship

Presented

, 19

Certificate of Diploma

In recognition of your completion of apprenticeship training, the Board of Trustees of the National Plastering Industry's Joint Apprenticeship Committee, on behalf of the Operative Plasterers' and Cement Masons' International Association and International Association of Wall and Ceiling Contractors, take satisfaction in providing you with this document.

This is to Certify that

Jerry W Crump

having completed and met the qualifications established by the National Plastering Industry's Joint Apprenticeship Trust Fund for Apprenticeship Training is eligible for Journeyman Plasterer.

Date of Award

June 1 1971

Josyph T. Power
Co-Chairman, Operative Plasterers' and Cement Masons'
International Association

J. E. Roushe Sr.
Co-Chairman, International Association of Wall and
Ceiling Contractors

agency issues a completion certificate which is recognized throughout the United States and Canada as proof of reaching journeyman status. (See certificate on Page 18.) In addition, the plasterer is awarded a *national* diploma. See page 19.

The graduated apprentice is now able to travel and work as a journeyman in the U.S. and Canada (subject to Canadian laws) under the membership of either the "Operative Plasterers and Cement Masons International Association of the United States and Canada," or the "Bricklayers, Masons and Plasterers International Union of America." These two organizations have interchangeable working cards which permit a journeyman to work in either organization's area of control. Over the years these two national unions have established geographical areas for each union. They train and exchange men to supply the needs of the contractors in any given area. Local unions have agreements with their contractors on supplying manpower (subject to federal regulations) for a given area. Inability to supply the manpower from the local market usually results in calling for men from other cities to come into the area to work.

Blueprint Reading

Essential to most of the work of the plasterer is an understanding of the working drawings (blueprints) which tell him what the architect requires in the various areas of the building. Working drawings are called *blueprints* because the print is blue with white lines. They are also, in many cases now, called *whiteprints* (white background with blue or black lines). The *working drawings* and the *specifications* (the written instructions concerning items or techniques that cannot be shown on the drawings) make up the complete set of instructions from the architect to the plasterer.

In general, the working drawings or blueprints will show what is to be done and where; the specifications will state what materials to use and how the work is to be done.

The plasterer in reading the prints must know the meaning of the various types of lines used, and the symbols, abbreviations and notations which are made to tell the reader what is required. The first step then in reading blueprints is to learn about these items. The next step is to learn about floor plans, elevations, cross sections and details. (The use of *Building Trades Blueprint Reading*, Parts 1 and 2, published by the American Technical Society, Chicago, Illinois, is recommended.)

To become proficient in reading blueprints requires considerable study and work in doing the actual drawing. Just as the plasterer apprentice must practice his trade to learn it, he must also practice drawing to be able to understand the drawings. This is why the writers of this book recommend to the reader the use of the books listed to help him work and understand the basic fundamentals of blueprint reading.

Practice in making freehand sketches of the various floor plans, elevations and details found in plans will be helpful later when this skill is required on the job in measuring various rooms and areas.

After the basics are mastered the reader should draw to scale the various geometrical figures (such as squares, rectangles, triangles, circles, etc.) so often found in the construction trades. This practice will be helpful later when various types of arches, ellipses, domes, lunettes and many other shapes must be developed full size on the job. Only by doing the actual drawing will the full understanding develop.

The formulas and construction of many of these figures will be covered in the chapter on ornamental plastering, Chapter 9. Refer to this and use reference books covering more detailed studies on mathematics to give the help needed.

A good background in basic mathematics is important to compute the areas and volumes encountered in plastering work. The use of a good practical mathematics book to review many of the problems and formulas will be helpful. *Appendix A*, at the back of the book, gives a review of basic math used in plastering.

Continuous study and work is required to develop the required skill in blueprint reading, mathematical calculation and sketching. During the apprenticeship training period the apprentice will be given 96 hours of drafting and blueprint reading plus 72 hours of mathematics (National Standards). To become really proficient and to qualify for advancement to foremanship, etc., the reader must devote additional time to these important phases of the trade.

Specifications

Depending upon the size of the job, the specifications may consist of a few sheets of paper or a few notations printed on the blueprints up to elaborate books covering hundreds of pages. Each trade will usually have its own section covering all the items and specific procedures involved in the job. It is always important to check the general specification section to learn what is expected from all the trades as well as the specific trade involved.

The specifications will list the types of materials and finishes to use in each area. Also covered will be such items as protecting other work from damage, cleaning up and removing rubbish, guarantee on workmanship and materials, working with other trades, use of scaffolds, time limit to complete the work, patching defects after other trades, temporary lighting and heating.

Read the specifications carefully and make notations on special items in the other trade sections that refer to plastering. For certain items the specifications sometimes list more information than the plans. Remember, in case of a dispute over what is to be done or how it is to be done, in most cases the specifications carry more weight than the plans.

The architect is always the final arbitrator of all disputes on what the plans or specifications say or mean. Therefore, the plasterer should study both the plans and specifications carefully before starting the job. Any item that is not clear should be questioned by a letter to the architect, and a clarification or specific explanation should be requested in writing so that no changes can be made later without agreement by both parties.

Safety

The observance of safe working conditions cannot be overemphasized. The careless plasterer is a dangerous worker — dangerous to both himself and to his fellow workers. Good working habits and careful consideration of safety rules and safety laws will, in the long run, pay dividends in no lost time due to injuries plus avoiding the pain and suffering accidents cause.

The National Safety Council's slogan, *Safety First*, is one that should be kept in mind at all times. Also remember that, "accidents don't happen—they are caused," is a truism that will prove itself over and over again on the job.

There are federal safety laws that control certain working conditions; also, each state has safety laws with

National Safety Council

departments to enforce the laws as well as to inform the trades on safety. Local communities also have various rules and regulations which control the safety of the worker and the public.

The plasterer, both apprentice and journeyman, should contact the safety departments of his state and local community. He should determine what the local safety laws are and should procure the counsel and materials available on safety in that area. These services are free and the agencies are anxious to serve the workers.

All violations of the safety laws or unsafe conditions should be reported to the job steward who is empowered under the contract between the union and the contractors to settle such matters on the job level with the foreman or contractor. The State Safety Inspector can also be called in for a ruling.

There is now a national trend to have all construction workers trained in "First Aid." The local Red Cross chapters usually offer these courses free to any groups that request this help. Each job should have at least one worker trained in first aid, so that when an accident occurs expert help will be on hand until medical aid is obtained. Many J.A.C. are now scheduling classes for their apprentices while unions are offering night classes for journeymen.

Each trade has its own dangers

from which the worker must protect himself. The plasterer must protect his skin from undue contact with the lime and other materials used in his work. Therefore the clothing of the plasterer is important in protecting him from injury.

National Safety Council

The way tools are used and the use of good equipment is also very important. Poor scaffolds may mean the loss of footing and result in a serious accident with pain and lost time for the careless worker.

The mechanical equipment the plasterer now uses requires special care in its operation. Air compressors can blow up if the safety pressure relief valves are inoperative. Plaster pump hoses may break if they become clogged with an obstruction. The proper care of the equipment will prevent this from happening. Spraying machines can blind a fellow worker if carelessly used or if goggles are not worn. Short circuits in the equipment can cause death under certain conditions.

Clothing

A white cap with a projecting bill or visor is worn by all plasterers. The visor is of great importance because it protects the eyes from possible falling mortar. When using mechanical spray equipment to apply plaster or other materials, safety goggles should be worn to protect the eyes.

If mortar, lime or other foreign materials should get into the eyes, vision could be completely impaired if the material is not washed out immediately with clean water. A drop of olive oil applied after thorough washing would be advisable. No matter how slight the injury to the eye may seem, a doctor should be seen as soon as possible.

Some plasterers use gloves to protect their hands, but in general gloves need not be worn. The important safety feature is to keep the hands clean and dry at all times. When working with chemicals use rubber gloves to protect the hands and goggles to protect the eyes.

Whether bib-style overalls or regu-

Fig. 1-8. Wear appropriate clothing and head gear for the job.

lar plasterers' pants with a shirt are worn, the color of the material is always white. White is the trademark of the plasterer. Pants and overalls made for the plasterer usually do not have front pockets or cuffs since these would tend to catch the falling mortar. A white shirt with long sleeves that fit snugly at the wrists is recommended. This precaution is taken to protect the plasterer's arms from the material being applied. Good, sturdy leather shoes with hard toe-caps should be worn to protect the feet from falling objects and from dampness. Never wear soft-soled shoes as you might step on a nail which would puncture the soft sole and enter the foot.

The hard hat is a must on many construction jobs. Any building, except in the house field, requires the protection of an approved steel or plastic hard hat. There is the constant danger of some workman dropping a tool or piece of material down from one of the floors above or from a scaffold. Fig. 1-8 shows plasterers wearing regular hard hats.

Tools and Equipment

It cannot be stressed too strongly that the greatest care should be taken in handling tools. If they are handled carelessly, serious accidents may result. The back of the trowel should be kept clean and the trowel should not be overloaded with mortar. An excessive amount of mortar

might cause some to drop or splash into the plasterer's eyes. Any mortar which is dropped from the tools used onto the hands, wrists, ankles or under the clothing should be removed immediately. Sores should be given special attention. If any rash occurs, a doctor should be seen.

Fig. 1-9 shows a plasterer mixing putty and plaster on the finishing board. Always mix the material using the toe of the trowel with a circular motion, pressing the trowel downward and inwards towards yourself rather than using the heel of the trowel and pushing towards your partner. Each plasterer should watch the action of his trowel so it will not sweep into the path of his partner's trowel with a good chance of getting

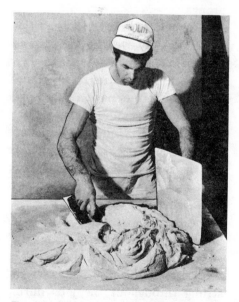

Fig. 1-9. Mixing putty and plaster.

a serious cut on his trowel hand.

When the finish material is being applied to a ceiling, it is better to let any surplus material dropping from the trowel fall on the floor rather than to try and catch it on the hawk because this may cause the material to splash and fly into the eyes when it hits the hawk.

Always work sharp-edged tools of any kind (except trowels, as described above) away from the body to avoid cuts if the tool should slip off the work. Files, knives and similar tools should have good wooden handles. When using snips, be careful to work away from the sharp edges of the metal being cut. Be particularly careful when cutting expanded metal lath, as this material when cut develops razor sharp prongs.

Mechanical plaster mixers are now used on nearly every job. A mixer is a dangerous tool; hands must be kept away from the rotating mixer blades. All mortar mixers have a guard over the charging or filling opening, but some plasterer tenders will remove this or lift it out of the way to clean out the mixer. Never run the mixer with this guard off or with the gear cover removed.

Electricity is the power supply for much of the mechanical equipment used on construction jobs today. Temporary lighting wires are found strung throughout the jobs during construction. Electricity is a very

dangerous power, there is nothing to show that the power is flowing through the wires except when a light is burning or a motor is turning. The electricity is there whether it is used or not. Great care must be taken not to touch any bare wire, or to create any condition where the current can flow through your body to a ground. Wet scaffolds, metal lath ceilings and wet ground are dangerous, as the wetness or mass of metal improves the grounding condition, permitting a greater flow of current to pass through the body. This causes severe burns and possible death.

Never attempt to touch a person who has live current flowing through him or you too may be killed. Try to remove the wire or equipment creating the problem by pushing or lifting it off using a dry piece of wood. Shut off the electricity immediately, if possible; this is the safest method.

Check all wires and equipment for bare spots, poor connections and for

National Safety Council

proper grounding before using them. Keep wire up off the ground and never operate electrical equipment in wet locations without the proper grounding conditions or instantaneous overload-shutoff devices.

Make sure all equipment has three-wire, grounded-type cords, using adequate size wire to carry the required load. Undersized wire may cause a fire due to overheating or may cause a motor burnout when operating under a heavy load.

Chemicals

The modern plasterer, in the course of his work, may come into contact with many chemicals. Some of them when used without the proper precautions can cause serious burns or loss of sight. Never use any material without first determining what dangers it may present and how to use it safely. Read all the manufacturer's directions and check with previous users to find out what problems may be encountered.

Wear rubber gloves and safety glasses plus the proper body covering to prevent chemicals from coming into direct contact with the skin. Keep a pail of clean water available to wash off any chemicals that might accidentally come in contact with the skin.

Scaffolds

Scaffolds must be built to support the load they are to carry. According to the National Safety Council, they should be designed to support at least four times the anticipated load of men and materials. This is necessary for safety because sometimes unexpected additional loads are placed upon the scaffold.

It is essential that those who erect scaffolding be familiar with the requirements of the safety codes or statutes of his own state as well as the national standards. The *Standard Safety Code for Building Construction* gives detailed requirements for the materials to be used and the manner of erection of scaffolds.

The plasterer uses four basic types of scaffolds and each of these types can be constructed of either wood or metal and sometimes a combination of the two materials. The basic types are as follows:

1. *Trestle scaffolds:* planks laid across trestles make safe scaffolds up to 10 feet high.
2. *Built-up scaffolds:* either wooden pole scaffolds or steel sectional scaffolds.
3. *Rolling scaffolds:* scaffolds on wheels that permit them to be moved.
4. *Hanging scaffolds:* suspended scaffolds supported by cables or metal straps. (Some large scaffolds in high ceiling areas are hung from cables to beams above to keep the floor area clear.)

A *swing stage scaffold* is also used

at times for some exterior work, but it is not as common as the others. A swing stage scaffold, such as the one shown in Fig. 1-10, is supported from above, using ropes and pulleys which permit it to be raised or lowered as needed. This type of scaffold is used on exterior work but is not used very often by the plasterer because it is dangerous and awkward to work on. Some of the new plastic matrix, exposed-aggregate jobs will sometimes be placed in small panels at various levels on the exterior of a building and the swing stage scaffold is the most practical one to use for this.

For *wooden scaffolds*, all supports and planks must be of sound lumber, free of large knots, cracks or split ends. All uprights must be crossbraced; ledgers and bearer planks must not be spaced more than 8 feet apart for use with 2″ x 10″ planks. New lumber standards will now reduce this plank to 1½″ x 9½″ net size. Allowances will have to be made for this by reducing ledger spacing.

Planks on inside scaffolds one stage high are normally spaced about 3 inches apart. This permits much of the mortar that falls during the plastering application to drop down be-

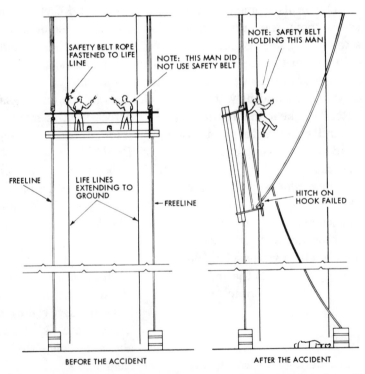

Fig. 1-10. Swing stage scaffold safety devices. (National Safety Council.)

low the scaffold, providing a cleaner and safer scaffold to work on.

On exterior scaffolds and for multiple stage scaffolds where men will be working above one another, the planks must be laid tight together to keep the mortar and tools from dropping on the workers underneath. Plank ends should extend at least 6 inches but not more than one foot beyond the bearer planks so they will not slip off. On steel scaffolds planks should have cleats nailed across their ends underneath so they cannot slide off the metal cross bars.

Exterior scaffolds for the plasterer should be at least three feet wide, and are more useful at four feet width, so as to permit a safe passage between the mortar board and the building.

On a narrow scaffold the plasterer's mortar board becomes a problem. It prevents free passage from one side of the scaffold to the other because the board, usually 3 feet square, fills up the whole space. A slightly wider scaffold (4 feet) permits passage and provides working room in front of the board between the building. Guard rails and toe boards should be used on all scaffolds over one stage in height. Not more than 14 inches of open space should be allowed between the scaffold and the walls.

Tie the scaffold to the building at every other staging from the bottom to the top and at every other upright

National Safety Council

scaffold pole for the length of the scaffold. Fig. 1-11 shows a typical wooden double pole, independent scaffold.

Scaffolds built on the ground must have the poles set on planks laid in solid contact with the ground so as to provide a firm unsinking footing. Nail the upright to these planks so they will not slide off later. This method of construction will prevent the scaffold from falling over due to sinking into wet ground caused by rain or other conditions.

To safely support the weight of the planks, men and materials, 4″ x 4″ poles are recommended for all wooden plasterers' scaffolds over one stage in height.

Most contractors stock only certain basic scaffold material in their storage yards. It is cheaper to use a 4″ x 4″ pole for a simple scaffold when you have it on hand anyway. It is useable for all scaffolds and because of its size can be nailed repeatedly without splitting.

Wooden scaffolds should not exceed 40 feet in height. Anything above this height should be built of steel for fire safety and strength. Various State and National Safety Regulations vary considerably in scaffolding. Local laws must be followed. Fig. 1-12, for example, shows a light trade, double pole scaffold recommended by the State of California safety orders. This is acceptable for plasterers. If the scaffold is

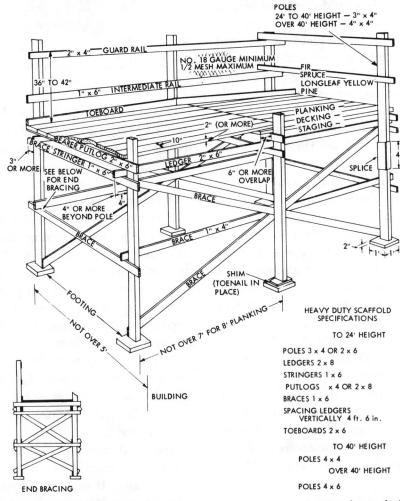

HEAVY DUTY SCAFFOLD
SPECIFICATIONS

TO 24' HEIGHT

POLES 3 x 4 OR 2 x 6
LEDGERS 2 x 8
STRINGERS 1 x 6
PUTLOGS   x 4 OR 2 x 8
BRACES 1 x 6
SPACING LEDGERS
    VERTICALLY 4 ft. 6 in.
TOEBOARDS 2 x 6

TO 40' HEIGHT

POLES 4 x 4

OVER 40' HEIGHT

POLES 4 x 6

Fig. 1-11. Typical double pole wood scaffold, showing safe construction for a single stage height. Multiple stage scaffolds require 4" x 4" poles and 2" x 8" bearer planks. (National Safety Council.)

over 32 feet high, 3″ x 4″ uprights are used. If heavy tools or materials are stored on the scaffold, 4″ x 4″ uprights spaced 7′-6″ apart are used for under 32 feet.

Steel scaffolds are now widely used for overall adaptability. They can be built to any height and are adaptable to all types of job conditions. These scaffolds can be purchased or rented and the supplying companies give technical services on needs and types best suited for each job or condition.

The type usually used for low heights is made up of prefabricated frames and cross braces, see Fig. 1-13. A factor of safety of not less than four times the load is required. Care must be used in the erection of the scaffold so that it rests on a firm base and is kept plumb and level as it is

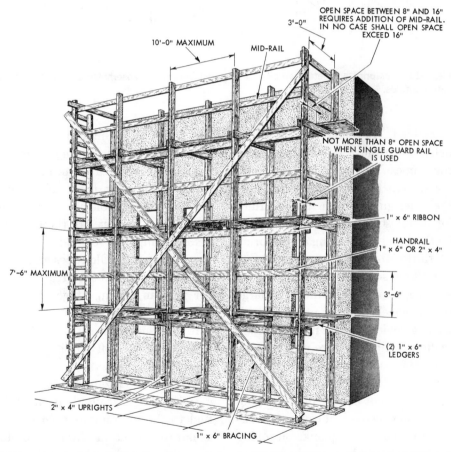

Fig. 1-12. A light trade, double pole scaffold must meet state code requirements. (State of California, "Construction Safety Orders.")

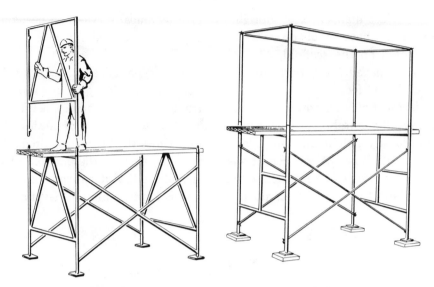

Fig. 1-13. Prefabricated metal frames and diagonal braces are assembled quickly to provide safe scaffolds.

assembled. It must be inspected daily. Care must be taken to keep the frames from injury or from rusting so that they do not lose part of their design strength. Safety rules for metal scaffolding, recommended by the Steel Scaffolding and Shoring Institute, are shown on page 33.

Rolling scaffolds, either steel or wooden, should be built with large strong wheels that are provided with locking devices. Never move a rolling scaffold while men are on the scaffold. Always clean the floor ahead of the move to be made so that the wheels will not be blocked by an obstruction which might cause the scaffold to tip over. All planks on rolling scaffolds should be securely fastened down so they cannot slide off while the scaffolds are moved.

Hanging and *swing stage scaffolds* depend upon the cables or block and tackle to support them. Never use frayed cable or cable clamps that are worn out. Inspect all these items before using the scaffold; your life depends upon it. Old ropes and worn blocks are perhaps the greatest danger in using swing stage scaffolds. Insist on good equipment, make sure the supports are securely fastened and tied off so they cannot slip or work loose.

When the plasterer has to work on a swing stage scaffold, it must be tied to the building at each end so as not to swing away from the building as the plasterer applies the mortar. Safety belts fastened to life lines

STEEL SCAFFOLDING SAFETY RULES

as Recommended by

STEEL SCAFFOLDING AND SHORING INSTITUTE

(SEE SEPARATE SHORING SAFETY RULES)

Following are some common sense rules designed to promote safety in the use of steel scaffolding. These rules are illustrative and suggestive only, and are intended to deal only with some of the many practices and conditions encountered in the use of scaffolding. The rules do not purport to be all-inclusive or to supplant or replace other additional safety and precautionary measures to cover usual or unusual conditions.

   I. **POST THESE SCAFFOLDING SAFETY RULES** in a conspicuous place and be sure that all persons who erect, dismantle or use scaffolding are aware of them.
  II. **FOLLOW LOCAL CODES, ORDINANCES** and regulations pertaining to scaffolding.
 III. **INSPECT ALL EQUIPMENT BEFORE USING** — Never use any equipment that is damaged or deteriorated in any way.
  IV. **KEEP ALL EQUIPMENT IN GOOD REPAIR.** Avoid using rusted equipment — the strength of rusted equipment is not known.
   V. **INSPECT ERECTED SCAFFOLDS REGULARLY** to be sure that they are maintained in safe condition.
  VI. **CONSULT YOUR SCAFFOLDING SUPPLIER WHEN IN DOUBT**—scaffolding is his business, **NEVER TAKE CHANCES.**

A. **PROVIDE ADEQUATE SILLS** for scaffold posts and use base plates.

B. **USE ADJUSTING SCREWS** instead of blocking to adjust to uneven grade conditions.

C. **PLUMB AND LEVEL ALL SCAFFOLDS** as the erection proceeds. Do not force braces to fit — level the scaffold until proper fit can be made easily.

D. **FASTEN ALL BRACES SECURELY.**

E. **DO NOT CLIMB CROSS BRACES.**

F. **ON WALL SCAFFOLDS PLACE AND MAINTAIN ANCHORS** securely between structure and scaffold at least every 30' of length and 25' of height.

G. **FREE STANDING SCAFFOLD TOWERS MUST BE RESTRAINED FROM TIPPING** by guying or other means.

H. **EQUIP ALL PLANKED OR STAGED AREAS** with proper guard rails, and add toeboards when required.

I. **POWER LINES NEAR SCAFFOLDS** are dangerous—use caution and consult the power service company for advice.

J. **DO NOT USE** ladders or makeshift devices on top of scaffolds to increase the height.

K. **DO NOT OVERLOAD SCAFFOLDS.**

L. **PLANKING:**
   1. Use only lumber that is properly inspected and graded as scaffold plank.
   2. Planking shall have at least 12" of overlap and extend 6" beyond center of support, or be cleated at both ends to prevent sliding off supports.
   3. Do not allow unsupported ends of plank to extend an unsafe distance beyond supports.
   4. Secure plank to scaffold when necessary.

M. **FOR ROLLING SCAFFOLD THE FOLLOWING ADDITIONAL RULES APPLY:**
   1. **DO NOT RIDE ROLLING SCAFFOLDS.**
   2. **REMOVE ALL MATERIAL AND EQUIPMENT** from platform before moving scaffold.
   3. **CASTER BRAKES MUST BE APPLIED** at all times when scaffolds are not being moved.
   4. **DO NOT ATTEMPT TO MOVE A ROLLING SCAFFOLD WITHOUT SUFFICIENT HELP** — watch out for holes in floor and overhead obstructions.
   5. **DO NOT EXTEND ADJUSTING SCREWS ON ROLLING SCAFFOLDS MORE THAN 12".**
   6. **USE HORIZONTAL DIAGONAL BRACING** near the bottom, top and at intermediate levels of 30'.
   7. **DO NOT USE BRACKETS ON ROLLING SCAFFOLDS** without consideration of overturning effect.
   8. **THE WORKING PLATFORM HEIGHT OF A ROLLING SCAFFOLD** must not exceed four times the smallest base dimension unless guyed or otherwise stabilized.

N. For "PUTLOGS" and "TRUSSES" the following additional rules apply:
   1. **DO NOT CANTILEVER OR EXTEND PUTLOGS/TRUSSES** as side brackets without thorough consideration for loads to be applied.
   2. **PUTLOGS/TRUSSES SHOULD EXTEND AT LEAST 6"** beyond point of support.
   3. **PLACE PROPER BRACING BETWEEN PUTLOGS/TRUSSES** when the span of putlog/truss is more than 12'.

should be used at all times. See Fig. 1-10. Use only safety approved equipment with the proper guard rails installed. Never overload the scaffold; keep it under its rated load capacity at all times.

Figure 1-14 shows another type of scaffold, a hanging scaffold, that can be used to work on the upper section of a building when it is not practical to set the scaffold on the ground. Notice the cantilevered steel beams from which the scaffold sections are hung. This unit can be moved along as the work progresses. The ends of the supporting beams must reach two-thirds of their length back onto the roof and be counterbalanced with sufficient weight to prevent the scaffold from tipping over.

Ladders

The ladders used in plastering are different from the ladders used by other trades. The hod carrier who supplies the mortar and other materials to the plasterer must climb up these ladders with a hodfull of mortar on his shoulder. The normal 12-inch spacing of the ladder cleats or rungs is too great to step up with such loads; therefore, the usual cleat

Fig. 1-14. Plasterer applying matrix coat for exposed aggregate, using hanging scaffold. (Ceram-Traz Corp.)

spacing for this type of ladder is 8 inch.

Hod carrier ladders are usually made from structural grade 2″ x 4″ lumber and 1″ x 4″ cleats with filler blocks nailed on the rails between the cleats to strengthen the cleat at this point.

It is very important that the ladders be properly placed and that they are securely fastened at both top and bottom. Figure 1-15 shows how the ladder is constructed and the proper horizontal distance out at the base. The distance from the foot of the ladder to the side of the building should be about ¼ the ladder length, as shown on Fig. 1-15. Sizes of side rails in relation to height and the required extension of the

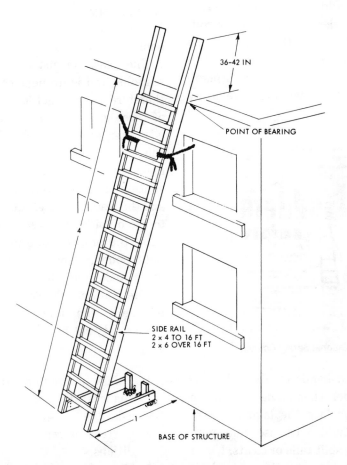

36–42 IN

POINT OF BEARING

SIDE RAIL
2 x 4 TO 16 FT
2 x 6 OVER 16 FT

BASE OF STRUCTURE

Fig. 1-15. Typical cleat ladder construction and formula for setting ladders against buildings or scaffolds. (National Safety Council.)

ladder above the scaffold or building (so there is hand support in getting off or on the ladder) are also given.

When the ladder is used outside and must be set on the ground, always set it on a plank and nail the side rails to this foundation. Fasten the side rails or the plank with stakes driven in the ground so the ladder cannot slip. Inside the building on concrete or wood floors tie or nail down the ladder for the same reason. When a ladder is used against a rolling scaffold, block the wheels of the scaffold opposite the ladder with pieces of 2″ or 4″ lumber or a plank so the scaffold cannot move away from the ladder.

National Safety Council

Use both hands on the side rails of the ladder when going up or down and always face the ladder. Never splice a ladder or use one with cracked or split rails or cleats. Never paint a ladder, as this may hide dangerous defects.

Building Enclosure

With year around constuction now an established practice, the enclosure of buildings to permit both interior and exterior work to continue even in freezing weather often creates a dangerous condition for the plasterer. Temporary heaters, if they are not properly vented, may give off obnoxious or deadly fumes. Salamanders, gas or gasoline heaters are very dangerous, as the carbon monoxide gases given off are odorless but lethal.

The plasterer usually works on scaffolds and is up near the ceiling; therefore he will get the full force of these gases first because the gas is lighter than air and will rise to the ceiling. All temporary heaters must be constructed with positive venting of the combustion chamber, and the resulting gases must be piped to the outside. The vent pipes must be of the proper size and have gas-tight points. "You only live once—don't make your life a short one."

The National Safety Council recommends the following top ten basic safety rules:

1. Follow instructions; don't take chances; if you don't know, ask.
2. Correct or report unsafe conditions.
3. Help keep everything clean and orderly.

4. Use the right tools and equipment for the job; use them safely.

5. Report all injuries; get first aid promptly.

6. Use, adjust and repair equipment only when authorized.

7. Use prescribed protective equipment; wear safe clothing, keep them in good condition.

8. Don't horseplay; avoid distracting others.

9. When lifting, bend your knees; get help for heavy loads.

10. Comply with all safety rules and signs.

National Safety Council

Checking on Your Knowledge

The following questions give you the opportunity to check up on yourself. If you have read the chapter carefully, you should be able to answer the questions. If you have any difficulty, read the chapter over once more so that you have the information well in mind before you go on with your reading.

DO YOU KNOW

1. How old is the plastering trade?

2. Name some of the important protections plastering adds to a building.

3. Can plastering add any decorative value to a building?

4. What basic differences exist between the early plastering materials and those used today?

5. Name the trades that developed from the original plasterer.

6. What great change is now taking place in this trade?

7. What is the basic difference between interior and exterior plastering?

8. Name the various levels of craftsmen in the plastering trade.

9. How does the architect tell the plasterer what he wants done in a building?

10. Name some of the important safety rules the plasterer should know and follow.

11. Name the four basic scaffolds used by the plasterer.

12. Why are some temporary heaters used to heat buildings dangerous?

Chapter 2

Tools of the Plastering Trade

In the beginning, the tools of a plasterer consisted of various shapes of wood, and his hands were primarily used to put the material on the surfaces. Through man's ingenuity and desire to do less strenous, more efficient and artistic work, he has developed many tools for his work. In the last few years, as in other trades, the use of machines has been developed to eliminate the slower and burdensome work of plastering.

Some of the tools the plasterer uses have more than one name. Different parts of the country may have different names for each tool. This situation causes confusion in teaching, and in practice it amounts to a serious fault. The plastering industry and manufacturers are standardizing the terms used in the plastering trade by using terms most easily recognized in naming tools and describing their uses.

The plasterer carries certain tools in his toolbox. Others are supplied by the contractor. This division of ownership of the tools has been worked out over a period of years with the interest of all in mind.

The small hand tools are the property of the journeyman plasterer. He carries them from job to job. The larger tools are placed on the job by the contractor. The plasterer, however, is responsible for the care of tools furnished him by the contractor for as long as they remain in his possession.

By agreement of the plasterer and the contractor, time is allowed at the end of each day to clean and put away tools used on the job.

Good tools, kept clean, are necessary to do a good job of plastering. From time to time a plasterer should check his tools for wear or warping, and he should repair or replace them as necessary. Knowing which tool to use and how to use it is necessary for a first-class job.

Basic Hand Tools

The plasterer brings most of the small tools which he is apt to use to the job site. They are small enough to be transported easily in the plasterer's *toolbag* or *box*. These are tools needed to do simple, plain browning and finishing operations and would be the starting kit for a beginning apprentice. Each of the tools that a plasterer usually carries will be discussed individually. Fig. 2-1 shows typical tools and toolbag of starting plaster apprentice.

Hawk. The hawk is used to hold and carry the various plaster mortars. See Fig. 2-1(E). It consists of a flat piece of aluminum or magnesium from 8″ to 14″ square with a detachable wooden handle centered and fixed to the underside. A soft sponge-rubber pad with a hole the diameter of the handle is placed around the handle, flat against the underside of the hawk to act as a cushion for the hand and help prevent callouses. Knowing the size and squareness of your hawk can be helpful in making approximate measurements.

Trowels. Trowels are used by the plasterer to apply, spread, shape and smooth the various materials used in plastering. The size of the trowel varies according to use and to the mechanic's preference as to the tool's feel and balance. Fig. 2-2(A) shows the common plastering trowel. (See also Fig. 2-1J). Common sizes are 10½″ x 4½″, 11″ x 4½″ and 11½″ x 4¾″. There are, of course, many other sizes available from which a plasterer can choose.

A trowel has three parts. They are called the *blade*, the *mounting* and the *handle*. Blades are made of tempered high-carbon or stainless steel. They must be perfectly straight for long performance. For certain materials that may stain with ordinary steel, care should be taken in selecting the quality and kind of steel used for the blade. Mountings are made of steel or aluminum. The mounting is the portion that connects the blade and handle. For long performance, care should be taken in selecting the quality of the rivets or pins used to fasten the blade to the mounting. Handles are made of

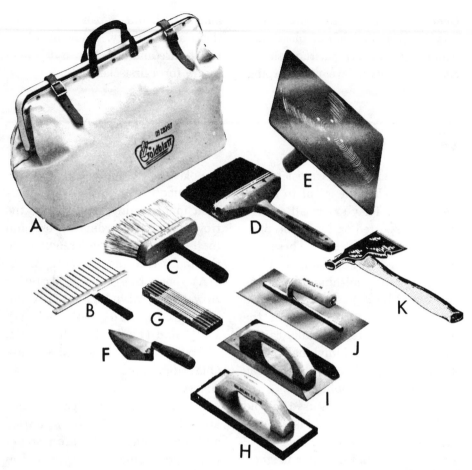

Fig. 2-1. Toolbag and tools of a starting plaster apprentice: (A) toolbag, (B) scratcher, (C) browning brush, (D) finishing brush, (E) hawk, (F) pointing trowel, (G) folding rule, (H) sponge rubber float, (I) angle float, (J) trowel, (K) half hatchet. (Goldblatt Tool Co.)

Fig. 2-2. Common trowels: (A) trowel, (B) margin trowel, (C) pointing trowel.

wood and are either curved (camel back) or straight. The mechanic will usually modify these to give him the proper balance and feel, so the tool will perform easily and smoothly. Fig. 2-3 illustrates how the handle is filed down to prevent blisters. The handle is fastened to the mounting with a locknut.

The plasterer generally carries at least three trowels in his tool kit. The newest and best one is used for finishing (applying the final coat of plaster). The next best trowel is used for applying the various mortars or first or second coats. The third is usually a well-worn trowel. It is used to gauge (mix) putty finishes and to do various scraping jobs, such as cleaning angles, grounds, etc.

Fig. 2-3. Filing down ridge on wood handle to prevent blisters.

Pointing Trowels. The pointing trowel is shaped like a triangle and is similar to a bricklayer's trowel. See Fig. 2-2(B) and Fig. 2-1(F). It can be bought in various sizes, running from 4½" to 7" for the blade. Blades should be of tempered steel with an integral forged post (mounting) and a handle (tang). It is important that the blade, post and tang be one piece made of forged metal. Post and tangs that are riveted to the blade soon break off from the blade.

The post of a pointing trowel is the part that rises up from the blade so as to place the handle up above the blade and provide room for the fingers. The tang is the part of the trowel that the wood handle slips over.

Pointing trowels are used for many purposes. They are small enough to be used in places where the larger trowels will not fit. They are used to clean tools and to handle small pats of material in many plastering operations. See Fig. 2-4.

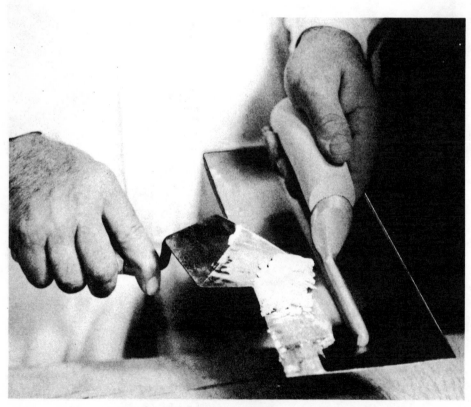

Fig. 2-4. Pointing trowel used for cleaning.

This tool is also helpful in ordinary cove miters.

Margin Trowel. The margin trowel is constructed similar to the pointing trowel, the difference being in the shape of the blade. See Fig. 2-2(C). The blade of the margin trowel come in sizes from 5″ x 1½″ to 8″ x 2″ with a square end. This trowel is used to apply and trowel-up material in narrow places, such as between moldings that are run closely together. The square end is better suited for troweling than is the pointing trowel. Like the pointing trowel, this trowel is used for cleaning tools. Many plasters put a notch in the sides of this trowel to aid in trimming back the base coat along corner beads. It is the tool most commonly carried by the plasterer in the special pocket of his work pants.

Angle Float. An angle float is a flat-surfaced tool with a flange along the two outer edges. See Fig. 2-5 left. A wooden handle is attached in the center between the flanges (such

Fig. 2-5. Left: Angle float for finished surface. Right: Angle paddle.

as 9″ x 4″ x ¾″ flanges). There are three types of material used to make angle floats. They are plexiglass, stainless steel and aluminum. The plasterer uses this tool to apply finishes to inside angles.

Angle Paddle. The paddle has a square, spade-like end that is tapered down to a thin wedge shape. It has a long round handle. See Fig. 2-5 right. Sizes are usually 2½″ to 3″ wide at the paddle end. The handle is usually ⅞″ thick and from 6″ to 8″ long from the edge of the paddle to the tip of the handle. Paddles are made of one of the following materials: wood, aluminum, rubber and plastic. Wooden paddles usually have a thin metal strip fixed into the front edge of the paddle to prevent wear. The paddle is used to clean out and, in most cases, to finish the angle or corner after it has been floated. The finished result,

however, is not comparable to the properly troweled out angle.

Brushes. The plasterer uses three main types of brushes. They are the browning brush, finishing brush and tool brush.

Browning Brush. The browning brush is used to throw water on the surface of applied mortar to provide "slip" or lubrication to the tools used to straighten the surface. Its prime requirement is that it holds a large amount of water. It is used often to wash tools. It is usually shaped like a Dutch Brush. Normal size is 6½″ x 2″ x 4″. See Fig. 2-6(A).

Finishing Brushes. There are two kinds of finishing brushes. They are the bristle type and the felt type. The finishing brush is used for the same purpose as the browning brush in that is is used to apply water to the plaster surface. In this operation,

Fig. 2-6. Basic brushes used by plastering apprentice include: (A) browning brush, (B) regular (top) and Dutch (bottom) finishing brushes, and (C) felt finishing brush. (Goldblatt Tool Co.)

however, the water is applied by direct contact with the surface. Its bristles or felt must be of the best quality to prevent the soft plaster surface from being scratched.

The best bristle brushes are made of 100 percent hog bristle set in three rows. See Fig. 2-6(B). The next best bristle brushes are made of pure nylon bristles. These outwear hog bristles, but do not hold much water. The sizes are 1¼″ thick by 8″ wide and 4⅛″ long. The finishing brush has a long handle so that it will stay dry when it is put in a bucket of water.

Some finishing brushes can be purchased with a strip of leather bound about the bristles where they are attached to the handle. This prevents water from running along the handle while it is being used. Fig. 2-7 illustrates one use of a finishing brush being used with a trowel.

The felt brush is used the same

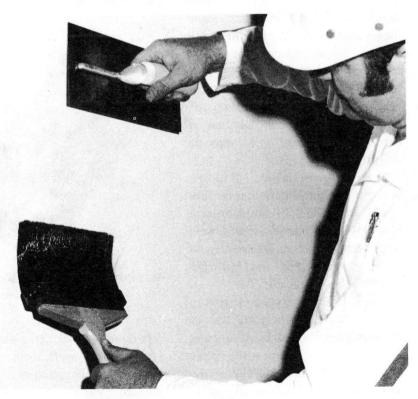

Fig. 2-7. Troweling finish surface.

way as the bristle brush. It is becoming widely used because it is inexpensive and makes complete contact with the surface while being used, whereas a bristle brush has a tendency to "finger" or spread apart. There are two kinds of felt brushes —with either one piece of felt folded double or two pieces of felt. These brushes are constructed with a handle and a channel 1¼″ x 7½″ with ⅛″ side walls in which the felt or felts fit. See Fig. 2-6C. A stiffener, usually made of aluminum, is placed between the felt ends and fastened to the channel. This gives the brush exactly the right flexibility. Handles are detachable and are made of wood or aluminum. One important thing to look for in this type of brush is its weight and the kind of fastener used to secure felt and stiffener to the channel. Because of corrosion, replacing felts may be difficult.

Scarifier or Scratcher. This is a tool with a convenient handle and flexible steel tines. It comes in sizes 6″ to 16″ wide. The Scarifier is used to scratch the surface of unset materials so that the next coat will have sufficient bond. See Fig. 2-8.

Floats. Many types and sizes of floats are used by the trade. As the name implies, the tool floats, or rides, over the surface of the work. Basically, the float is a straightening device used by the plasterer to level off the humps and fill in the hollows

Fig. 2-8. Scarifier or scratcher.

Fig. 2-9. Sponge rubber float. (Goldblatt Tool Co.)

left by other tools. It is also used to compact the work into a smooth dense mass or produce a textured surface. Various types of texture floats produce different surface treatments. They are constructed with handle, back and pad. See Fig. 2-9. The handle is secured to the back usually with two screws. The pad is glued to the back. This type of float has three measurements: length, width and thickness of pad, such as 8″ x 5″ x 1″, 9″ x 4″ x ½″. The important part, of course, is the pad. The selection will be determined by the texture required. The selection of pads are many. For example: cork, rubber, sponge rubber, foam plastic and carpet are used. The most commonly used are sponge rubber and the foam plastic.

Fig. 2-10. Half hatchet.

Half Hatchet. The half hatchet is a heavy-duty combination hatchet and hammer. See Fig. 2-10. It is used to chop off hardened plaster, and it is used to do other heavy work.

Hand Tools for Advanced Work

The term "advanced work" is not to be literally translated to mean that an apprentice plasterer must have these tools in order to advance from apprentice plasterer to journeyman. These tools are additional tools that most journeymen have. The plastering apprentice will add these to his tools as he goes along (not necessarily in the same order as they are described here) to help him improve his work and to give him independence by not having to depend on the journeyman to supply them.

Elastic Knives. Elastic knives are made of a highly flexible carbon steel, ground and polished to a mirror-like finish. See Fig. 2-11. The handles are molded of shatter-proof

Fig. 2-11. Elastic knife. (Goldblatt Tool Co.)

plastic and fastened with brass "hammer" heads. There are two sizes of knives: either 5″ or 6″. An elastic knife is used at stop beads, expansion beads, grounds, etc., to speedily trim back the browning (before it sets). The plasterer will cut a notch on each front corner of the blade to the depth he wishes the brown coat to be.

Steel Square. The plasterer uses a steel square for mold making and for squaring corners and checking his tools for square. See Fig. 2-12. The steel square most commonly used has a 12″ blade and 8″ tongue. A rust-proof steel square is the best for the plasterer to use, because he is constantly working with wet materials. Some plasterers prefer a larger steel square such as an 18″ x 12″ square. Three parts of a steel square

bear names: the *tongue* is the shorter arm, the *blade* is the longer of the two arms, and the *heel* is the meeting place of the two arms.

Six-Foot Folding Rule. The six-foot folding rule (or zig-zag rule) is the standard folding rule sold to the trades for general use. See Fig. 2-13. It folds up in 6″ sections and can be opened up to 6 ft. in length. Wood and aluminum are the two common materials from which the rule is made. Some of the more expensive 6″ folding rules have a graduated brass slide on the first section. It extends a full 6″ and is very handy for measuring inside dimensions.

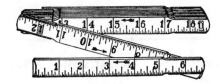

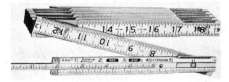

Fig. 2-13. Folding rules. (Lufkin Rule Co.)

Spring-Loaded Steel Tape Measure. This tool (also called a push-pull rule) is carried by many plasterers in place of the six-foot folding rule because it is compact and usually has a clip which he can use to attach it to his belt. See Fig. 2-14. It is similar to the steel tape in design. The tape is held in a metal case and

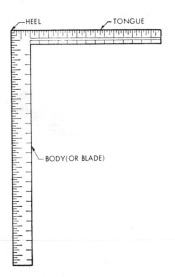

Fig. 2-12. Steel square.

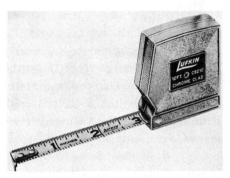

Fig. 2-14. Spring-loaded steel tape in metal case. (Lufkin Rule Co.)

Fig. 2-16. Chalk line and reel.

will coil itself back into the case after being extended. It comes in various lengths: 6 ft., 8 ft., 10 ft. and 12 ft. The most commonly used is the 10 ft. length.

Hand Saw. The hand saw is usually about 20″ to 24″ long and of the same pattern as the coarse-toothed rough carpenter's saw. See Fig. 2-15. The teeth must be large in order to prevent the wet plaster from filling them up and to prevent the saw from binding while it is being used. About 6 teeth to the inch is recommended.

Chalk Line and Box. A chalk

line is a length of string that has been thoroughly filled with chalk dust or plastering color additives. See Fig. 2-16. It is used to strike a straight guide line on work. Chalk lines are sold in 50′ balls, in 1 lb. skeins and 100′ shanks. The chalked string is held taut close to the work and is used either as a measuring guide or plucked with fingers to make a line. In long distances care should be taken to avoid a sag in the line.

The common chalk line and box carried by the plasterer serves a dual purpose. It is a combination chalk

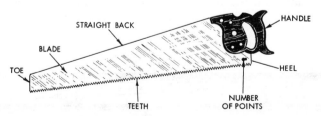

Fig. 2-15. Straight back hand saw.

line and plumb line. The box is shaped with a point on one end and its winding arm has a lock pin to hold the line taut when using it as a plumb line.

Plumb Bob. The plumb bob is a weight of lead, steel or brass attached to a line and used by lathers, plasterers, etc. See Fig. 2-17. It is used to indicate and check vertical direction to make sure that the vertical surfaces are straight and true. When a plasterer refers to a "plumb wall", he means that the wall surface is true vertically. Plumb bobs are classified by weight rather than by size. The common weight of a plumb bob runs from 4 to 16 oz.

Fig. 2-17. Plumb bob.

Level. A level, or spirit level, is used to establish a horizontal or vertical line. See Fig. 2-18. It indicates whether any part of a horizontal surface is higher than another or if a

vertical surface is not true. The essential part of the level consists of a slightly curved glass to be nearly filled with alcohol or ether. The tube of liquid contains a movable air bubble. When the bubble is centered, it indicates that the surface is level or straight. Centering the air bubble is easy, since most levels have lines indicated on the glass tube. When the air bubble is located between these lines the surface is level. Slight tilting of the level at either end will cause the bubble to move away from the center, indicating a line that is not horizontal or level or that a vertical line is not straight or true. The glass tube is protected by a metal or wood casing of some length. The common lengths of levels used by the plastering trade are 18″, 24″ and 48″ and may be made of wood, aluminum or magnesium. The metal levels are preferred since they will not warp or wear as wood might.

Most levels today contain more than one tube of liquid. The more expensive ones contain many tubes so that they can be used regardless of the way they are picked up or set against any surface. Most levels also contain two types of tubes filled with

Fig. 2-18. Levels are available in either aluminum or wood.

liquid. One is used to determine how level or horizontal a surface is as described in the preceding paragraph, and another, usually at the ends of the level to determine verticality of the surface. The tube determining horizontal straightness will be parallel to the length of the level. The tube determining how plumb a vertical surface is will be parallel to the width of the level. Fig. 2-19 shows a vertical surface being checked.

Tin Snips. There are two types of snips used in the plastering trade. One is the *regular snips* which come in various sizes and are numbered; such as #10 is 12½″ long, #8 is

Fig. 2-19. Level used to check vertical surface.

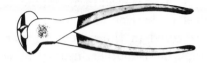

Fig. 2-21. Nippers.

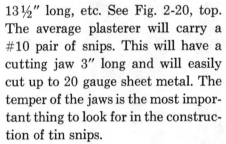

Fig. 2-20. Tin snips. Top, regular snips; bottom, aviation style snips.

13½″ long, etc. See Fig. 2-20, top. The average plasterer will carry a #10 pair of snips. This will have a cutting jaw 3″ long and will easily cut up to 20 gauge sheet metal. The temper of the jaws is the most important thing to look for in the construction of tin snips.

The other type is the *aviation style snips*. These come in three styles: one for left cutting, one for right cutting and one for straight and combination cutting. Fig. 2-20, bottom, illustrates an aviation snip for straight or combination cutting. The advantage of aviation type snips is the spring-loaded leverage which takes the effort out of cutting.

Tin snips are used in the plastering trade to cut metal lath, corner beads and making cornice molds.

Nippers. Nippers (also called end cutting pliers) are small pincers used for holding, breaking or cutting. See Fig. 2-21. Two types of nippers

are used. One has solid jaws, the other has removable jaws which can be replaced after wear. The plasterer uses nippers to erect plaster casts when they are tied up with wire and plaster wads. Nippers are also used for general work, such as nail pulling and miscellaneous lathing work left undone by the lather. The nippers are often carried by the plasterer when scratching metal lath. The sizes are 7″ long by 1⅜″ jaws to 8″ x 1⅜″.

Angle Plane. The angle plane is a strongly built tool. It is constructed of a handle and a channel base to which 7 blades are attached, perpendicular to the channel and at different angles to each other. See Fig. 2-22. Each blade is 1¼″ deep and the overall size is 10″ long by 4″ wide. Its handle is the same as that of a float. It is used for preparing the surfaces (after brown coat has set) for a finish coat, by knocking down high spots, cleaning angles and scraping. The angle plane is much easier to use for this work than an old trowel.

Pencil or Marking Chalk. The pencil is used by the plasterer for any

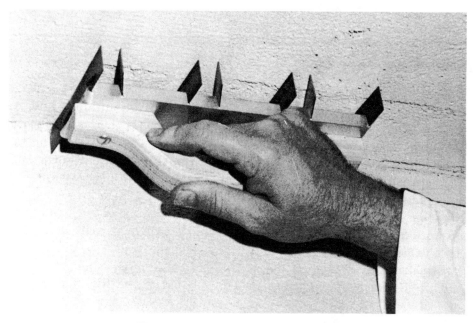

Fig. 2-22. Plasterer using an angle plane.

Fig. 2-23. Tool brushes.

marking or writing that will be done on a job. This is a common lead pencil of any kind. Marking chalk is used, as the name implies, for marking. Level points, for example, are marked with marking chalk.

Tool Brush. Tool brushes are used to apply water to small places and to brush out miters, ornamental work, etc. Two sizes are generally used. They are 1″ round sash type and 1½″ flat, long-handled type. See Fig. 2-23. A good quality paint brush is ideal as a tool brush.

Specialized Hand Tools

These tools are designed to enable the plasterer to apply materials to surfaces where his ordinary trowels will not fit. They also enable the plasterer to perform a specific job more efficiently. Special individual textured finishes are also accomplished with the use of special texturing float or brush.

Pool Trowel. The pool trowel is constructed with blade, mounting and handle. The blade comes in sizes 10″ x 3″ to 18″ x 5″. Both ends of the blade are rounded, and its mounting has a short shank which gives it more flexibility than the regular trowel. See Fig. 2-24. The handle is curved (camel back). This trowel can be used for troweling various curved surfaces, such as can be found in swimming pools, domes and curved light coves.

Midget Trowel. Midget trowels are constructed like the regular

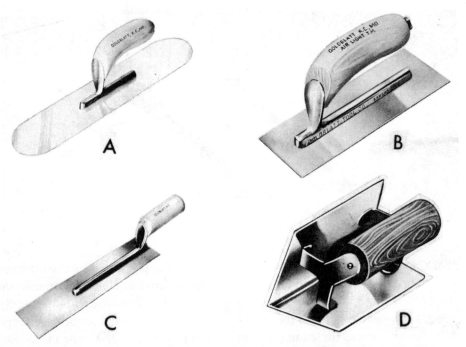

Fig. 2-24. Specialized trowels: (A) pool trowel, (B) midget trowel, (C) pipe trowel, (D) inside corner trowel.

trowel but are smaller, ranging from 7½" x 3" to 8" x 3¾". The handles are all curved (camel back). See Fig. 2-24(B). They are used for window sills, closets and small areas.

Pipe Trowel. The pipe trowel is constructed like the regular trowel, having a blade, mounting and handle. However, it differs in size and proportion (10½" x 3") with the handle offset and the blade directly in front of the handle. See Fig. 2-24(C). This long nose allows it to

be used behind fixtures and between pipes. Fig. 2-25 illustrates the usage.

Inside Corner Trowel. The inside corner trowel comes in two sizes: 6" x 2½" x 2½" and 6" x 4" x 4". Fig. 2-24(D). It is used, as the name, implies, to finish inside corners.

Outside Corner Trowel. This trowel is constructed the same as the inside corner trowel except in reverse. Its size is 6" long, 2½" wide. This is an ideal tool for freehanding arrisses for exposed aggregates.

Fig. 2-25. Plasterer using a pipe trowel.

Wide Blade Angle Plow. Wide angle plows are made of a single piece of highly flexible stainless steel bent to form two surface blades at a 90° angle. They have an offset handle mounted between and in front of the blades. The blades are 4″ wide at the ends, tapering to 5″ at the center, and are each 4¼″ long. See Fig. 2-26, top. This tool saves time in trimming angles when browning, by floating it through the angles instead of cutting the angles at two planes.

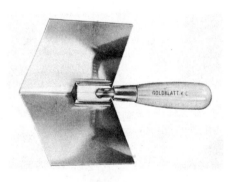

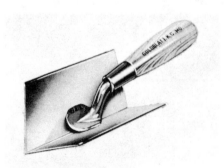

Fig. 2-26. Plows. Top, wide blade angle plow; bottom, angle plow.

Angle Plow. The angle plow is much like the wide angle plow except that it is made of a stiff stainless steel. The blades are 2½″ wide and are not tapered. See Fig. 2-26, bottom. It is used primarily to apply finishes to inside angles. See Fig. 2-27.

The difference between the *corner trowel* and the *angle plow* is manly in the handle. The inside corner trowel is a more loosely controlled tool, so it slides through the angle with little pressure and is easier to use on unset material. The angle plow, however, is used to apply pressure and finish setting materials, such as putty coat materials, in place of using the angle float.

Lathing Hatchet. The lathing hatchet comes with a permanent blade or a replaceable type of blade. See Fig. 2-28. It is a lightweight tool having a very thin, sharp, long blade. It is very highly tempered. The tempering provides a long lasting, sharp cutting edge. The head of the hatchet is usually cross-hatched with grooves in order to prevent it from slipping off the nailhead when striking it. The handle has a thick-grip — a feature that means comfort when the hatchet is used for long periods without interruption and the head is set at the most comfortable angle for overhead work. The replaceable blade lath hatchet usually is an unbreakable one-piece construction forged of finest steel and is provided with a ny-

Fig. 2-27. Plasterer using an angle plow.

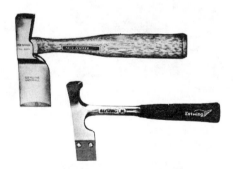

Fig. 2-28. Lathing hatchets. (Goldblatt Tool Co.)

lon-vinyl cushion grip. The end of the blade is easily replaced by removing two round head screws. The lath hatchet is used to cut and nail rocklath or to nail metal lath.

Steel Tape. The steel tape is used to measure distances longer than is possible with the six-foot folding rule. Two sizes are in general use: 50′ and 100′ lengths. See Fig. 2-29. The 50′ steel tape is the popular size used by most plasterers. The tape itself is made of spring steel $\frac{3}{8}$″ wide and it winds up into a metal case. The marking for inches and feet should be stamped into the metal for permanence. The best types have rust-resistant chrome or enamel finishes.

Cement Stucco Dash Brush. This brush is 5″ x 2¼″ block with 4 rows x 8 rows bassine fiber 5¼″ long. See Fig. 2-30. The cement stucco

Fig. 2-29. Steel tape. (Lufkin Rule Co.)

Texture Brushes

These brushes are used, as the name implies, for texturing.

Rice Brush or Stippling Brush. The rice brush is constructed of a round handle 6½″ long and a base 5½″ x 5½″ to which the bristles are attached. See Fig. 2-31. This brush is most commonly used for acoustic plasters. The material is applied to the surface and straightened. This brush is then used to punch the surface evenly to create a textured surface.

Fig. 2-31. Rice brush or stippling brush.

Fig. 2-30. Cement stucco dash brush.

dash brush is used by dipping the brush into the material and throwing it on to the surface with various techniques.

Wire Texture Brush. This brush is a finger-grip brush having a 3″ x 1″ wood base with 2½″ long wire bristlse attached. It is primarily used to texture cement by dragging it through the applied finishes before it is set. See Fig. 2-32.

58

Fig. 2-32. Scoring the surface, using a wire brush.

Fig. 2-33. Plasterer's rubber sponge. (Goldblatt Tool Co.)

Plasterer's Rubber Sponge. The plasterer's rubber sponge is oval shaped to fit the hand and is made of pure sponge rubber. The size is 7" x 4" x 2". See Fig. 2-33. This tool is ideal for texture floating curved surfaces.

Texture Float. This refers to the uncommon float, such as a carpet or cork float.

Ornamental Tools

These are additional tools necessary to perform ornamental work such as cornice and staff work.

Small Tools. Sometimes these are called *modeling tools* or *ornamental tools*. Small tools are used in pointing the joints when erecting cast ornamental work. They are used to finish the final intersection of a miter in cornice work. They are also used for any work that requires a fine, small-sized tool. See Fig. 2-34.

Two general shapes are in common use by the plasterer. These are leaf and square, and trowel and square. Each of these types has a long handle with a blade on the end. As the name implies, the leaf and square has a pointed, leaflike shape on one end and a square shaped blade on the other. See Fig. 2-35(A) The trowel and square has a pointed blade offset from the handle about ¾" on one end and a square blade on the opposite end. See Fig. 2-35(B) The offset is constructed so that the blade can be laid flat on the surface while providing room for the fingers when

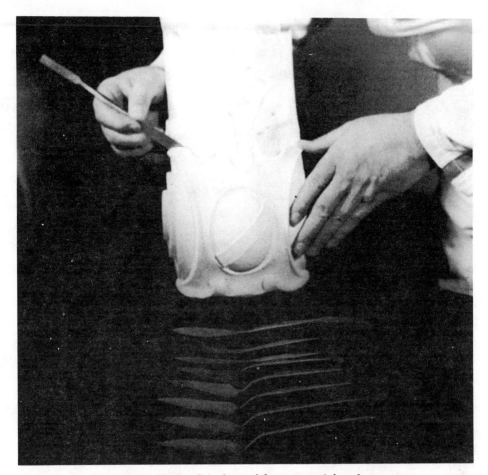

Fig. 2-34. Small tools used for ornamental work.

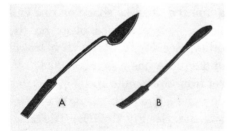

Fig. 2-35. Small, modeling, or ornamental tools: (A) trowel and square, (B) leaf and square.

grasping the handle. The offset also makes it possible to use the tool in recessed places. These small tools are made of the best spring steel and are forged and ground by hand. Consequently, the tools are very light and flexible. The tool size is measured by the width of the square blade: ½″, ⅝″, ¾″, ⅞″ for leaf and square. The sizes for trowel and square are $\frac{3}{8}$", $\frac{1}{2}$", $\frac{5}{8}$", $\frac{3}{4}$" and 1". The overall length runs from 8" to 12".

Miter Rods or Joint Rods. Miter rods are pieces of metal from 2" to 24" long, 4" wide and $\frac{1}{16}$" thick. See Fig. 2-36. The shorter side of the sheet of metal is cut at a 45° angle. The longer side is ground to a knife-like edge. Miter rods can be purchased in the following lengths: 2", 4", 6", 8", etc., through 24". Joint rods should be made of the highest grade of steel and should be perfectly straight. Miter rods are used to form and shape joints and miters in cornice work. The soft plaster, placed at the intersection of two cornices, is drawn with this tool until the two cornices intersect or join completely as a finish miter. Because cornices vary in size, and therefore, produce different sized miter joints, many different lengths of miter rods are required. Three sizes are generally carried: they are the 6", 12" and 18" miter rod. Smaller or larger sizes than those mentioned are used only in very special work. Another good use for the miter rod is the sharpening of inside angles.

Rubber Gloves. The plasterer uses rubber gloves when applying cornice plaster and similar work. They protect his hands from lime burns, permitting him to manually feed the cornice mold or "stick ornament" (apply ornamental work). The thick "gauntlet" type of glove is preferred because it wears well and

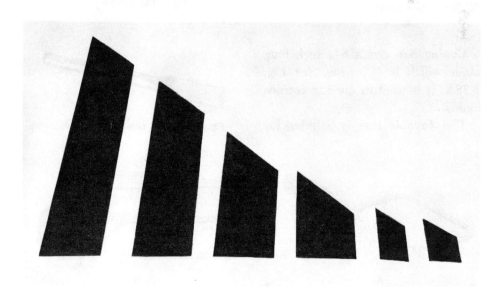

Fig. 2-36. Joint rods.

can be quickly removed from the hand with a jerking motion.

Bevel Square. The bevel square is a combination square and 45° angle. The shape is triangular. The square corner of the triangle forms a 90° angle, the other two corners form 45° angles. See Fig. 2-37. The plasterer uses the bevel square to miter cuts and also to mark off 45° or 90° lines as he might need them.

Fig. 2-37. Bevel square.

Coping Saw has a 6⅜ inch long blade which is ⅛″ wide. See Fig. 2-38A. It is used in making cornice molds.

The *keyhole saw* is popular for mold-making work. See Fig. 2-38B. Some mechanics prefer the coping saw to the keyhole saw. The coping saw is better for contour cutting, but its fine blade will not stand much abuse when used for other work.

Cold Chisels. These are available with blades from ¼″ to 1¼″ wide and from 5″ to 18″ in length. See Fig. 2-39. They are made to cut metal such as nails.

Wood Chisels. These are available in widths from ⅛″ to 2″. Blade length may vary from three to six inches. See Fig. 2-40.

Claw Hammer. The common carpenter's claw hammer is used by the plasterer to pull out nails and for all forms of mold, template, bracing and similar work. Hammers are sold by weight, and the 1 lb. size is about right for the plasterer's use. See Fig. 2-41.

Fig. 2-39. Cold chisel.

A

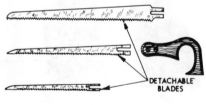

B

Fig. 2-38. Special-purpose saws: (A) coping saw, (B) keyhole saw with detachable blades.

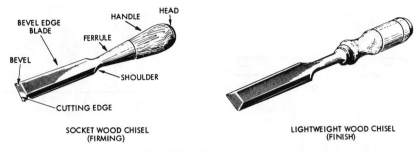

SOCKET WOOD CHISEL
(FIRMING)

LIGHTWEIGHT WOOD CHISEL
(FINISH)

Fig. 2-40. Wood chisels.

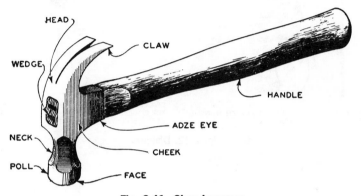

Fig. 2-41. Claw hammer.

Dividers. These are available in lengths from six to eight inches. This tool is used for layout work in making a cornice mold. It is also used to step off or accurately repeat several measurements. See Fig. 2-42.

Prick Punch. The prick punch is a round rod having one end sharpened to a fine point. See Fig. 2-43. It is used to punch holes through sheet metal for making profiles and also to outline profile patterns through blueprints onto sheet metal when making cornice molds.

Small Plane. The small plane (also called a block plane) is used by the plasterer to cut both wood and plaster. See Fig. 2-44. When fitting and sticking ornament, one section of material may be slightly large; the plane is used to cut it down so that it will fit.

Files. Files are used by the plasterer to produce profiles. (Profiles are patterns of a design used as a guide in forming cornice work.) The common shapes of files used are the rattail round, half round, flat, square

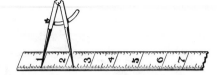

TO SET DIVIDERS HOLD BOTH POINTS ON THE MEASURING LINES OF THE RULE.

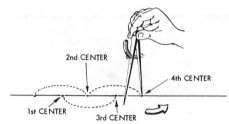

2nd CENTER

4th CENTER

1st CENTER

3rd CENTER

DIVIDERS ARE USED TO STEP OFF A MEASUREMENT SEVERAL TIMES ACCURATELY

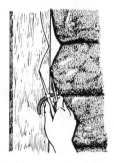

DIVIDERS MAY BE USED TO SCRIBE A LINE TO MATCH AN IRREGULAR SURFACE MASONRY OR WOODWORK.

DIVIDERS ARE USED FOR SCRIBING CIRCLES OR AN ARC. ALSO FOR COMBINATIONS OF CIRCLES AND ARCS FOR MAKING LAYOUTS FOR CURVED DESIGNS, ETC.

Fig. 2-42. Dividers.

Fig. 2-43. Prick punch.

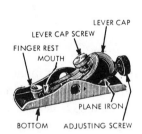

LEVER CAP

LEVER CAP SCREW

FINGER REST

MOUTH

PLANE IRON

BOTTOM

ADJUSTING SCREW

Fig. 2-44. Small plane.

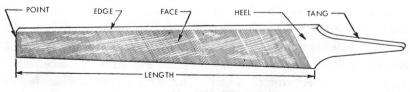

POINT EDGE FACE HEEL TANG

LENGTH

Fig. 2-45. File.

and triangle or saw file. Fig. 2-45 shows a flat file. File teeth should be fine enough to work on thin metal without scratching.

Flat and square files should have one edge smooth in order to be able to file one surface or corner without damaging the other. The flat file is also used by the plasterer to keep his other tools in good condition.

A file, when in use, should always be equipped with a handle. The handle fits over the tang.

Drywall Tools

The following tools are primarily designed for Drywall. Drywall has in the past been somewhat neglected by the plasterers in general, but is now growing to be one of the fine skills of a Journeyman Plasterer.

Drywall Joint Tool. The drywall joint tool is constructed like the regular plastering trowel except for the shape of the blade, which is made of flexible steel with approximately $\frac{3}{16}''$ concave bow. See Fig. 2-46. The plasterer uses this tool to spread the mud over the tape joints. Sizes are 8" x $4\frac{1}{2}''$ to 16" x $4\frac{1}{2}''$ with curved (camel back) handles or straight handles.

Flexible Corner Tool. The flexible corner tool is constructed the same as the wide blade angle plow. The only difference is in the degree of angle formed. The flexible corner

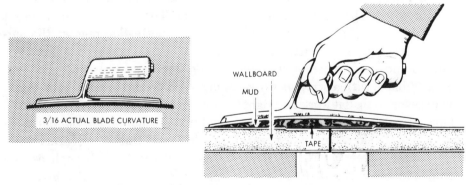

Fig. 2-46. Drywall joint tool. (Goldblatt Tool Co.)

Fig. 2-47. Flexible corner tool. (Goldblatt Tool Co.)

Fig. 2-48. Plasterer's knives: left, concave knife; right, straight knife. (Goldblatt Tool Co.)

Fig. 2-49. Trio finishing tool. (Goldblatt Tool Co.)

tool is formed to 103° angle and flexes to 90° for a tighter and more perfect corner. See Fig. 2-47.

Knives. Knives used by plasterers are of two types. One has a straight edge and the other a slightly concave shearing edge bowed in .050″ to allow build-up of material in the center while removing it at both ends. Fig. 2-48 shows both types. These knives are used by the plasterer to cover nail heads and little nicks in the boards. Blades are 4″, 5″ and 6″ wide.

Trio Finishing Tool. This tool has three interchangeable and replaceable blades: (1) an 11″ x 4″ blade with concave edge that shears, feathers and finishes joints; (2) an 11″ x 4″ blade with a straight working edge for doing flat work and tight skimcoating; and (3) an 8″ x 4″ concave blade for skimcoating or shearing excessive freshly applied joint compound. See Fig. 2-49. Each blade fits securely and quickly into the lightweight blade clamps by means of two small brass wing nuts. Joint tape can be slipped into the blade clamp to shim blades for almost any desired "bow." The handle is offset to allow better leverage at the surface.

Corner Tape Creaser. The corner tape creaser makes both inside and outside corners. This sturdy, lightweight aluminum tool hangs on a roll of tape while doing corner work. Just pull the tape through as you need it and cut it off. See Fig. 2-50. There are no wrinkles in tape to get

Fig. 2-50. Corner tape creaser. (Goldblatt Tool Co.)

air bubbles and show later after painting.

Mud Holder. The mud holder is a hawk 12″ x 6″ with a lip on one side 1½″ x 12″, and a handle fastened to the underside. See Fig. 2-51. The plasterer uses this to carry *mud* to his work. (Mud is a special plaster with additives used to control setting and hardness. It is used to fix joints in drywall and to cover nail holes. Sometimes it is used to patch plaster cracks.)

Sanders. There are three types of sanders: the hand, the long handled and the corner sander. The *long handle sander* is as the name implies. It is constructed of a metal base to which a 9″ x 4″ rubber face is glued.

Fig. 2-51. Mud holder. (Goldblatt Tool Co.)

It has two spring clamps on the back side, one at each end. The clamps hold the sandpaper to the face. On the back of the base and in the center is a universal swivel device which holds the long handle. With this tool a plasterer can sand anywhere in a room without a bench. The *hand sander* Fig. 2-52B) is basically the same except the handle is permanently mounted to the back of the base and there is no universal joint. The *corner sander* (Fig. 2-52C is constructed with the base and face pad shaped at 90° forming two sides. Its size is 4″ x 4½″ for each sanding face. Its handle is 5″ long attached to an adjustable swivel at the back side where both bases meet.

Utility Knife. The utility knife is used for cutting board tape, etc. It is made of metal, has two sides and a single-edge razor blade held between the two sides. The size is about 5″ long by 1″ wide and ¾″ thick. See Fig. 2-53.

Banjo Tapers. There are two types of banjo tapers: the dry taper and the wet taper. The *dry taper* has a shape circular on one end and tapered at the other. It is constructed like a case with one side hinged to the other, and has a handle on one side and another handle on top. There are two latches to hold the hinged side closed. A roll of tape is placed inside the case at the round end. The tape is then drawn to and past the tapered end. The tape may

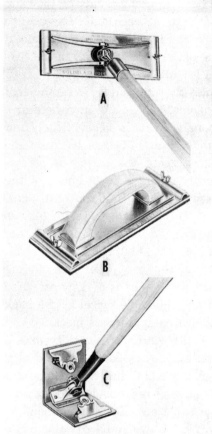

Fig. 2-53. Utility knife. (Goldblatt Tool Co.)

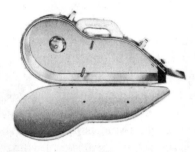

Fig. 2-54. Wet tape banjo taper. (Goldblatt Tool Co.)

Fig. 2-52. Three types of sanders: (A) long-handled sander, (B) hand sander, (C) corner sander. (Goldblatt Tool Co.)

be threaded two ways: one way for ceilings and one way for walls. The dry taper usually will have a portion of the hinged side missing at the round end so the tape roll can be checked to see how much is remaining.

The *wet taper* is the same except

it is in a completely enclosable case, so that the mud may be stored inside the case with the roll. See Fig. 2-54.

To use the banjo taper the plasterer guides it along the seams releasing the tape as he moves it.

Wallboard Stripper. The wallboard stripper is constructed with a handle and two steel wheels attached to an adjustable slide which is fastened to the handle. See Fig. 2-55. The two steel wheels on this tool perforate both sides of a drywall

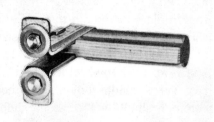

Fig. 2-55. Wallboard stripper. (Goldblatt Tool Co.)

Fig. 2-56. Mashers: left, round head drywall masher; right, square head drywall masher. (Goldblatt Tool Co.)

board so that you can snap off the trimmed edge cleanly and neatly. The comfortable handle guides the wheels accurately. Cuts as wide as 4½″ can be made with the tool. A scale along the top edge of the tool makes setting easy. Just loosen a thumbscrew. Strippers come in sizes for ⅜″, ½″ boards and ½″, ⅝″ boards.

Drywall Mud Masher. This is the perfect lightweight tool for mixing small amounts of the joint cement evenly. Shaped much like the familiar potato masher, it is made of steel wire, a wooden handle, and has a 5″ diameter round head or a 5″ square head. Its overall length is 24″. Fig. 2-56 shows two types in general use. Use of this tool is limited to small work.

Contractor Supplied Tools and Equipment

The contractor furnishes all of the larger tools used on the job, such as the straight edges (rods), feather-edges, darbies and slickers.

The plasterer's tender or helper, who mixes the plastering materials and brings them to the plasterer, uses many tools that are also supplied by the contractor. Among these tools that the plasterer's tender uses are scaffolds, ladders, mortar boxes, mortar boards and stands, etc.

The contractor also furnishes mechanical tools or machine equipment, such as plaster mixer, texturing guns, mechanical pump which applies plaster, etc, as well as all the required lumber, nails, drop cloths, plaster, temporary heaters and any other equipment or material needed to do the job. The tools and equipment supplied will vary from job to job.

Rod. A *rod* (Fig. 2-57A) is the trade name for what is also known as a straightedge. This tool is used to straighten the face of walls and ceilings. The lengths most commonly used by the trade are 5 ft., 6 ft., 7 ft. and 8 ft. In order to apply screeds, the plasterer will sometimes use a rod up to 20' in length. Wooden rods are generally 6" wide x 1¼" thick. The metal rod is made of magnesium and is hollow. Its size is generally 4½" wide x 1" thick, and comes in 6 ft. to 8 ft. lengths.

Featheredge. The featheredge (Fig. 2-57D) is used to straighten angles (corners) in the finish coat. It may also be used to produce putty coat screeds for cor-

nice, and many plasterers use this tool in place of the rod to straighten the walls and ceilings. The basic sizes of the featheredge are 4' to 8' long and from 4½" to 6" wide. There are two types made today. The most common shape is the wedge shape. Normally it is 1" thick and tapers to ¼". It is constructed of magnesium, aluminum or wood. The metal featheredge will hold its straightness longer, but is not as easily repaired as it becomes worn. The most common featheredge found on the job today is made of aluminum.

The other type is the magnesium featheredge. See Fig. 2-58. This new design is particularly helpful for straightening work behind the plas-

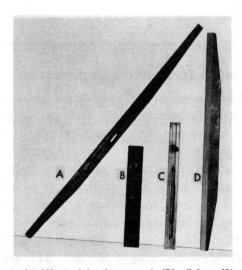

Fig. 2-57. Contractor's tools: (A) straightedge or rod, (B) slicker, (C) darby, (D) featheredge.

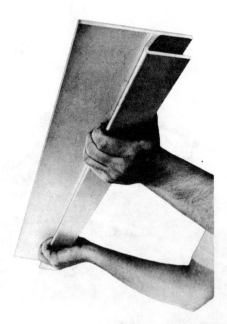

Fig. 2-58. Magnesium featheredge. (Goldblatt Tool Co.)

tering machine. Its feature is the finger-tip control. Its design is like an "s" shape, having one curve of the "s" 4½" wide and the other curve of the "s" 1½" wide.

Darby. The darby (Fig. 2-57C) consists of a long, thin flat blade usually ½" thick, 3½" wide and 42" to 45" long, made of mahogany or aluminum or a combination of both. Handles are attached to one surface of the blade. One handle is made of wood and is round-shaped like the hawk handle. The other handle is wood about ¾" thick, 1½" wide and placed away from the round handle. The tool is held with both hands and is used to float over freshly rodded

brown mortar. This action compacts and smooths the mortar into a flat, dense surface. Darbies are also used to smooth out acoustic materials that are applied over a previous coat of brown or are used to smooth up putty coat screeds and to straighten putty coat that has been applied over an uneven brown coat. See Fig. 2-59.

Slicker. The slicker (Fig. 2-57B) is used in place of the darby and is preferred by some plasterers because it can be bent slightly while it is being floated over the mortar and can be handled at a better angle to the surface. This feature enables the plasterer to cut off the higher points of the surface. The slicker is made of

71

Fig. 2-59. Darbying on acoustical plaster.

wood bevel siding like that often used to sheath the exterior of a house. There are also metal slickers available from manufacturers. See Fig. 2-60. A magnesium slicker with one end bent for a better finger grip is also available. Sizes of a slicker are usually 3 ft. to 4 ft. long and 4¾" to 8" wide. It does not have handles, but is held by the thicker edge.

Scaffolds and Other Equipment. The contractor also has to send to the job tools and equipment that will be used by the plasterer's tender to mix and supply the mortar. See Fig. 2-61 for an illustration of many of these. These consist of mortar boxes, putty boxes, water barrel, water pails, shovels (No. 2 square end), mixing hoe, screens (for sand and putty), floor scraper, hod, hod stand, water hose, wheelbarrow, scaffold, ladders, various sizes of adjustable scaffold horses, scaffold planks (various lengths), canvas covers, finishing boards, high and low stands for mortar or finishing boards, masking tape, polyethylene covering, electrical wiring for equipment and for lighting.

Finishing boards are used as a surface upon which to mix plaster (mor-

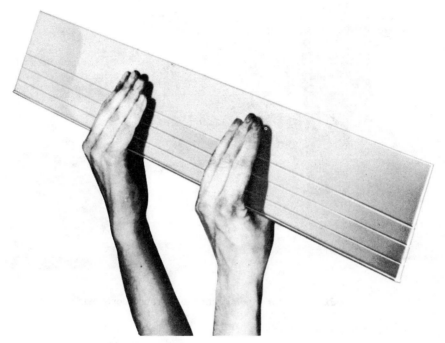

Fig. 2-60. Lightweight plasterer's slicker or shingle. (Goldblatt Tool Co.)

tar). They are from 5 ft. to 6 ft. square and are supported by a stand which is about 3 ft. high. Finishing boards have, by arrangement between the contractor and journeyman, been limited as to size. The height to which the boards are elevated is calculated to afford ease of movement. The board material most popular is waterproof plywood of ¾″ thickness. This kind of board has a smooth top that wears well and facilitates the work.

Scaffolds are erected to the proper height for the work to be done. As a rule of thumb, the distance from the plank to ceiling should be 6 ft. and the distance from the edge of the nearest plank to the wall should be no more than 14″. Best results will be obtained when the equipment is in good condition and all of the materials are at hand. (See Chapter 1 for more information on scaffolds.)

Water Level. A water level (Fig. 2-62) is used wherever extreme accuracy is required. It is very useful on such work as ornamental plastering. It consists of a good quality tubing, usually made of rubber, resembling a garden hose with pyrex glass tubes attached at both ends. The

Fig. 2-61. Contractor's tools needed to mix and supply mortar:

(A) putty screen
(B) metal scaffold jack
(C) putty board
(D) water hose
(E) board stand
(F) scaffold horse
(G) wheelbarrow
(H) hod
(I) floor scraper

(J) mixing hoe
(K) no. 2 shovel
(L) mortar box
(M) water barrel
(N) beam strips
(O) cornice strips
(P) 14-quart pail
(Q) scaffold planks

Fig. 2-62. Water level.

tubing usually has a ⅜″ or ½″ inside diameter and is available in 50′, 75′ and 100′ lengths.

Stilts. In many parts of the country stilts are being used by the plasterer in place of low scaffolding. There are basically two types.

The first type, unadjustable stilts, are available in heights of 16″, 18″ and 20″. They are made of sturdy, lightweight aluminum alloys, and are designed with a platform on which the plasterer stands. The platform has an extension with a strap which is used to tie the stilts to the plasterer's legs just below the knees. There is usually a strap adapter which is attached to the standing platform. This is used to tie the feet securely to the platform. Some plasterers bolt a pair of shoes to the platform. These stilts use mechanical ankles, which

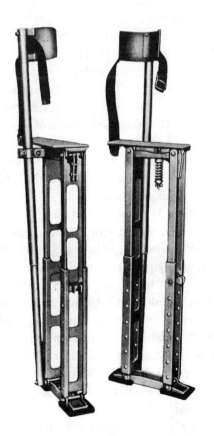
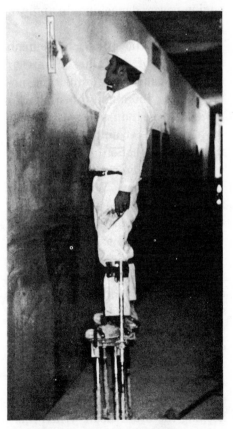

Fig. 2-63. Adjustable stilts. (Goldblatt Tool Co.)

75

actually walk as the wearer walks, relax as he relaxes. They have a special spring attached under the platform to help leg muscles offset fatigue.

The second type is the adjustable stilts. See Fig. 2-63. These are made the same as the unadjustable stilts except they can be adjusted from heights of 18" to 30" at 2" intervals or steps.

Machine Equipment

Keeping step with construction techniques, the plastering industry has developed various electrical or engine-powered machines to apply or mix materials.

Plaster Mortar Mixer. The main advantage of the plaster mortar mixer is the fact that it mixes the material more thoroughly in a shorter period of time than would be possible by hand. For certain types of material the blade action of the mixer is essential to the production of the right consistency. Where a manufacturer requires a certain length of time for his material to be mixed, this type of machine will be required.

Most mortar mixers have rotating paddles enclosed in a drum-shaped container. The paddles are made of steel with a leading edge and an extension made of hard rubber which can be replaced when worn. The purpose of the rubber blades is to allow the paddles to compress against the sides of the drum while turning, thus keeping the drum clean. When mixing has been completed, the drum can be tilted so that the mortar can be poured easily into a wheelbarrow or mortar box or hopper of a pumping machine. Fig. 2-64 shows the plaster mortar mixer with the drum in a tilted position. The power unit can be either an electric motor or a gasoline engine. Mixer sizes are measured by cubic feet. Sizes range from 3½ to 10 cubic feet. Most of these mixers are mounted on an axle with two wheels and a telescopic tongue, making it adaptable for hauling to the job like a trailer.

Lime Mixers and Mud Mixers. Mixers are similar to kitchen mixers in construction, although they are larger. They come in sizes of "one bag" or "three bags". They are constructed of a drum which holds the material to be mixed and a steel shaft to which several impellers are fastened. See Fig. 2-65, left. The impellers for lime are shaped so that the material will be drawn down into the water. The mud mixer's impellers (see Fig. 2-65, right) are shaped so as to blend the mixes without draw-

Fig. 2-64. Plaster mortar mixer. (Gilson Brothers Co.)

Fig. 2-65. Mixers: left, lime mixer; right, mud mixer. (Goldblatt Tool Co.)

ing too much air into the mix.

The mixer has an electric motor which clamps onto the top of the drum. The impeller shaft is connected to the motor when ready to be used.

Plastering Machine. The plastering machine (Fig. 2-66) will handle all types of material and its volume output is larger. The material is forced from the hopper through the hose by horizontal or vertical pumps. It can be either electric or gasoline powered.

Glitter Guns. Glitter guns are

Fig. 2-66. Plastering machine. (Essick Div. of A-T-O Inc.)

used to apply glitter (dry colored particles) to wet acoustic plaster. There are two types available. One is a glitter gun that can be used with any air compressor that can produce 15 to 30 lbs. pressure. It is constructed like a paint sprayer. See Fig. 2-67. It comes with a 10″ curved nozzle to apply glitter to the ceiling and 2″ straight nozzle for walls. An

Fig. 2-67. Glitter gun.

adjustable air valve on the gun gives fine control of spray.

The second type of hand glitter gun is also designed to apply glitter to walls and ceilings and needs no compressor. It is manually driven by turning a crank. This revolves a wheel that throws the dry glitter out by centrifugal force.

Texturing Machines. Texturing machines can be classified into two types: one type is *gravity fed*, the other has the *material fed by means of a pump.*

Basically the principle of texturing is the same for both. The material is forced through an orifice by means of air pressure. There are usually two sizes of air nozzles: one is $\frac{1}{16}''$, the other $\frac{9}{64}''$. The sizes of the orifices are usually $\frac{7}{32}''$, $\frac{1}{4}''$, $\frac{5}{16}''$, $\frac{3}{8}''$ and $\frac{7}{16}''$. The air pressure is controlled by a valve on the compressor and is usually registered in PSI (pounds per square inch) by a visible pressure gauge at the compressor. The sizes of the air hose are usually $\frac{1}{4}''$ or $\frac{1}{2}''$.

The *gravity texture machine* has a hopper connected to a pistol grip, trigger control unit. See Fig. 2-68. The hopper is either plastic or aluminum. It holds about $1\frac{3}{4}$ gallons of material and is designed in such a way that the material will not spill when texturing ceilings. It also has an air compressor and hose and a motor to run the compressor.

The pump fed texturing machine.

Fig. 2-68. Gravity fed texturing machine. (Goldblatt Tool Co.)

This machine consists of a hopper which will hold about 35 gallons of material, a rotor/stator pump, 5 h.p. engine, air compressor and hose, material hose (usually $\frac{3}{4}''$) and a pattern nozzle. The pattern nozzle is a lightweight metal pipe to which the material supply hose and the air hose are attached. This is what the plasterer holds and directs toward the surface to be textured. Fig. 2-69 illustrates two varieties of the pump-fed texturing machine.

Along with this type, there are machines available which have a mixer combined with pump.

Fig. 2-69. Pump-fed texturing machines: top, machine supplying both material and air, and requiring an air compressor; bottom, pattern pump, which pumps only the material and is more mobile than other types. (Goldblatt Tool Co.)

The plasterer uses both the gravity and the pump-fed texturing machines for applying the final texture in sand finishes, acoustic finishes and stucco finishes. The pump type is also used for certain types of fireproofing.

Catalyst Additive Gun. The catalyst additive gun is used for special veneers where a catalyst is required. It is constructed of a hopper for the material and a tank to hold the catalyst. There is a material supply hose, a catalyst supply hose, and an air hose connected to a special type nozzle. See Fig. 2-70. The catalyst is added to the material at the nozzle just before it is blown through the orifice onto the wall.

Note: A catalyst is used to accelerate the set of the material. The whole concept of veneer plaster is the thin coat of plaster that is applied; the full strength of this material is in the quickness that the material sets hard. Some types require a catalyst that is sprayed on the base just before applying the thin coat of plaster. This machine mixes the catalyst with the material as it leaves the nozzle. The application varies with the manufacturer.

Fireproofing Machine. The fireproofing machine is a machine used by the plasterer to spray special insulating fibers directly onto steel beams and metal decking. Fig. 2-71 illustrates such a machine. This machine breaks the dry material up with rotating fingers and feeds it into a hose by means of a rotor-stator. The material is then blown through the hose. Water is added by an atomizing nozzle (Fig. 2-72) at the end of the hose. Different nozzles are used, depending upon the type of material being used. The water is supplied by a regular water hose connected to a

Fig. 2-70. Catalyst additive gun. (National Gypsum Co.)

Fig. 2-71. Fireproofing machine. (The Universal Insulating Machine Co.)

Fig. 2-72. Atomizing nozzle. (The Universal Insulating Machine Co.)

pump or water booster (Fig. 2-73). This is used to provide a constant volume of water at a constant pressure, giving a smoother stream. The nozzles are precision made with very fine holes around the outside ring which form a fine water ring around the fibers as they are blown out of the nozzle.

In multi-storied fireproofing jobs, savings often can be made by leaving the fiber spray machine in one location rather than moving it up as the job progresses. A high-pressure

Fig. 2-73. Water booster. (The Universal Insulating Machine Co.)

blower supplies the air power to make this possible. Usually fiber hose or aluminum tubing can be run up the elevator shaft or the stairwell from the machine to the floor being sprayed.

Aggregate Gun. The aggregate gun is used to apply aggregates to a surface. It consists of a hopper attached to a short round pipe. See Fig. 2-74. This pipe is connected to a

blower which is turned by an electric motor. This machine operates by the aggregates feeding into the round pipe by gravity and then being forced out the muzzle end by air from the blower.

Aggregate Seeder Gun. The aggregate seeder gun is similar to the aggregate gun and is used to apply aggregates to a surface. It consists of a hopper attached to an electric

Fig. 2-74. Aggregate gun (Cement Enamel Development, Inc.)

motor which has an oscillating gear attachment. The machine operates by having the aggregates feed into the pan by gravity and the sponge oscillating enough to push the aggregates onto the surface.

Foldstir Mixer. The foldstir is used to mix drywall mud, acoustic and exposed aggregate materials. It lifts, folds and blends the materials so they are thoroughly lump-free. It has a ½″ diameter shaft 30″ long.

The paddle is 9¾" wide and 6⅜" deep and is permanently fastened to the shaft. See Fig. 2-75. The chuck end will fit any ½" or larger slow speed drill (about 500 RPM shaft speed). The foldstir can be used in a 5 gallon to a 55 gallon round container.

Jiffler Mixers. Jiffler mixers are used to mix drywall mud and exposed aggregate materials. The jiffler mixer is constructed of scientifically designed 5-fin dispersal wheels, called the upper wheel and the lower wheel. This pair of wheels is attached to a ⅜" x 24" long shaft by a screw to allow for replacement. See Fig. 2-76. The upper wheel cuts the material downward and the lower wheel cuts the material upward, thus creating a counter flow to disperse lumps quickly. The jiffler mixers can be used with a ⅜" or ½" drill (450 to 1200 RPM).

Jiffy Mud and Resin Mixers. This type mixer is used by the plasterer to mix drywall joint cement or epoxies; used properly, material won't splash out of the container. It will produce a smooth, lump-free, bubble-free mix. These paddles should be used with a ¼" or ½" drill with recommended speed of 450 to 650 rpm.

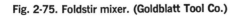

Fig. 2-75. Foldstir mixer. (Goldblatt Tool Co.)   Fig. 2-76. Jiffler mixers (Goldblatt Tool Co.)

Fig. 2-77. Jiffy mud and resin mixers. (Goldblatt Tool Co.)

The two vertical blades keep the mixing action inside the cage of the tool. See Fig. 2-77. The "sealed ring" design prevents puncturing or damaging of the mixing container.

Checking on Your Knowledge

The following questions give you the opportunity to check up on yourself. If you have read the chapter carefully, you should be able to answer the questions. If you have any difficulty, read the chapter over once more so that you have the information well in mind before you go on with your reading.

DO YOU KNOW

1. What are the two basic tools of the plasterer?
2. Name three types of trowels used by a plasterer?
3. What is a rod and what is it used for?
4. What is the difference between a float and an angle float?
5. What is a featheredge?
6. What is a slicker?
7. How many types of brushes are used by a plasterer?
8. What is a plumb bob?
9. Name two kinds of levels used by a plasterer.
10. What is a scarifier?
11. What are some tools designed for dry-wall?
12. What tools are usually furnished by the contractor?
13. Name some of the equipment used to mix plaster or related materials.
14. How does the fireproofing machine differ from the plastering machine?
15. What is an aggregate gun?

Materials of the Plastering Trade

This chapter covers the materials and explains their uses. Some of the history of them is covered but the emphasis is on their present use. The plasterer is judged on the performance of the finished job; therefore he must know the materials that can be used to produce the desired results. He must also know how the materials are best handled and the problems that must be avoided. His skill is wasted if the materials do not provide satisfaction on the job.

Any job is only as good as the materials and talent that go into using it. Talent, to a large extent, is knowledge intelligently applied. In plastering, as in most trades, a basic knowledge of the materials of the trade is essential. A plasterer would certainly be handicapped if he did not know the advantages and disadvantages of the materials he uses every day.

The accomplished plasterer realizes the importance placed upon understanding the nature and usefulness of materials. They are the key to superior work. It is essential that the journeyman stay abreast of new materials that are being added to his trade. The apprentice cannot advance in his career unless he knows how to work with the various materials.

Cementing Materials

The three main cementing materials used in plaster mortars are gypsum, lime, and Portland cement. Gypsum and lime have been used since ancient times. Portland cement is a relatively recent discovery. Gypsum and lime are made from natural rocks which are processed into several forms to produce the various kinds of cementing materials desired. Portland cement is a manufactured material that contains lime, gypsum, and other substances. Portland cement can be manufactured to meet a number of specific uses by varying the ingredients that go into its manufacture.

Gypsum

This is one of the common substances used for plastering. Gypsum occurs as a white or yellowish rock that is soft enough to be scratched with a fingernail. It is found in almost every country. In some deposits the gypsum may be dug from close to the surface while in others deep mines are needed. Fig. 3-1 shows the mining of gypsum.

Gypsum is calcium sulphate ($CaSO_4$ $2H_2O$). Some of the natural deposits are so pure that little processing is required. It should be noted that about 20 percent of gypsum is

Fig. 3-1. Mining gypsum rock. (United States Gypsum Co.)

water (H_2O) by weight or about 50 percent by volume. The water is contained in the crystals and is part of their chemical structure. This *water of crystallization* does not change until it reaches a temperature of 212° F, when it changes into water vapor and leaves the gypsum. It is this water of crystallization that explains why gypsum is so fire resistant. When the plaster becomes intensely hot the water becomes steam and retards the spread of flame.

Gypsum rock must be processed before it becomes suitable for plastering. The mined rock is first passed through a dryer which removes the surface moisture. Then a hammer mill pulverizes the rock and delivers it to vibrating screens which remove large pieces for further crushing. The pulverized raw rock is known as *land plaster* from its use in farming as a soil corrective.

After being crushed to the desired fineness the rock is sent to the calcining process. There are two methods of calcining the rock, the kettle method and the kiln method. In the *kettle method* the rock is put into large kettles and heated until it starts to boil. After enough water has gone out of the gypsum the boiling ceases. The calciner then dumps the kettles into cooling pits. At this point about ¾ of the water of crystallization has been removed and the processed gypsum is called *plaster of Paris*.

In the *kiln method* the crushed gypsum rock is fed into the upper end of a rotary kiln. This is a long revolving cylinder that is slightly tilted. As the kiln revolves the rock gradually slides toward the lower end. Heat from pulverized coal, oil, or gas is generated at the lower end so the rock is gradually heated as it moves along the kiln. By the time the gypsum is discharged at the lower end of the kiln it has been changed into plaster of Paris by partial removal of the water of crystallization. The rotary kiln process is a continuous one where raw rock is constantly added at one end while the calcined gypsum is being removed at the other. In the calcining process the gypsum which was chemically CaSO4 $2H_2O$ becomes hemi-hydrate of calcium sulphate which is $CaSO_4$ $\frac{1}{2}H_2O$.

After calcining, the plaster of Paris is cooled and stored in bins until it is sent to mills for grinding to the fineness needed. Some is sent through another calcining treatment to remove even more of the water of crystallization. This produces an even better product needed for certain types of plaster.

After being pulverized by the grinding mills, the plaster goes to the mixing department where it is tested for setting time, hardness, workability, and other characteristics. Then various materials are added to the plaster and the proportioned materials are carefully blended and bagged. All processing in the mill is scientifically controlled and handled by the latest types of conveyors, kilns, mills, and other materials-handling machinery. This results in a uniform product that can be depended on to meet the specifications set for it.

When calcined gypsum or plaster of Paris is mixed with water the reverse reaction occurs; the plaster takes up water and "sets" into the rock-like condition of its original state. It changes back to $CaSO_4$ $2H_2O$. This binds any material mixed with it into the hardened mass. Fig. 3-2 illustrates the cycle.

Gypsum Products

Unfibered gypsum is the neat product. (Neat plaster has no aggregate, or filler, added.) This sometimes is called compound or cement plaster. Water and aggregate must be added to it. It is used for scratch, brown, and leveling coats. There are three types: *regular* for use with sand aggregate and hand application; *LW* for use with lightweight aggregate and hand application; and *machine application* for use with either sand or lightweight aggregate.

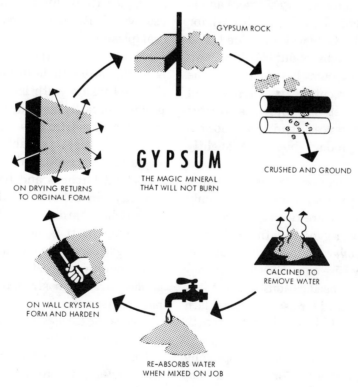

GYPSUM ROCK

GYPSUM

THE MAGIC MINERAL
THAT WILL NOT BURN

CRUSHED AND GROUND

CALCINED TO
REMOVE WATER

RE-ABSORBS WATER
WHEN MIXED ON JOB

ON WALL CRYSTALS
FORM AND HARDEN

ON DRYING RETURNS
TO ORGINAL FORM

Fig. 3-2. Cycle of gypsum. (United States Gypsum Co.)

Fibered gypsum is a neat gypsum with cattle hair or sisal fiber added. It is also called compound or cement plaster. The fibers are added to the plaster so it will hold together better and to reduce dropping. When used with metal lath the fibers help prevent too much plaster being pushed through the openings. It is not recommended for machine application, as the fibers hinder the flow through the nozzle.

Wood fibered gypsum is a neat gypsum with selected wood fibers added at the mill. The wood fiber plaster can be applied to all standard lath and masonry surfaces and is recommended as a scratch coat for metal lath. Only water is added. The plaster weighs about ¼ less than sanded gypsum basecoat and has over 50 percent greater strength and hardness. It also provides greater fire resistance than the normal sanded product but at a slightly higher cost.

Perlite gypsum plaster is a factory prepared plaster that requires only water. There are two types: *regular*, for use over gypsum lath (rocklath); and *masonry*, for use on high-suction masonry base only. Check the manufacturer's specifications to see if it is recommended for use over metal lath when the finish coat is to be smooth troweled.

Bond coat gypsum is a specially prepared gypsum for use as a base coat on monolithic concrete surfaces which are smooth and dense and have insufficient suction and key for ordinary plaster bases. Only water is added to this plaster. It adheres well because it does not expand during setting and therefore does not disturb the knitting of the gypsum crystals to the concrete.

Gauging plaster is a special material composed of screened particles of hand-picked gypsum processed to regulate its set at definite time intervals. Three settings are available: slow set, medium set, and quick set. This is mixed with slaked lime to make finishing plaster coat. It is also available with perlite fines added for use over lightweight aggregate base coat plaster.

"Structo-gauge" (gauging plaster) is a high-strength finishing plaster designed for use with lime putty to produce an easily applied finish of extreme hardness. Quick-set and slow-set formulas are available. This is suggested for walls and ceilings in household kitchens and bathrooms, hallways in schools, and similar places where hardness and durability are required. It is not to be used where excessive or continued moisture conditions exist. It must be applied over high strength gypsum base coats.

Keenes or Keenes cement is a gypsum for special uses. It is doubly calcined, and almost all the water from the gypsum rock has been removed. There are two types: regular (slow setting); and quick setting. Keenes is a high-strength, white plaster used with slaked lime and is the only gypsum plaster that can be *retempered.* (Retempered means to remix material that has stiffened or started to set. Water is added, if needed, and the material is remixed to bring it to the proper working plasticity.) Keenes is used most commonly with lime and sand for a float or sprayed finish.

Molding plaster is a selection of calcined gypsum ground very fine. It has a very fine powdery form which brings out details in ornamental trim, cornices, and cast work. For running cornices slaked lime is added for plasticity and as a lubricant for the template.

Casting plaster is even finer than moulding plaster, giving smoother castings used for plaques and art statuary. It is highly plastic and has great surface hardness and strength. Only water is needed for application.

"*Hydrocal*" is a special molding plaster that develops two to three times the strength and hardness of regular moulding plaster.

"*Hydrostone*" is another special plaster that develops six to eight times the strength of regular plaster. It has compressive strength of 11,000 psi, and is used for industrial casting and pattern making.

Lime

Lime, calcium oxide (CaO) or hy-drated calcium oxide ($Ca(OH)_2$), is one of the most common minerals in the world. It is very active chemically and combines easily with other elements. In some forms it is extremely caustic and in all forms it is highly alkaline. While the mineral lime is present in many combinations, only the combination with carbon dioxide which forms limestone (calcium carbonate, $CaCO_3$) is important as a source of plastering lime.

Lime for plastering or other use is obtained by quarrying limestone, which is a rock made up mostly of calcium carbonate. In order to change limestone into a lime suitable for plastering, it is crushed, screened, selected, washed, and graded. Fig. 3-3 shows how lime is removed from an open pit quarry. The selected stone is then placed in kilns where it is heated up to 2,500°F. This process drives off the moisture and also removes certain gases from the stone, especially carbon dioxide. This is a calcining process similar to that used for gypsum. It is also referred to as *lime burning.* Fig. 3-4 shows rotary kilns in which lime is burned.

The product of the calcining or burning is *quicklime,* chemically calcium oxide (CaO). Calcining lime is a process that is nearly as old as the use of fire. It is likely that it was discovered by ancient man when he built a fire on limestone rock.

Quicklime is a very caustic material. When it comes in contact with

Fig. 3-3. Dolomitic lime plant and quarry. (Ohio Lime Co.)

Fig. 3-4. Rotary kilns in which lime is burned. (Marblehead Lime Co.)

water a violent reaction occurs that is hot enough to boil the water. It can also cause severe burns on the skin. In the past quicklime was the material the plasterer received on the job and he had to use it to make plaster. The first step in using quicklime is to slake or hydrate it. This is done by adding enough water to it so that the oxide becomes a hydroxide ($Ca(OH)_2$). During the slaking process the caustic quicklime becomes very hot and is hazardous to work with.

Today the lime manufacturers slake the lime as part of the process of producing lime for mortar. The slaking is done in large tanks where water is added to convert the quicklime to *hydrated lime* without saturating it with water. The hydrated lime is a dry powder with just enough water added to supply the chemical reaction. Hydration is usually a continuous process and is done in equipment similar to that used in calcining. After the hydrating process the lime is pulverized and bagged. Hydrated lime when received by the plasterer still requires soaking with water.

The properties of the lime depend largely on the limestone from which it is produced. Calcitic limestone is about 85 percent calcium carbonate and makes high calcium lime. Dolomitic limestone consists of about 40 percent magnesium carbonate and 60 percent calcium carbonate. The magnesium content produces a unique property in dolomitic limes which improves the workability or spread of the finish lime.

When regular lime of the dolomitic type was used without proper soaking, failures sometimes occurred. To avoid this, a type of lime called *autoclave, double hydrated,* or *pressure lime* was developed. In preparing this, autoclaves or pressure kettles are used to hydrate the magnesium in the lime up to 92 percent of its content.

In standard lime not all the magnesium is hydrated either in its processing or when it is soaked on the job. If it is used as a finish coat of more than usual thickness, there is the possibility that the unhydrated magnesium will hydrate in a few years because of the gradual absorption of moisture from the air. As the magnesium hydrates it expands and bulges are produced that separate the finish coat from the base coat. Autoclave lime overcomes this trouble and the time for soaking is eliminated.

Lime hardens through a process of recarbonation. This is a slow process in which the hydrated lime absorbs carbon dioxide (CO_2) from the air to form calcium carbonate ($CaCO_3$). Consequently the strength of lime plaster gradually increases over a long period of time. Years ago lime combined with sand and hair was used for all plaster coats. Today this is no longer true and gypsum has largely replaced lime. The slow hardening of lime and the possibility of shrinkage are the reasons for the change to gypsum. Lime is presently being used only in finish coats.

Lime Products

There are several kinds of lime on the market. The one used by the plasterer is the finishing lime, which is the best grade. Mason's lime used in making mortars for masonry is not as select a quality. Limes for many special purposes are being marketed, but they do not make a satisfactory putty for plaster use.

Hydrated or regular finish lime was hydrated during manufacturing, so that only water and a soaking time of 12 to 16 hours, or overnight, is required. It is used for trowel or texture finishes.

Autoclave lime does not require soaking because it is double-hydrated by the manufacturer. It is mixed with water and used immediately. It is applied by trowel or textured finishes and can be added to Portland cement plasters to improve their spreadability.

Fibered lime is an autoclave lime with reinforcing fibers added. It is added to cement for all types of lime-cement plasters and stucco work. The fibers help hold the plaster in place during application.

Lime-colored sand finishes are basically an autoclave lime with silica sand and colors mixed in at the mill. Only water is required to produce textured finishes in colors.

Mason's lime is almost as fine and pure as plastering finish limes. However, it is prepared to meet the requirements of bricklaying and is not recommended for plastering use.

Portland Cement

In 1824, Joseph Aspdin, a plasterer and bricklayer of Leeds, England, was experimenting to produce a mortar that would harden under water. He achieved this by burning limestone and clay together in his kitchen stove. The gray powder was called Portland cement because of its resemblance to stone quarried on the Isle of Portland.

Since its invention, Portland cement has replaced almost all other cements, both natural and artificial. This is due to low cost and superiority. Portland cement has become a standardized product of high quality and uniformity, regardless of where it is made. It is manufactured all over the United States and Canada and in most other countries.

Modern cements are manufactured in the same basic manner that Joseph Aspdin used. Some form of lime (which may be obtained from limestone, marble, chalk, slag, oyster shells or coral) is mixed with certain kinds of clay; this is ground and then calcined at temperatures around 2,700°F. The ingredients lime, silica, aluminum oxide, and iron oxide combine chemically and form lumps or clinkers. The clinkers are pulverized and a small amount of gypsum is added to control the setting properties. Fig. 3-5 illustrates the process of manufacturing Portland cement.

This cement is a combination of calcium silicates and aluminates. Water starts a complex reaction yielding crystalline substances in an amorphous gel which sets as a hard mass.

Cement hardens as a result of hydration of the materials in it. While setting takes place in a short time, the cement requires several days for the hydration to become complete. As a result, cement continues to increase in hardness for about a month and in some cases for years.

Cement Products

Regular Portland cement is the most commonly used by the plas-

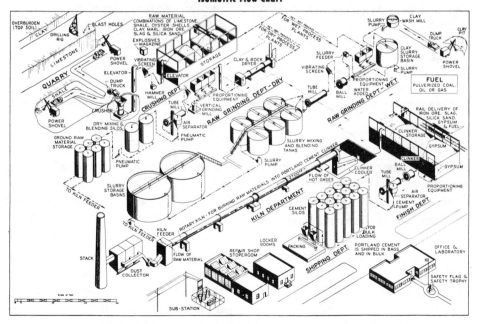

MANUFACTURE OF PORTLAND CEMENT
Isometric Flow Chart

Fig. 3-5. Drawing of typical cement manufacturing.

terer. It is used for scratch, brown, and finish coats for exterior work, or where dampness is present, or where an extreme hardness is required. Aggregate and water must be added.

High early-strength cement is a specially processed cement that reaches high strength in about half the time of regular cement. This is important where time is a factor, as succeeding coats can be applied sooner. This develops the same strength in three days that the regular cement does in seven. Aggregates and water must be added.

Air-entrained cement is a special formula that holds air in the mix. This improves workability, durability, and resistance to efflorescence (the white crystals that may form on the surface). Aggregates and water must be added to make mortar. Do not use special additives with this ce-

ment unless they have been approved by the additive manufacturer.

White cement is a regular type cement made of selected materials that are white instead of the more common gray ones. It is especially useful where colorings or exposed aggregates are used in the finish coats. Water and aggregate must be added.

Mortar cement is used primarily by the bricklayer. It is a regular cement that has a portion of hydrated lime added in manufacture. This is sometimes used as a plasticising ingredient in place of lime in Portland cement base coat mortars.

Prepared exterior finishes are convenient mixes of white cement, silica sand, and colors. They are bagged by the manufacturer and are used for textured finishes over Portland cement bases. They require only the addition of water.

Aggregates

Most cementing materials if used without aggregate would shrink to such an extent that cracks would develop. To avoid this and to make the mortar more economical, some kind of aggregate is always mixed with the cementing material. The amount and kind of aggregate used affects the mortar that is produced and the plastering skills and methods used to apply it.

Aggregate for plastering must be clean, sharp, strong, and of proper gradation. Materials lacking these characteristics may not make satisfactory mortars, as all of these charteristics have an influence on the final result.

Two types of aggregate are widely used today. One consists of various kinds of sand and the other of the more recent lightweight materials.

Sands

Several kinds of sand are used in plastering. Most sands are obtained from natural deposits such as beaches, river bottoms, lakes, and banks. Some is obtained by crushing rock to the desired fineness. In many plants sand is washed and screened to obtain the cleanliness or size wanted. Plastering sand must be small enough to pass through a No. 4 (4 meshes an inch) sieve and coarse enough so that less than 5 percent will pass through a No. 100 sieve.

Cleanliness is important in plastering sand, as the presence of clay or other material may cause poor mortar, streaking, and unevenness of the surface. Sand can be tested for cleanliness by placing a representative handful in a jar and filling it with clean water. The jar is shaken vigorously to mix the sand and water and then allowed to stand. The sand will settle to the bottom while silt, clay, and other material will stay

DIRTY SAND   CLEAN SAND

WATER

SAND

Fig. 3-6. Bottle test for dirty sand.

on top or in the water. While this is a simple test, it can show the presence of material that would make the sand unfit for use in plastering. See Fig. 3-6.

Moisture in Sand. Most sand contains varying amounts of moisture (water), which must be considered when mixing plaster. It is essential to remember that the amount of water used in making plaster mortar is important. If a great deal of water is in the sand to be used, it means that less water should be added to make the mixture sufficiently pliable.

The moisture content of sand has been standardized to some extent as follows:

Dry Sand. Dry sand, which flows freely, is seldom available for plasterwork. It has no appreciable moisture content.

Damp Sand. Damp sand feels slightly damp to the touch but leaves very little moisture on the hands. Such sand usually contains about ¼ gallon of water per cubic foot.

Wet Sand. Wet sand, which is the kind most usually available, feels wet and leaves a little moisture on the hands after being handled. Such sand contains about ½ gallon of moisture per cubic foot.

Very Wet Sand. This sand is dripping wet and leaves more moisture on the hands than wet sand. Very wet sand contains about ¾ gallon of moisture per cubic foot. If the sand is composed mostly of very fine

particles, it may contain as much as 1¼ gallons per cubic foot.

To learn how to feel sand to determine its moisture content a simple experiment can be performed.

Spread about ⅔ of a sack of sand in a thin layer on paper, canvas, or a dry floor inside a building and let it dry. It should be stirred now and then to make sure all surface moisture disappears. When the sand feels dry and flows freely, it is ready for use.

Measure out 3 gallons of the dry sand, place one gallon in each of three pans. Then measure 5 oz. of water to one pan, 12 oz to the second, and 20 oz to the third. Mix each pan thoroughly. (Note: 1 cup equals 8 oz; 1 pint equals 16 oz.) The one with 5 oz. of water is damp sand, the second is wet, and the third is very wet. This will teach you the appearance and feel of the moisture content of sand.

Normally, on the job, the plasterer takes a handful of sand and squeezes it. If water emerges through his fingers, it is *very wet*. If he squeezes the sand, releases it and a large amount of sand clings to his hand, this is *wet* sand. A small amount of sand remaining on the hand is *damp* sand. *Dry* sand squeezed in the hand will sift through.

Silica sand or white sand is a sand available for plastering. This sand comes in various sizes, each carefully screened for uniformity. This special sand is used for texture finishes, exposed aggregate mixes, and some veneer plaster bases.

Mason's sand is similar to plaster sand except that the grain size is smaller. It is used primarily by the mason or bricklayer but may be used in some veneer type plaster mixes.

Lightweight Aggregates

Lightweight aggregate materials are important in plastering. They weigh only 1/10 as much as the sand they replace and they produce a better insulating and sound-absorbing quality, are cleaner and of more uniform plasticity.

Lightweight aggregates are made from natural deposits found in the West and Southwest United States. The deposits are dug and given a processing which includes a heat treatment. This treatment causes water trapped in the porous rock to expand into steam and explode the rock into many tiny hollow cells. Grading and sizing for market are part of the processing.

The main quality of lightweight aggregate is lightness. This reduces the weight added to a building by the plastering, and lighter supporting framework can be used. Much less weight is handled by the plasterer, a fact that he appreciates at the end of a working day.

Lightweight aggregates have superior qualities for insulation and sound absorption. Their insulating qualities add greatly to the fire resistance of plaster made from them. As the aggregates have been heated to about 2,000°F in processing, they are not affected by ordinary heat or cold. Lightweight aggregate plaster is resilient and resists cracking.

Lightweight aggregates are uniformly graded, cleaned, and packaged. Proportioning is simplified as well as handling and storing.

Three types of lightweight aggregates are in common use. They are vermiculite, perlite, and pumice. The first two are available in all parts of the country. Pumice is somewhat heavier and is not as widely distributed. Several other kinds of lightweight materials such as expanded shale, processed slag, and other by-products from industries are used as aggregates in local areas.

Vermiculite. This consists of silica, magnesium, aluminum oxides and some other minerals. Combined water is about 5 to 9 percent by weight. The ore resembles mica and is made up of very thin layers (or laminations), about a million to an inch. There is a tiny amount of moisture between each layer. When the ore is heated to about 2,000°F the moisture changes to steam with explosive force, which causes the layers to separate and move apart. The moisture content then drops to about one percent of the weight while the flakes of ore have expanded about 12 times. The expanded vermiculite contains thousands of dead air spaces that act as insulators. The shiny surfaces of the layers also act as efficient reflectors for repelling heat.

When the vermiculite ore is heated it expands in a way that causes the pieces to elongate in a peculiar worm-like way which accounts for the name, meaning "worm-like."

The expanded ore is graded by size and weight and then packaged for various uses. In building, vermiculite is used for general plaster, acoustic plaster, insulating plaster, and lightweight concrete. It is used loose as insulation in walls and ceilings, and for soil modification in gardening.

The vermiculite plastering aggregate weighs about 7 to 12 lbs a cubic foot compared with 100 lbs for sand. It has 3½ times more insulating value than sand. When used in a gypsum mortar the insulating ability makes the plaster much more fire resistant, so lightweight plasters are often used for fireproofing in steel construction. While vermiculite will melt at about 2,500°F, it will not burn.

Structural steel tends to bend under high heat and can cause extreme damage to burning buildings. When the steel is encased in concrete or plaster it is considered to be fireproof. Experiments show that as the thickness of lightweight aggregate plaster increases so does the fire-resisting ability. A ceiling with 1 inch thickness of vermiculite (or perlite) plaster on metal lath gives a 4-hour fire-resisting ability. Table 3-1 gives some of the properties of vermiculite, along with fire resistance properties.

Vermiculite plaster aggregate is used in place of sand when making mortars. It comes in bags of 4 cubic feet that weigh about 25 lbs.

Vermiculite fines are added to lime finishes when used over lightweight brown coats.

Perlite. This is a mineral from the western part of the United States where ancient volcanoes threw off lava. Perlite ore is glassy and has many rounded granules that have a resemblance to pearls. See Fig. 3-7. Perlite consists of silica and alumina compounds which contain from 4 to 6 percent water. Its properties are similar to those of vermiculite.

The perlite ore is heated rapidly to more than 2,000°, which forces the water content to expand into steam, causing the perlite to become sponge-like in texture. This expanded perlite is literally a mass of air cells. After processing, the perlite is graded for size and then bagged. It is used for plaster aggregate, concrete aggregate, loose insulation, in the same manner as vermiculite. Chemically, perlite is similar to pumice, but there is a difference in physical characteristics.

Perlite weighs from 7 to 12 lbs. per cubic foot and adds a minimum of weight to the building. This material

TABLE 3-1 VERMICULITE INSULATION & PLASTER COVERAGE

VERMICULITE FILL INSULATION	VERMICULITE PLASTER
One 4-cubic foot bag covers 26 sq. ft. 2" thick 17 sq. ft. 3" thick 14 sq. ft. 3 3/8" thick 9 sq. ft. 5 5/8" thick Based on joists spaced 16" on center.	6 bags Vermiculite plaster aggregate 9 bags Gypsum plaster Covers 100 sq. yds. 1/2" thick over gypsum lath. Materials required over masonry and metal lath bases will depend on the evenness of the wall and the depth of the key.
VERMICULITE PLASTER FINISH (over vermiculite base coat) 1 bag (2 cu. ft.) Vermiculite Finish Aggregate 2 bags (100# ea.) Unfibred gypsum plaster or gauging plaster 16 gals. (Approx.) Water Covers about 40 yds. This is a smooth finish made from gypsum and vermiculite aggregate — the same basic material used in the base coat.	**VERMICULITE ACOUSITICAL PLASTIC** 1 bag Vermiculite Acoustical 10 gals. Water (Approx.) Covers about 4 sq. yds. 1/2" thick. Sound Absorption — Vermiculite acoustic has a noise reduction coefficient of .65 for a 1/2" thickness; that is 65% of noise is absorbed.

MASONRY WALL INSULATION

BLOCK WALLS:	CAVITY WALLS:	1:16 MIX INSULATION
One 4 cu. ft. bag will fill cores of about 25 standard 8 in. block or 22 sq. ft. of wall area.	One 4 cu. ft. bag will fill an area of about 48 sq. ft. 1 in. thick.	1 bag Portland cement, 4 bags Vermiculite aggregate and 48 gals. water covers approx. 76 sq. ft.; 2 1/2" thick.

DESCRIPTION OF CONSTRUCTION			Fire Resistance
WOOD JOIST FLOORS	Metal Lath Ceiling	3/4" Vermiculite Plaster	1 3/4 hrs.
	Gypsum Lath Ceiling	(1) 1/2" Vermiculite Plaster on 3/8" plain gypsum lath. Twenty guage, 1" wire mesh nailed to joists through lath.	1 1/2 hrs.
		(2) 1/2" Vermiculite Plaster on 3/8" perforated gypsum lath.	1 hr.
WOOD STUD PARTITIONS	Metal Lath on Both Sides of Studs	3/4" Vermiculite Plaster	1 hr.
	Gypsum Lath on Both Sides of Studs	1/2" Vermiculite Plaster on 3/8" perforated lath. (Bearing)	1 1/4 hrs.
		(Non-Bearing)	1 1/2 hrs.

COURTESY, VERMICULITE INSTITUTE

Fig. 3-7. Crude, crushed and expanded perlite. (Perlite Institute, Inc.)

Aggregate Size	Passed by Screen	Retained on Screen	
A	$1/4''$	$3/32''$	 ◈
B	$1/2''$	$1/4''$	 ⬡
C	$7/8''$	$1/2''$	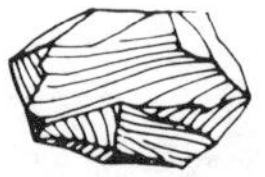 ⬣
D	$1^3/8''$	$7/8''$	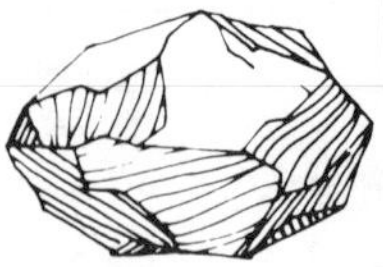 ⬣

Fig. 3-8. Aggregate sizing. (General Stone and Materials Corp.)

Note: The headings "passed by screen" and "retained on screen" are misleading but correct. The screen is placed outside the box. On a ¼" screen, the term "passed by screen" means the rock ¼" and larger is passed over the screen into the box while the ³⁄₃₂" and smaller sizes are retained or dropped through the screen. This method grades the stones by sizes gradually separating each size as it drops through or over the screen of that size.

TABLE 3-2   EXPOSED AGGREGATE. COLOR CLASSIFICATION AND GEOGRAPHICAL ORIGIN

State of Origin	Opaque White	Translucent White	Off White	Yellow Buff	Brown	Green	Pink	Red	Gray-Black
Colorado	Rocky Mountain Quartz	Colo. Milky Quartz Suprema Milky Quartz Devils Head Quartz Snowy Crystal Quartz		Platte Valley Granite	Iroquois Brown (Jasper)	Ice Green Quartz Suprema Green Quartz Suprema Gami Green Diablo Green Mountain Green	Agate Quartz Pink Feldspar Flamingo Quartz Suprema Royal Spar Royal Ruby Quartz Suprema Pink Granite	Idaho Springs Red Granite Suprema Sunset Red Granite	Suprema Lido Quartz Suprema Paxi-Gray Granite Suprema Light Gray Granite Black Obsidian
Georgia	Polar White Quartz								
Maryland				Parley's Buff (Quartzite) Pebble Jewels (Quartzite)		Cardiff Green (Serpentine)			
New Hampshire	Lyndeboro White Quartz	Raymond White Quartz	Topaz White Quartz T & M White Granite					Ruby Red Quartzite	T & M Black Granite
New Jersey						Royal Green (Epidote)			
New York						T & M. Green Granite		T & M. Red Granite	
North Carolina		Arctic Quartz Eldorado Quartz	Mt. Airy White Granite	Cherokee Quartz					
South Carolina			White Pebbles (Quartzite) Hartford Quartz	Amberlite Quartz Sunburst Quartz Marlboro Quartz Marlboro Buff Pebbles (Quartzite)	Brown Pebbles (Quartzite) Chocolate Pebbles (Quartzite)		Coral Gray Granite Superior Pink Granite		Smoky Quartz Twilight Granite
Texas		Crystal Quartz		Tan Quartzite Colortone Quartz		Moss Green Serpentine Antique Green Serpentine	Apache Quartz Town Mt. Granite	Baldwin Red Granite	Stewart Blue Granite Gillespie Black Granite
Virginia							Garnet Quartz		

COURTESY, GENERAL STONE AND MATERIAL CORP.

also has excellent insulating properties, fire resistance, and low water absorption. Perlite plaster has three times the insulating effectiveness of sand.

Pumice.   This is another material formed by volcanoes. It could be called foamed lava or volcanic glass froth. Pumice is filled with minute cavities produced when the pumice was exploded from the volcano and the water in it turned to steam, blowing holes or bubbles all through the molten rock.

In many cases pumice can be used with just crushing and grading. In order to become as light as other light-weight aggregates, however, pumice is heated to explode it into a still more porous structure. Even with this it is still heavier than the others —which makes its use largely local because of the higher cost of transporting it. Pumice is the oldest of the lightweight building materials and was used by the Romans in constructing many of their large works.

Architectural Aggregates. These are used by the plasterer for exposed aggregate finishes. "Exposed aggregate" refers to any surface that has the aggregate (actual stones) exposed for final appearance. These aggregates are made from quartz, granite, marble, and other rocks that are crushed and graded by size and color. Fig. 3-8 lists aggregate sizes.

The colors are those of the natural rock and cover a wide range. Table 3-2 gives a breakdown of natural colored aggregates and their geographical origin. Recently, colored glass aggregate has been used. This is available in three sizes from ¼″ up to ¾″ and in numerous vivid colors that are much brighter than the natural aggregates. Marblecrete is classified as one of the exposed aggregates. It is possible to get dry pre-mixed material in one bag for marblecrete.

Water

Water performs two functions in the plaster mix. One is to make the mixture workable by making it semi-liquid. The other is to dissolve the cementing material or binder so that it will act as an adhesive with the aggregate.

The amount of water to use in a plaster mix depends on the materials being used and the method of applying them. It has been found with gypsum plasters that when just enough water is used to complete hydration, maximum strength results.

Too much water may make the mortar too thin and there will be excessive dropping and waste when it

is applied. The correct amount of water is important when one considers the suction ability of bases. Some bases absorb more water than others and more water may be necessary in the mortars used on them. For example, a gypsum block base would require more water in the mix for easier application (because of its high suction) than gypsum lath.

Water should be clean and pure. Organic or other material in the water may affect the quality of the plaster by causing a change in the rate of set or by discoloring the mortar. Unclean water will often lessen the strength of the plaster.

Admixtures

A plaster mix contains an aggregate, a cementing material (binder), and water. These are the basic ingredients. Quite often they are not the only ones used but are supplemented by materials added during manufacture or at the job site. Any ingredient, other than the three basic ones, is called an *admixture* or *ad mix*.

These materials are added to change or modify the characteristics of the plaster mix in some way. For example, Portland cement mortars are "harsh" (not easily spread) and set slowly. Pure lime mortars tend to be weak. Pure gypsum mortars have fast and irregular setting times. Consequently, admixtures are often added to them to affect the setting time either by accelerating it or retarding it, depending on the effect desired. Other admixtures are used to control the strength and the color.

Setting Time Admixtures. These are substances that affect the setting time. "Setting time" was described earlier as the "period of time necessary for mortar to harden or to become rigid." Some mixtures harden very slowly and may need some additional ingredient that will speed up the setting time. This is called an *accelerator*. Other mixtures set so rapidly that the mortar cannot be applied fast enough to do a good job of smoothing or finishing. These need to be slowed down and admixtures that do this are called *retarders*.

Accelerators are added to cause faster setting of the mortars. Gypsum mortars generally set quickly and do not need to be speeded up. However, there are local conditions of water, weather, or aggregate that may slow down the setting time.

One of the most common and powerful accelerators is gypsum which has been allowed to set and then is ground for use. Set gypsum forms needle-like crystals that accelerate the early setting action when mixed with mortar. It is added to gypsum or lime mortars to speed up their setting time. Aluminum sulphate solution is a good accelerator for gypsum, or a commercial one can be purchased that carries directions for its use.

Many prepared cementing materials can be purchased with accelerators included. Keenes cement sets at a slow rate. Sulphate of potash is sometimes added by the manufacturer to adjust the setting time.

Portland cement mortars set slowly enough to warrant the use of an accelerator. Calcium chloride is sometimes used for this purpose. Caution should be used with this chemical as the chloride can corrode metal lath or electrical conduit, causing expansion and cracking of the plaster. Soda, dissolved in hot water, then cooled and screened into a cement mixture, will also hasten the set. The best method to speed up the setting of cement is to use heated water and aggregates.

Retarders are added to slow down the setting time of plaster mixtures that set too fast. Normally, pure gypsum sets too quickly for plastering. On the other hand, Portland cement sets so slowly that a retarder is seldom needed. A retarder is generally added during manufacture of gypsum plasters so they will not set in less than four hours.

The amount of retarder used will depend on various factors, such as the period of set fixed by the manufacturer, the type of retarder, and the plasterer's time needed to mix and apply the material.

In general, retarders slow down the absorption of water by the gypsum. It is the taking up of water in chemical combination that causes the hardening of gypsum plaster.

Many types of retarders are used by the trade. Certain sections of the country prefer definite types. Some of these are cream of tartar, gelatin, glue, ammonia, zinc sulphate, dextrine, gums, soap, starch, and animal or vegetable oil. The best and safest method is to use the product sold by the plaster manufacturer as a retarder.

It is sometimes useful to retard cement when doing exposed aggregate work. An oil solution called "Sonatard" is often used.

Strengthening Admixtures. These are used to increase the strength of mortars. For many years fiber or hair was used and in fact these are still being used in some places. The necessity for their use, however, has decreased because new plastering mediums and methods make the addition of fiber unnecessary.

There are three kinds of fiber which were generally used in the past and are occasionally used today: sisal fiber, animal hair, and wood fiber. Wood fiber is used in some plasters to add bulk and to aid in getting better coverage. Fibered plasters today are used more in special applications than for general strengthening.

Fiber and hair were used mostly when lime was the primary plastering material. Pure lime plaster sets very slowly, and fiber or hair was added to give the plaster more strength and cohesiveness.

Gypsum plasters are in general use today and these set hard in two to three hours. Consequently, they do not require fiber reinforcement. The only time when fiber or hair is necessary is when plaster is used as a scratch or base coat over metal or wood lath. The fiber or hair is added to the plaster mixture to prevent an excess from going through the keys of the lath and dropping on the floor behind. Fibered plaster formulated for this purpose can be purchased from manufacturers. Unfibered gypsum plaster is used for the brown and finish coats.

Fiber or hair is also used with some Portland cement work. Portland cement plaster is primarily used as exterior stucco and is applied in three coats. Fiber or hair is used as a reinforcing agent for the first or scratch coat applied on metal lath. The hair or fiber helps form plaster keys that hold the mortar in place. Asbestos is another fiber sometimes used to improve cohesion.

Water Repellent Additives. These are sometimes used to increase the water repellency in normal cement mixes. Ammonium stearate, aluminum stearate, or butyl stearate emulsions, sold under various trade names, are added to cement mortars for this purpose.

Color Admixtures. These have enjoyed various periods of vogue. Tinted plaster has again come into favor in the last few years because it saves the initial expense of painting new walls. The demand for low cost housing has increased the use of tinted plaster. Color is often used in the matrix (the mortar in which the aggregate is embedded) for exposed aggregates to increase the decorative value.

Lime has a tendency to bleach color. Consequently, colors used must be able to resist this action. Mineral colors are fast and will not fade. Other colors may fade quickly. Limeproof mineral pigments include amber or sienna (raw or burnt) for shades of brown, red iron oxide for shades of red, black iron oxide for gray through black, yellow iron oxide for yellow, and chromium oxide for green. These may be mixed to produce the shades desired.

There are several manufacturers that supply properly ground mineral colors for plaster and cement. These colors are usually packaged in five pound packages but are available in larger quantities. The plasterer adds them to the plaster or cement finishes either by weight or by volume, depending on the product.

Bonding Agents

"Paint-on" lath is a manufactured paint-type material that consists of a resin emulsion or synthetic latex. It is colored to assist in application, so that skips or misses can be seen. They are alkali-resistant and may be cleaned off tools with water. Fig. 3-9 illustrates the application of paint-on lath.

The plasterer uses these materials

Fig. 3-9. Paint on lath. (Standard Dry Wall Products, Inc.)

to bond gypsum, lime, or Portland cement mortars to any clean, sound surface. Paint-on lath can be applied by brush, roller, or spray. It dries quickly to a thin, flexible film that provides a permanent weld for the mortar. Many plasterers find this material very helpful in repair or patch work.

There are several brands of this material on the market and each has its advantages and limitations. Pack- aging is the same as paint products in cans of pint, quart, gallon, and five gallon sizes.

There are other bonding agents on the market but they are not as widely available. Most are for specific uses and are distributed under franchises. They differ from the "paint on" type as they are mixed into cementitious material, inorganic fibers, or aggregates.

Acoustic Materials

The word *acoustic* means, in its simplest sense, "related to hearing." The science of acoustics, therefore, treats of sound and noise in reference to hearing.

Sound goes outward from its

source in a series of waves. When the sound waves hit a surface such as a wall or ceiling, some of the waves are absorbed by the solid material of the wall or ceiling. Some of the waves travel through this material to the air on the other side. Other waves bounce off the surface and cause echoes.

All materials have some capacity to absorb sound waves. The amount of waves that can be absorbed varies considerably among different materials. Those with smooth, hard surfaces often reflect sound waves and the sound is reverberated. Some materials allow sound waves to pass through them and come out the other side. Usually all of these things occur at the same time, but the degree of each varies with the material.

Porous, rough, or soft surfaces tend to absorb sound waves more than they transmit or reflect them. The absorbed waves die out and do not carry the sound to other places. Acoustic materials are ones that tend to absorb sound waves.

Plaster excels other wall and surface treatments in reducing the amount of noise (sound waves) that can travel through it to another place. It also keeps noise from coming through in the opposite direction. Plaster is a good sound insulator because it absorbs sound waves.

Acoustical plasters are specially designed to absorb sound waves to a maximum degree. Some of the in-

gredients used in making acoustical plasters are rock wool, perlite, asbestos, pumice, and vermiculite. Gypsum, lime or Keenes cement is used as the binder. Many of these products on the market also use a foaming agent that acts to create a porous or spongelike body to the plaster. The foaming agents cause tiny bubbles to form within the mixture, leaving air pockets in the plaster. These air pockets absorb sound waves.

There are five compositions used in making acoustical plaster.

1. Lime and Keenes cement is used as the binder with a lightweight aggregate filler, a foaming agent and fiber for hand textured, and no fiber for the machine textured. It is applied to a plaster base coat ½ inch thick.

2. Gypsum is used as a binder with lightweight aggregate as a filler and a foaming agent. It is usually hand textured to ½ inch thickness over plaster base coat.

3. Lime is used as the binder with vermiculite or perlite as its filler. It is hand or machine textured to ½ inch thickness over plaster base coat.

4. A clay-adhesive type is used which has clay and a special bonding agent such as acrylic liquid plastic with vermiculite or perlite as the filler. It is usually machine applied to any sound surface such as concrete.

5. Combinations of 1, 2, 3, 4 are used and are usually machine applied and may be applied to any

sound‧surface.

These acoustical plasters are packaged in bags and only water is added. Most of the gypsum and lime manufacturers have acoustical plasters on the market. They are usually the ½ inch coat type. The bags are marked with the amount of coverage at ½ inch thickness; the normal coverage is about 4 sq. yds.

The bonding agent types are made by companies in and out of the plastering industries. These materials usually cover 200 to 300 sq. ft. per package.

Another type of sound absorbing medium is acoustic tile. The tiles are preformed and generally square. They are either nailed or glued in place with a mastic. There are many kinds of these on the market. They are made of different materials and present different appearances, but all will absorb sound readily.

Contact Fireproofing Materials

These materials are designed to be applied directly to steel beams, colums, or decking and are not considered decorative or abrasion-exposed materials. There are two types. One uses a bonding adhesive, usually of

Fig. 3-10. Fire Shield plaster being sprayed on. (National Gypsum Co.)

a latex type, and the other uses water to dampen it at the nozzle. Fig. 3-10 illustrates the application using a bonding adhesive.

Materials consist of asbestos fiber, mineral fiber, lightweight aggregates and inorganic binders. Each manufacturer has his own formula of these ingredients. They are usually sold under a franchise. The plasterer uses this material with the manufacturer's recommended machine to apply fire protection to the under structure of steel buildings. Materials are commonly packaged in 50 lb. bags.

Insulating Materials

These are franchised and require specialized machinery and experienced applicators. The materials are classed as urethane and polystyrene. Specially trained plasterers use these materials for insulating walls and roofs and for plaster bases. They are usually sprayed but may be poured in some instances. The materials expand about 30 times over their original volume after they have been sprayed. They form a rigid mass that is very light and has millions of tiny air cells.

Exposed Aggregates

Use of exposed aggregate is a growing field within the plastering industry. It offers the architect a material comparable to precast ones at a much lower cost and with weight additions as low as 1 to 2 lbs. a sq. ft. troweled on, and 3 to 4 lbs. in the seeded method. Because it is applied on the job, it can be applied to any contour, angle, or curve. Jointed stone patterns, murals, and other decorative effects can be formed with simplicity. There are a number of manufacturers supplying materials and each has his own formula and methods. Mostly they work through franchised dealers.

Exposed aggregates can be classified into two types. One uses a matrix that the aggregate is seeded or forced into. See Fig. 3-11, top. This is called the *stone-seeded* exposed aggregate. The other mixes the aggregate with a binding agent which is then troweled onto the surface. The aggregate is exposed after the binder dries to a

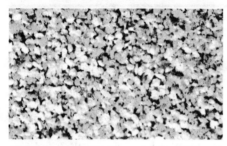

Fig. 3-11. Emulsion mix. (Cement Enamel Development, Inc.)

clear finish. See Fig. 3-11, bottom. This is called the *troweled-type* exposed aggregate.

Binder

The binder or cementing agent used may be the epoxy type or the latex type. The *epoxy type* uses a two component system. One is an epoxy resin that hardens when mixed with a catalyst which is commonly called the hardener. The two substances are mixed just before application. The epoxy materials are factory-prepared and are packaged in 5 or 6 gallon containers in various colors for use in either the troweled or seeded methods.

The *latex type* may be a latex, lastric, or acrylic binder. Binders are designed to be mixed with cement and silica sand to form a matrix for the seeding aggregate. The material is liquid and is used in place of water in mixing the mortar. Packaging is in 5 or 6 gallon containers and 55 gallon drums. Colors are added on the job by using special color additives from the manufacturer.

There are trowel applied lastric or

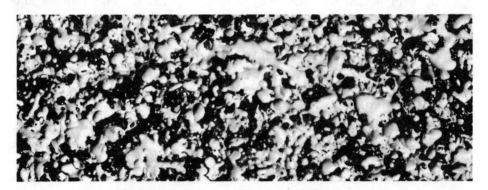

Fig. 3-12. Special textured finish. (Cement Enamel Development, Inc.)

acrylic types that are factory mixes of marble, quartz, or ceramic aggregate and the lastric or acrylic binder. They come in many colors and in 5, 6, or 10 gallon containers.

The manufacturers of these specialized plastering materials usually have materials composed of the same basic formulas that are used without the exposed aggregate to create special finishes and textures that are resistant to severe weather. Fig. 3-12 illustrates a special textured finish.

The plasterer uses all these types of materials to put decorative finishes over any sound surface such as plywood, asbestos board, blocks, or concrete. The manufacturer's directions should be followed in using them.

Veneer Plaster

The industry is grouping a number of systems such as "rapid plaster," "thin coat plaster," and "one coat plaster" under the general heading of *veneer plaster*. Basically all the systems are designed to eliminate or reduce the brown coat used in conventional systems. The two types of veneer plastering are a one-coat system and a two-coat system.

Two Coat System

The *two coat system* uses a standard rocklath or gypsum board for the base. A thin coat of specially processed gypsum and wood fibers is applied. When this thin coat is mixed with sand and water it will develop a compressive strength of at least 1,800 and up to 3,000 psi. This type of material can be either hand or machine applied. The coat must set very quickly or it will dry before setting, due to its thinness. Some coats of this type use a catalyst to speed the setting. In fact, some machines that are used to apply a thin coat mix in the catalyst at the nozzle.

Any standard finish is used over the base coat with either hand or machine application.

One Coat System

The *one coat system* may consist of a bonding agent and a standard finish with extra amounts of high strength gypsum gauging added. Or it may consist of specially prepared gypsum with a plasticizing ingredient to enable the plaster to trowel to a smooth finish.

These types of extremely strong plasters are also used with embedded radiant heat systems. The thin coat radiates the heat rapidly from the cables without building up excessively high temperatures and thus reduces the danger of calcination over a long period.

Veneer plastering tape is a fiberglass mesh 2 ½ inches wide. This material is used to reinforce joints and angles before applying veneer plaster.

Joint cement used in drywall systems is an air-drying type. It is manufactured in powder form requiring only water. It is packaged in bags of 5 or 25 lbs. This is used to cover nail heads and to apply joint tape over the joints in the base material. Joint cement is also available ready-mixed in 5 gal. containers, though it is more costly than powdered. This material has several advantages over powder. No job mixing is required, it has better coverage, more plasticity, extra smooth finish with less shrinkage, and less sanding is necessary.

Dry wall tape is used on dry walls and is a thin, strong paper 2 inches wide in rolls of 50, 100, and 500 lineal feet. It is manufactured either plain or with tiny perforations.

Checking on Your Knowledge

The following questions give you the opportunity to check up on yourself. If you have read the chapter carefully, you should be able to answer the questions. If you have any difficulty, read the chapter over once more so that you have the information well in mind before you go on with your reading.

DO YOU KNOW

1. What is one of the most common substances used for plastering?
2. What does the term *calcining process* mean?
3. Why is gypsum a good fire retardant?
4. What does cement basically consist of?
5. What is Keenes cement?
6. Name two types of lime manufactured for plastering.
7. Name three lightweight aggregates used by the plastering trade, and some of the advantages of using them.
8. What is gauging plaster?
9. What is used by the plasterer to hold back the setting action of his materials?
10. What is used by the plasterer to hasten the setting action of his materials?
11. What sand is best for mixing with plaster?
12. What does water do for a mix?
14. What mineral pigments can be used by a plasterer for coloring his materials
15. What are the usual ingredients found in acoustic plaster?
16. What are exposed aggregates?
17. What is veneer plaster?
18. What materials are used for drywall?

Mixing Materials

There are three things we must consider before we begin mixing materials. First, the kind of material we are going to mix; second, to what backing the material will be applied; and third, the method we will use to mix the material.

Normally, the mixing of most plaster materials is done by the plasterer's tender. However, as we mentioned earlier, the responsibility of the over-all plastering is that of the plasterer; so it is very important for a plasterer to know how to mix the materials he intends to use.

It is important to keep tools used in mixing materials clean at all times. In mixing plaster mortar do not wash the tools in water intended for use in the mixture itself. It is important not

only to keep the tools used in box mixing clean, but it is also important to keep the paddles or spinners and the mechanical plaster mixer clean.

As you finish mixing each batch, clean paddle or spinner type blades by spinning them in a drum of water set alongside your mixing containers.

Clean the plaster mixer whenever it is to remain idle for any length of time. Use a water hose and spray clean the sides and blades. The mixing blades should be either disengaged by a clutch or turned off. The dirty water is then dumped out before starting a new batch. If the mixer runs continuously, it need not be cleaned until it is stopped at some normal period of delay, such as lunch time.

Methods of Mixing Materials

The oldest and occasionally still used method of mixing material is the box method. This method requires more time and individual effort than mechanical mixing. Materials that can normally be mixed by hand are the gypsum and cement mortars used for scratch and brown coats. Fig. 4-1 illustrates the hoe and box needed for box mixing. When mortar is mixed this way, the cementing material and aggregate should be mixed dry until a uniform color is achieved. In the case of gypsum mortars, sometimes a small amount of retarder is sprinkled over the sides and bottom of the box before adding materials. By doing this the box can be kept cleaner for succeeding batches. When using mill mixed gypsum aggregate materials, the dry mixing is not necessary.

Box Mixing

To dry mix, first place half the amount of aggregate required for one

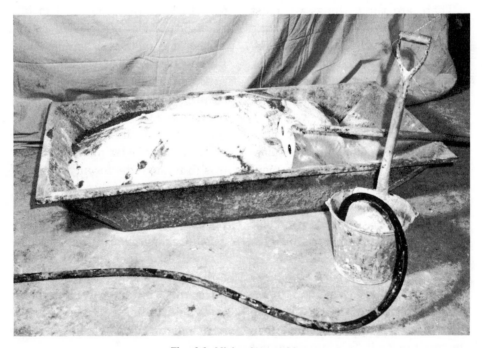

Fig. 4-1. Mixing box and hoe.

bag of cement or plaster in one end of the box, spread the first bag of cement or plaster over the sand, then lay the balance of the aggregate over the cement or plaster. If any dry admixtures are needed, spread the amount required for a bag of cement or plaster over the top of the sand. Repeat this until all the required material is in the box. This is called charging the box. Now with a mortar hoe (a hoe with two holes in the blade) start at one end of the box and pull the hoe in short choppy strokes toward yourself until you have completely hoed all the material in this manner. Then with the hoe pull as much material as you can to one end of the box, stacking it carefully. This will allow room to add water.

Now add some water and begin to pull the dry material into the water with short chopping strokes. Make sure the hoe cuts to the bottom of the box. Continually add water as needed; and as you proceed further out into the box make your strokes longer so as to bring the hoe farther through the wetted material. Keep chopping with the hoe until all the dry material has been wetted and pulled to the end of the box. Now repeat this operation, pulling the mortar to the opposite end of the box. Do not add any more water than is needed to bring the mix to a soft, plastic mass.

Mixing should be continued until the materials are thoroughly combined. A uniform color usually indicates that a good blend has been achieved. Do not over-mix; to do so causes mortar to begin setting sooner than it should. With experience comes the ability to know when the mortar is ready by the feel of the hoe as it is pulled through the mortar.

Machine Mixing

This method of mixing is used for gypsum, cement, acoustic plaster, lime, some veneers, and some exposed aggregate matrixes; it is required when using a plaster gun. Fig. 4-2 illustrates machine mixing.

Machine mixing of the mortar is

Fig. 4-2. Uniform consistency is easily achieved with a mechanical mixer. (United States Gypsum Co.)

119

recommended because of the uniformity it provides. For strong mortars the gypsum or cement must cover each particle of aggregate and this is done more sucessfully by the thorough mixing provided by the mechanical plaster mixer.

When mixing with a plaster mixer, ingredients should be added while the machine's blades are turning. First, add the required amount of water, then add half the required aggregate, then any additives that may be required. Add the cement or plaster, then the remainder of the aggregate. Let the mixer run about one more minute then dump the batch into a box, wheelbarrow, or plastering gun hopper.

Most plaster mixers are designed to operate at top capacity when the mortar is about 2 inches, at most, above the blades. When the mixer is charged higher than this, the proper mixing action fails to take place. Instead of blending materials, the mixer simply folds them over and over, resulting in excessively dry mix on top and too wet mix underneath. Thus, the first part dumped will be soupy and weak. Obviously, this is bad mixing; in addition, too large a batch may be dumped before the plaster is properly mixed because it will look as though it is mixed on the top.

If too little water was put in the mixer at the start, more can be added. If too much water was put in, dry material can be added to bring the batch to the proper consistency. However, in the case of mortars containing lightweight aggregates, the mortar sometimes looks dry even though sufficient water has been added. If mixing continues without adding any more water, the mix will become wetter; this is because surface tension on the lightweight aggregate is held to a certain point and then is lost, releasing some of the water it has held to itself. It is recommended that these mixtures should not be mixed in a plaster mixer for longer than three minutes after all the materials have been added. It is well to note that any excessive agitation of gypsum mortars will hasten the setting time.

Many of the acoustical and cement based matrixes for exposed aggregates require the use of a plaster mixer. Their special manufacturer's specifications regarding the amount of water, liquid binder, sand (if called for) and the length of time to mix must be followed exactly.

Paddle Type Mixing

An electric drill with specially designed paddles or spinners is used for paddle type mixing. This method is ideal for mixing small quantities of gypsum, lime, or cement materials, and is the recommended method for acrylic and epoxy emulsions as well as drywall materials. Fig. 4-3 illustrates a paddle type mixer.

Fig. 4-3. Paddle or spinner type mixer for smaller container.

The containers used to hold the materials are round and range in size from 5 to 25 gallons. The containers that are the easiest to keep clean are made of plastic or rubber.

To mix materials that are supplied in a dry state, such as lime or gypsum, the required amount of liquid or water is first added to the container. Dry materials are slowly added in proper proportions. They are spin-mixed intermittently as the materials are added. When the batch reaches the proper consistency required for smooth application, no more dry material is added. Continue spinning the material by moving the spinners up and down and

121

around to be sure the entire batch has been thoroughly mixed. When using factory prepared emulsions, the material is put into the container and the required amount of catalyst is added. The material is then spin-mixed for the length of time specified by the manufacturer. If the material does not require any additive it is simply put into a container and spun until it reaches a workable condition.

Mixing Precautions

Most plasters are mixed according to definite proportions of water, aggregate and cementing material. The hardness and durability of the plaster surface will depend upon the accuracy with which the correct proportions are followed. Too much water will cause a very fluid plaster which would be hard to apply and which would cause unnecessary dropping from the plasterer's trowel. Such waste can be expensive. Too much water will also cause a series of small holes to develop in the finished mortar coat. When the mortar sets and excess amounts of water remain as drops, the water evaporates and leaves pockmarks in the plaster. Any holes in the base coat plaster weaken the plaster job since maximum strength depends upon a uniform thickness of base coats.

Too much aggregate in a plaster mixture can also weaken the plaster wall. A certain proportion of binder is needed to hold the aggregate together. When too much aggregate is present in a mix, and not enough binder to unite the mixture, some of the aggregate particles will tend to crumble off.

Sometimes mortar mixtures are made "poor" or "rich". A plasterer calls a mixture *poor* when the mortar mixture contains less cementing material than normal. A plasterer calls a mixture *rich* when the mortar mixture contains a large proportion of cementing material.

Dirty sand, water, tools, or mortar box will affect the speed or rate of set of the plaster mortar mixture. Plaster mortar that has set or hardened cannot be retempered or used; therefore, the rate of set must be watched closely. Lime, however hardens so slowly that it can be left to age and used later.

Organic matter such as leaves and other kinds of vegetation, or foreign matter of any kind, may prevent the set of mortar altogether. When such a condition is encountered, the sand should be tested. Usually this component, and no other will be the culprit.

If the sand is found to be clean when tested, the fault may be traced to excessive amount of retarder accidently added at the mill. Such plaster can be made to set by the addition of a commercial accelerator. This should be added during the mixing. Commercial accelerators

tact. See Fig. 4-4. In this way a good bonding surface is provided for the finish coat. If lime mortar is not floated (and compacted), it will crack because of shrinkage of the putty.

To strengthen the mortar, it is good practice to add either Portland cement or Keenes cement to the lime mixture. This admixture strengthens the mortar and prevents the cracking caused by shrinkage and by extreme suction.

Admixtures. These may be added to the mortar in the proportion of 10 to 15 percent of the lime content. Hair or fiber is added to help hold the mortar together when it is to be used as a scratch coat over wood or metal lath. Smaller amounts of the hair or fiber may be used in the brown mortar.

Hair or fiber must be clean and unmatted. Use about 1 lb. of hair or fiber to 100 lbs. of dry lime for scratch coats and about half this amount for brown mortar. If hair is used, pull it apart and add it to the water while mixing or spread it over the sand and chop it into the mix with the hoe. Mortar should be uniform in color and soft enough to spread easily.

Box Mixing. First mix the dry sand and lime together until the dry mixture is consistent throughout. When the color seems uniform, one can usually be sure that the sand and lime are well mixed. Then add the water and hoe the mixture until the proper workability is achieved.

If a mechanical mixer is used, place water in the mixer first. Then add half of the sand and then all of the lime. Then add the remainder of the sand. Mix the mortar from 1 to 3 minutes, no more, depending upon the speed of the mixer. Then dump all of the mixture at once.

Lime mortar may be made ahead of time and left to age. When it is needed, it can be softened by hoeing and by adding enough water to replace that lost by evaporation. Such aged mortar will spread better and be more plastic than mortar freshly made.

It was common practice in years past to make and store all of the mortar required for a given job at least one month prior to use. This practice insured working mortar that was well soaked and properly cured.

Unfibered Gypsum or Fibered Gypsum Mortar

Proportions vary for these depending upon the base on which the plaster is to be placed. For all scratch coats over metal lath, gypsum lath, insulation, styrofoam, and also over surfaces of low suction ability, the mix is proportioned, when using sand, in ratio of 2 to 1: 2 parts of sand to 1 part of gypsum. The amounts are gauged by weight or proportioned in the fixed ratio of 14 #2 shovels of sand per 100 lb. bag of plaster, and 7 to 9 gals. of water. Table 4-2 gives measures for mixing gypsum. When scratching metal lath by hand, fibered gypsum is generally used because the fibers aid the mortar to hold together better and prevent too much of the mortar from going through the metal lath.

When using lightweight aggregates, either perlite or vermiculite, for scratch coats, the proportion of mix is 2 cu. ft. of aggregate to 100 lb. bag of fibered or unfibered gypsum. The amount of water for perlite is 8-10 gals. and for vermiculite is 9-11 gallons. Table 4-3 gives measures for mixing gypsum with lightweight aggregates.

Scratch mortar for metal lath should be mixed a little stiffer (with less water) than other plaster mixes.

When doubling or browning over

TABLE 4-2   GUIDE CHART: FIBERED AND UNFIBERED GYPSUM   (APPROXIMATE AVERAGE)

GYPSUM	SAND RATIO	SAND AMOUNT	WATER*	SUCTION	COAT
100 LBS.	2 TO 1	14 NO. 2 SHOVELS	7 TO 9 GALS.	LOW AVERAGE	SCRATCH COAT OVER METAL LATH OR BROWN OVER ROCK LATH, ETC.
100 LBS.	2 1/2 TO 1	18 NO. 2 SHOVELS	7 1/2 TO 9 1/2 GALS.	MEDIUM	BROWN COAT PREVIOUS SCRATCH
100 LBS.	3 TO 1	21 NO. 2 SHOVELS	8 TO 10 GALS.	MEDIUM	BROWN COAT MASONRY SURFACES: CEMENT BLOCK CINDERBLOCK BRICK, CLAY TILE
100 LBS.	4 TO 1	28 NO. 2 SHOVELS	9 TO 11 GALS.	VERY GOOD	BROWN COAT GYPSUM TILE, SOFT BRICK, ETC.

*U.S. MEASURE

TABLE 4-3   GUIDE CHART: FIBERED AND UNFIBERED GYPSUM WITH LIGHT WEIGHT AGGREGATES (APPROXIMATE AVERAGE)

GYPSUM	PERLITE		VERMICULITE		SUCTION	WATER*	COAT
	RATIO	CUBIC FT.	RATIO	CUBIC FT.			
100 LBS.	2 TO 1	2	———	———	8 TO 10 GALS.	LOW OR AVERAGE	SCRATCH COAT OR BROWN OVER ROCKLATH WITH AVERAGE SUCTION
100 LBS.	———	———	2 TO 1	2	9 TO 11 GALS.	LOW OR AVERAGE	SCRATCH COAT OR BROWN OVER ROCKLATH WITH AVERAGE SUCTION
100 LBS.	2 1/2 TO 1	2 1/2 TO 1	———	———	9 TO 11 GALS.	MEDIUM	BROWN COAT OVER PREVIOUS SCRATCH
100 LBS.	———	———	2 1/2 TO 1	2 1/2	10 TO 12 GALS.	MEDIUM	BROWN COAT OVER MASONRY, CINDER, CEM. BLOCK, ETC.
100 LBS.	3 TO 1	3	———	———	10 TO 12 GALS.	MEDIUM	BROWN COAT OVER MASONRY, CINDER, CEM. BLOCK, ETC.
100 LBS.	4 TO 1	4	———	———	11 TO 12 GALS.	VERY GOOD	BROWN COAT OVER MASONRY, GYPSUMTILE, SOFT BRICK, ETC.
100 LBS.	———	———	4 TO 1	4	13 1/2 TO 15 GALS.	VERY GOOD	BROWN COAT OVER MASONRY, GYPSUMTILE, SOFT BRICK, ETC.

*U.S. MEASURE

previously scratched gypsum mortar the proportion is 2½ parts sand to one part gypsum. This would be equal to about 18 shovels of sand (using a #2 shovel) to 100 lb. bag of gypsum; the amount of water is 7½ to 9½ gallons. When using lightweight aggregates for doubling or browning over previously scratched gypsum mortar, the proportion of mix is 2½ cu. ft. of aggregate to 100 lb. bag of fibered or unfibered gypsum. The amount of water for perlite is 9 to 11 gals., and for vermiculite 10 to 12 gallons.

For a brown coat over most masonry surfaces of average suction, such as brick, clay tile, cement block, or cinder block, a mix of 3 to 1 is used: 3 parts sand to 1 part gypsum (fibered or unfibered). When using sand this would be 21 shovels to a bag (100 lbs.) of gypsum with 8 to 10 gals. of water. When using lightweight aggregate it would be 3 cu. ft. of aggregate to a bag of gypsum. Perlite requires 10-12 gals. of water; vermiculite, 12 to 14 gallons.

Due to the great suction of gypsum tile, some soft brick and clay tile, a greater amount of sand must be added to prevent shrinkage cracks in the brown coat. In fact, the ratio may be as high as 4 parts sand to 1 part gypsum; this would be 28 shovels of sand to a bag (100 lbs.) of gypsum with 9 to 11 gals. of water. When using lightweight aggregate this would be 4 cu. ft. to a bag of gypsum.

Perlite requires 11-12 gallons of water; vermiculite, 13½ to 15 gallons.

The plaster mix for rocklath and insulation lath is the same as that for metal lath scratch coat: 2 parts sand to one part gypsum when using sand or 2 cu. ft. (½ bag) of lightweight aggregate per bag (100 lbs.) of gypsum.

Wood Fibered Gypsum. Wood fibered gypsum normally requires only the addition of water. When it is used over masonry or gypsum tile or as a brown coat over scratched metal lath, 1 cu. ft. (7 No. 2 shovels) of sand per 100 lbs. of wood fibered gypsum should be added.

Perlite Aggregate. Perlite aggregate and gypsum (mill mixed) require the addition of water only (approximately 6 gallons), but it should be noted that the manufacturers make two types: one for masonry and regular that is used over gypsum lath.

Bond Coat Gypsum. This requires the addition of water only (approximately 8 gallons).

Keenes Cement Mortar

Keenes cement is another product of gypsum rock that is ground, calcined, and processed by the manufacturers in a special way. The result is a cement that produces a mortar with special qualities. It forms a very hard, durable surface and it can be retempered or reused. None of it need be wasted.

Keenes cement is mixed with lime in order to increase its plasticity since it is, by itself, hard to work. It is not only used as a cementing material for scratch and base coats, but it is used as a very fine finish coat as well. For scratch mortar and brown mortar, Keenes *regular* cement is used. This cement has a normal setting time of 4 hours. However, it can be retempered or remixed as often as necessary to use it up.

Proportions. For scratch coats over wood or metal lath, use 1 bag of Keenes cement (100 lbs.) and 700 lbs. of sand (42 No. 2 shovels) plus 160 lbs. of lime putty (9 No. 2 shovels). If dry hydrated lime is used, 80 lbs. will equal the required putty. see Table 4-4.

It should be noted that when dry lime is soaked into a putty it will be approximately twice as heavy as when dry. One 50 lb. bag of dry lime when made into a putty will equal 1 cu. ft. and will weigh about 100 lbs.

For brown coat mortar, change these proportions somewhat. To 1 bag of Keenes cement (100 lbs.) and 1,000 pounds of sand (60 No. 2 shovels), add 160 lbs. of lime putty. If dry hydrated lime is used, 80 lbs. will equal the required putty. See Table 4-4.

This mortar should not be used over gypsum lath or any of the fiberboard insulation laths. As a brown mortar it can be used over all masonry bases and any previously scratched surface.

Admixtures. For scratch coats over wood or metal lath, add 2 lbs. of hair or fiber to increase the rigidity of the scratch coat and brown coat. Since Keenes cement is manufactured so the time of set is highly accelerated, no additions of this sort are necessary.

Mixing. Mix dry Keenes cement thoroughly with an equal amount of sand, then add water and mix again. While this mixture is soaking, mix the balance of the sand with the required amount of putty and hair. Wet this mixture as much as may be necessary to obtain a good blend. Now mix the two batches together, making one pile or box of mortar. Add water, if necessary, to bring the mass to the proper consistency.

The mixing of separate piles of mortar is done to insure complete

TABLE 4-4 GUIDE CHART: KEENES CEMENT MORTAR (APPROXIMATE AVERAGE)

KEENES CEMENT	HYDRATED DRY LIME	LIME PUTTY	SAND	WATER*	COAT
100 LBS.	80 LBS.	9 NO. 2 SHOVELS	42 NO. 2 SHOVELS	13 - 14 GALS.	SCRATCH COAT
100 LBS.	80 LBS.	9 NO. 2 SHOVELS	60 NO. 2 SHOVELS	15 GALS.	BROWN COAT

*U.S. MEASURE

and easy blending of the various ingredients and to prevent the formation of lumps. If dry Keenes cement were added to a wet lime putty, the Keenes cement would form lumps. When the cement and sand are mixed dry and then wetted, lumping is avoided.

Machine mixing will save much time and labor and yield a mortar that is thoroughly combined. Use the same steps for machine mixing as for hand mixing.

Hydrated Lime

Hydrated Lime is soaked in tubs from twelve to twenty-four hours to complete hydration and to form a putty plastic enough to be used to advantage. For the regular or stand-

Fig. 4-5. Lime, putty tank, and screen in preparation for soaking lime.

ard hydrated lime, the common practice is to fill a clean mortar box or putty tank with the required amount of water. Two 14 quart pails of water per 50 lb. bag of lime will be about right. Figure 4-5 shows the lime, putty tank and screen, ready for the lime to be soaked.

After the water is placed in the tank, slowly sift the lime into the

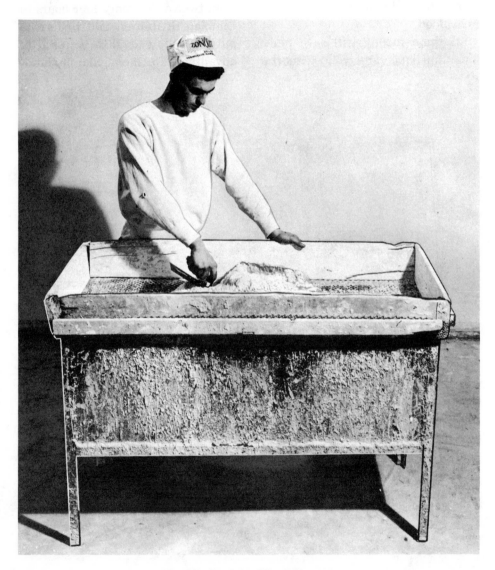

Fig. 4-6. Screening lime into water.

water. The lime is sifted by screening it. See Fig. 4-6. When all of the lime has been sifted into the water, it should come just up to the surface. Slow soaking is promoted by screening the lime and this helps prevent the formation of lumps.

Some plasterers prefer to hoe the putty shortly after it is soaked to break up any lumps and to form a smooth paste. See Fig. 4-7. After soaking overnight, the putty will be fairly stiff. Usually it is necessary to screen the soaked lime through a

Fig. 4-7. Hoeing lime putty to break up lumps.

screen made of ¼" mesh, screen or metal lath. This will break up any lumps that would make the lime difficult to mix with gauging gypsum and difficult to spread with a trowel.

If too little water is used, the lime will be improperly soaked and may be *burned*. The term "burning" refers to the small pebble-like lumps of unsoaked lime. In the finish coat, these lumps shrink and form small indentations similar to pockmarks.

When a plaster mixer is used, there is usually no need to screen the lime the next day, provided it was given a sufficient amount of time in the mixer before dumping. Generally 5 to 10 minutes is good.

When using a limestir, paddle, or spinner type mixer, screening the next day is not necessary and the time is shortened considerably. Be-

cause of their design these mixers homogenously combine the water with the lime.

Autoclave Lime

Pressure or autoclave lime is soaked somewhat differently than standard or regular lime. Most manufacturers recommend using only a small amount of water. The lime is hoed into the water and worked until a stiff paste is formed. Then water is added until the putty is of the right consistency. Too much water "drowns" this type of lime. The only other difference is that it need not be left to soak overnight, as does the regular lime.

If warm water is used when mixing the lime, it will produce a more plastic putty and will make the regular lime usable sooner.

Finish Coats

The plastered area in a room takes up about 75 percent of the visible surface area. Consequently, the mixing and application of the finish coat is of extreme importance in producing a hard, durable surface finish.

The finish coat must also have the kind of surface appropriate to a particular style of architecture. Above all, the finish coat must be workable

and plastic enough for the plasterer's use.

Many kinds of materials are used in finishing mixtures. Among them are some of the materials discussed earlier. Finishing coats use lime and gypsum products. The lime used is the hydrated or partially hydrated type soaked into a putty. Hydrated limes, used as a finish material, are two types: regular lime and auto-clave lime. The gypsum products used are gauging plaster and Keenes cement.

Lime putty forms the basis for many finishing mixtures. One of the finishes using lime putty is the finish called putty coat.

Putty Coat

Putty coat finish is the most common type of finish used today by the plastering industry. Some of the other names used are whitecoat, hardcoat, creamcoat and smooth coat. It can be worked to a smooth, straight, hard surface and it is well suited to all types of surface treatments. One of its great advantages is ease of repair. This finish can be cut full of holes, scratched, dented, and otherwise marred, yet a plasterer can with little trouble patch the work so that repair cannot be detected.

Proportions. For work of a general character the putty coat is mixed with gauging plaster in the proportions of 3 to 1 (75 percent putty to 25 percent gauging plaster). This produces a good hard surface. If a very hard surface is required, the proportions may be changed to increase the amount of plaster used.

The use of too much gauging plaster, however, will produce a mixture that is hard to spread. Forty percent is about the maximum amount of plaster that could be used while preserving an easy working material.

The plasterer will often use the condition of the brown coat as a basis for adding a higher proportion of gauging plaster. The conditions are determined by the degree of suction the brown coat has. The less suction there is, the stronger or harder the surface must become in order to finish properly. This is why some material companies have designed special gauging plaster, such as "Structo-gauge," to be used over veneer plastering bases.

Admixtures. These are used in a putty coat mixture. Lime, as stated previously, sets very slowly. As a result, some agent is needed in the putty coat mixture to accelerate the setting time. Gauging plaster, the other basic ingredient, accelerates the setting time.

There are two kinds of gauging plaster: quick-set gauging plaster and slow-set gauging plaster. Quick-set gauging plaster is used in most parts of the country. The plasterer, however, wants to be able to control the set of the finish material to suit his job conditions. For this reason various amounts of retarder are added, the amount depending upon the job. Retarder may be needed as an admixture to provide ample time for applying the putty coat.

Mixing Putty coat mixing is done by the plasterer on a finishing board. This is a large, smooth-topped board about 5 ft. square. It is placed on a stand to bring it about waist

Fig. 4-8. Tools and materials ready for mixing putty coat.

high. This height permits the plasterer to mix the materials at a position affording ease of movement. Fig. 4-8 illustrates the correct height of the stand. Tools and materials are ready for mixing process.

To prepare or gauge a batch of putty coat, first place about the amount of lime putty needed on the board. This should equal 75 percent of the total amount of finish coat needed to do the job (in other words, a 3 to 1 mix). Form this putty into a ring as large as the board will permit. Place a small dab or lump of lime putty in the center of the ring. See Fig. 4-9.

Over this putty place as much re-

Fig. 4-9. Putty ring ready for retarder and water.

tarder as will be needed to retard the set of the plaster that is to be used in this gauging. (Commercial retarders usually have the correct amounts of retarder to use indicated on the package.) Use a trowel to mix the retarder and putty together and add a dash of water to help these materials combine thoroughly.

Clean water is now poured into the ring. Use as much water as you think will dissolve the retarder completely and which will also absorb the gauging plaster to be added later. Be sure that the retarder is dissolved, however. Watch out for lumps, for when lumps are left in the gauging they may burn through the finished paint later. In some sections of the country plaster with retarder previously

added at the mill is used. For plaster of this character no retarder is added to the gauging.

With water and the retarder combined, the gauge is ready for the plasterer. Sift or add the plaster slowly letting it soak into the water without lumping. See Fig. 4-10. Fill the ring with plaster until all of the water is absorbed.

When a putty coat is used over lightweight aggregates, vermiculite, perlite fines or fine silica sand should be added in the proportion of $\frac{1}{2}$ to 1 cubic foot per 100 lbs. of gauging plaster. The plasterer usually has this added to the lime when it is being readied, but sometimes he will choose to add this while he is gauging. He does this by putting a little

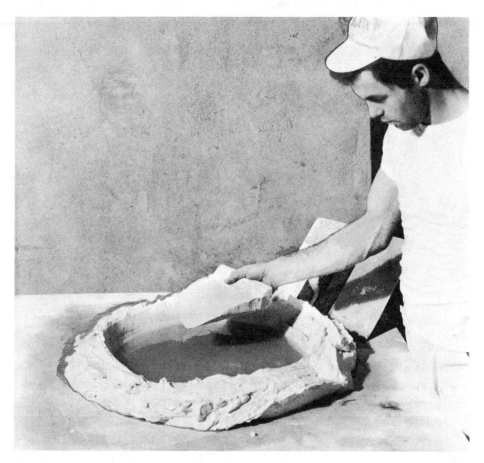

Fig. 4-10. Sifting plaster into water.

more water into the ring. After all the required gauging has been sifted, he sprinkles the lightweight fines into the remaining water and over the lime.

Let this mixture stand for about one minute to thoroughly saturate the plaster. Caution should be used when gauging so as not to splash any of the material into your eyes or into your partner's eyes. This danger ex-

ists because of the position a plasterer takes with his face directly over the mix, leaving his eyes susceptible to injury. Now, using the hawk and trowel, start combining the putty and plaster. Begin in the center and break up the stiff, soaked plaster, gradually mixing it with the putty. Begin by using the inner portion of the putty ring. Then start pushing in the outer edge of the putty ring. Mix

TABLE 4-5 GUIDE CHART: PUTTY COAT (APPROXIMATE AVERAGE)

LIME PUTTY	WATER*	GAUGING	SUCTION	HARDNESS
12 NO. 2 SHOVELS	2 1/2 GALS.	25 LBS.	MEDIUM	AVERAGE
12 NO. 2 SHOVELS	5 - 7 GALS.	50 - 70 LBS.	LOW	EXTREME
12 NO. 2 SHOVELS	1 1/2 GALS.	15 LBS.	HIGH	LOW

*U.S. MEASURE

the mass together until it is thoroughly combined. Gauged material should be quite stiff, otherwise it will run off the hawk before it can be applied.

The gauging should begin without delay as the setting action of the plaster begins as soon as it is wetted, and the plaster is usable only for a period controlled by action of the retarder.

Once the material has set or become hard it cannot be reused. The batch must not be disturbed after it has been mixed except as it is lifted from the hawk in the process of application. Any further mixing or moving of the gauge will speed up the setting. Table 4-5 gives mixtures for different types of putty coats.

Sand Finish

Sand finish is one of the oldest forms of finishing materials used in the plastering industry. The mixture for sand finish is very similar to the mixture for regular lime mortar used as base coat material. The difference lies in the fact that sand finish uses a very fine grade of sand which is screened for uniformity of size of the sand particles. A white or light colored sand is preferred.

After sand finish has been applied, it is not troweled. A cork, sponge, or felt covered float is used instead. The floating of this finish promotes a rough, though even, surface that has the general appearance of sandpaper.

Proportions. The proportions for sand finish are 3 to 1 (three parts of sand to one part of lime). Aging of the mortar will increase its plasticity. In many cases, the sand used is a screened product sold by material dealers for this purpose. For a pure white finish, silica sand is used. An important thing to remember is that sand of unusual fineness tends to produce minute cracks and a weak surface.

Sand finish will crack because of shrinkage if applied in too thick a coat, or to fill a hollow. The procedure recommended is to do the finishing the day after the work is

TABLE 4-6 GUIDE CHART: SAND FINISH (APPROXIMATE AVERAGE)

AMT. PUTTY	DRY LIME	SAND		GYPSUM	WATER*
		SHOVELS	BAGS		
12 NO. 2 SHOVELS	2 BAGS (50 LBS. EA.)	42 NO. 2	3	15 LBS.	15-16 GALS.

*U.S. MEASURE

browned. This will insure suction that is uniform, but not excessive. Table 4-6 gives proportions for a sand finish.

Admixture. Keenes cement, which is a slow setting plaster, or retarded gauging plaster is used by many plasterers. These may amount to 10 to 15 percent of the lime content. Admixtures are usually added if the background is dry or the finish must be applied more thickly than ordinarily is done.

Quick or slow setting gauging plaster may be added just before the sand finish is to be applied. It should not be added at any other time. Keenes cement, used as an admixture, may be added at any time prior to application of the sand finish.

The mixing of gauging plaster is done the same way as putty coat, except less gypsum is required.

Colored sand finish is popular in many parts of the country. Color must be added first to the dry sand, then both color and sand are screened together. This screening breaks up any lumps present in the color. Next, combine the colored sand with the dry lime and screen again. Do not use more than 10 percent of color in proportion to lime. Too much color will weaken the finish coat.

Mixing. Sand finish can be made with the least amount of work if the materials are screened and mixed at the same time. Use a clean mortar box and a good screen. A ⅛″ mesh wire cloth will be about the right size for this work. Place a section of pipe or roller that will span the width of the mortar box and place the screen on this.

Beside the box, build up a pile of sand and lime proportioned to the mix required. Turn this over once or twice with a shovel to start the mixing. Now place a few shovels of the mix in the screen and roll the screen back and forth on the roller. See Fig. 4-11. This procedure screens the material and also mixes it. Dry sand, of course, is easier to screen than wet sand.

Water can be put into the box before the screening and mixing process is begun. Putting water into the box first will allow the screened and mixed material to soak and will cut down on the dust raised by the screening.

When the required amount of material is screened the mixture is hoed

Fig. 4-11. Screening sand and lime mixture into box.

until it is well mixed. Then it should be allowed to soak for at least one night. For material of superior spreading quality and good floating ability, let the sand finish age for about a week, because sand finish improves with age.

It is not necessary to mix sand finish by hand. It can be done as well with a mixer and, of course, it will take less time. Whether hand mixed or machine mixed, the quality of sand finish depends upon the screening of the sand.

Sand finish of superior spreading quality and good floating ability can be made for immediate use by mixing aged lime putty or autoclave lime with the screened sand in a mortar box or in a mixer.

Texture or grain of the finished surface is controlled by the looseness of the sand used. Therefore, the screen used will determine the texture of the finished product.

Keenes Cement Finish

Keenes cement is used whenever a dense, smooth surface is required because it produces a harder than average finish. For use in kitchens and for wainscoting it is especially advantageous. Because it is a gypsum product, however, Keenes cement is an easy prey to dampness. Consequently, it should not be used around bathtubs and showers.

Proportions. The proportions of Keenes cement finish vary because the amount of Keenes cement that is mixed with the lime putty will determine the hardness of the finish coat. One part putty to one part Keenes (1 to 1) is about as weak a mix as would be used by the trade. The ratio of putty to Keenes is varied to suit the job, and in many cases it is closer to 3 parts Keenes to 1 part putty (3 to 1). Table 4-7 gives proportions for Keenes cement.

Admixtures. Silica sand may be added to the batch as a hardening agent and as a protection against shrinkage cracks. Five or six scoops of silica sand or fine marble dust mixed into the gauging will do the trick. This material will also help to break up any lumps of putty that may be present.

Do not add gypsum plaster of any kind to Keenes cement. To add it is to defeat the purpose for which this material was formulated. It weakens the cement.

Once Keenes cement is applied and has set it cannot be retempered. But as long as the mortar remains on the board it may be wetted again and then remixed.

No retarder is needed since it is a relatively slow setting material. You can purchase Keenes cement that sets in two, four, or six hours. The two and four hour sets are used for finishing work and for casting work; the six hour set for scagliola work (imitation marble).

Mixing. Keenes cement can be applied *neat* (undiluted). That is, all Keenes cement and no lime putty. However, this is seldom done except in special cases. The cement is mixed with water and made into a smooth, soft paste. It is then troweled on the

TABLE 4-7 GUIDE CHART: KEENES CEMENT FINISH (APPROXIMATE AVERAGE)

AMT. PUTTY	WATER*	KEENES CEMENT	HARDNESS
6 NO. 2 SHOVELS	7 GALS.	100 LBS.	MEDIUM
6 NO. 2 SHOVELS	14 GALS.	200 LBS.	HIGH

*U.S. MEASURE

Fig. 4-12. Putty ring filled with water ready to receive Keenes cement.

wall or ceiling and worked into a fine hard surface.

For the average job, Keenes is soaked into the water, which is retained in a ring of lime putty, just as in putty coat work. See Fig. 4-12. Mixing of the outer ring with water and Keenes cement, is accomplished in the same way as for the putty coat.

Mixing Portland Cement Materials

The need for a versatile product which would withstand the effects of weather, dampness, and changing temperatures was met with the production of Portland cement. Descriptions of its abilities and its versatility have been given. Here you will learn to proportion and mix Portland cement and to determine whether admixtures are to be used.

141

Portland Cement Stucco and Plaster

Portland cement is used primarily as an ingredient in mortar mixtures that are to be used on exterior wall surfaces. When mortar containing Portland cement is used on exteriors it is called Portland cement or stucco.

Portland cement is also used as an ingredient in mortars used for interior plasterwork jobs. When it is used for interior work Portland cement mortar is called Portland cement plaster.

Portland cement plaster or stucco are modern wall coverings which satisfy all requirements of durability and appearance when properly applied. These products actually are Portland cement concrete. They resist fire. They have the plasticity which the plasterer wants. They harden slowly enough to permit texturing. They are durable; age increases their strength and density.

Portland cement plaster is the best interior finish where walls, partitions, and ceilings are subject to rough use or extreme moisture conditions. It is used in basements, steam rooms, vapor baths, toilet rooms, swimming pools, garages, storerooms, elevator shafts, and other places subjected to hard service.

Portland cement mortars set in about two to three hours. Many plasterers prefer using a cement called high early strength cement, because it has characteristics of early strength and better workability than other cements. It is possible to apply succeeding coats at closer intervals when this cement is used. It attains the same strength in three days that regular cement does in seven.

Specifications for cement, grading of aggregates, use of plasticizing agents, water requirements, and the use of mineral pigments and colored cement for Portland cement plaster are the same as for Portland cement stucco. The information regarding the proportions, admixtures, and mixing of Portland cement mortar will apply to both plaster and to stucco, even though the word "stucco" is apt to be used.

Proportions. Each coat of stucco should be not richer than 3 parts of damp, loose aggregate to 1 part of Portland cement. In other words, a mix of 3 to 1 is recommended Too rich a mix is the cause for most of the cracks and other Portland cement failures. The Portland Cement Association recommends that where the aggregate is well graded with a good proportion of coarse particles, a 3½ to 1 or 4 to 1 mix may be found satisfactory.

Table 4-8 gives proportions for a Portland cement mortar.

The proportions given are proper for all scratch and brown coats ap-

TABLE 4-8 GUIDE CHART: PORTLAND CEMENT MORTAR (APPROXIMATE AVERAGE)

| PORTLAND CEMENT | SAND AMOUNT | PLASTICIZING AGENTS | | WATER* |
		AUTOCLAVE LIME	MASON CEMENT	
94 LBS.	18 NO. 2 SHOVELS	1/2 BAG	———	9 GAL.
94 LBS.	18 NO. 2 SHOVELS	———	1/2 BAG	9 GAL.

*U.S. MEASURE

lied, regardless of the base or the purpose for which the mortar is intended. The ratio of 3 to 1 is maintained for each scratch and brown coat applied in order that the rate of expansion and contraction may be kept uniform while providing a mixture of sufficient body to be spread with hawk and trowel.

The amount of water needed to wet the mixture of sand and cement will seem of little importance to many plasterers, yet it is one of the most important factors in producing a strong and dense mortar.

Use only water that is fit to drink. It should be free of oil, vegetable material, and other impurities. Just enough water should be used to wet the cement and sand and bring it to the proper consistency. The amount of water used must be determined by personal experiment.

Cement and sand that have been wet with the proper amount of water will, at first, seem too dry. Only by thorough mixing will the water succeed in wetting each particle of sand and cement. A common mistake

many mechanics make is to use too much water at first and then attempt to dry up the batch by adding dry cement or sand. This upsets the proportions and wastes time.

About six gallons of water to one bag of cement will produce a good mortar. If aggregate such as sand is used, the amount of moisture it contains will vary. This variation is an important fact that should be considered when proportioning and mixing Portland cement stucco or plaster. If the sand is very wet, add a smaller amount of water and mix and then add more at the end to bring it to the right consistency. Ordinary sand, as it comes from the dealer, can be considered as "damp".

For cement work the type and size of the sand particles is very important. All sand or aggregate used must be clean, well washed, and screened to produce a balanced grading of the sand grains. Sand should be screened and graded to yield a product that combines both fine and coarse particles. This will produce a mortar that is dense and without voids, and

TABLE 4-9 GUIDE CHART: PORTLAND CEMENT MORTAR WITH LIGHTWEIGHT
AGGREGATE (APPROXIMATE AVERAGE)

| PORTLAND CEMENT | PERLITE | VERMICULITE | PLASTICIZING AGENTS | | WATER* |
			AUTOCLAVE LIME	MASON CEMENT	
1 BAG 94 LBS.	3/4 BAG 3 CU. FT.	———	1/2 BAG	———	10 GALS.
1 BAG 94 LBS.	3/4 BAG 3 CU. FT.	———	———	1/2 BAG	10 GALS.
1 BAG 94 LBS.	———	3/4 BAG 3 CU. FT.	1/2 BAG	———	11 GALS.
1 BAG 94 LBS.	———	3/4 BAG 3 CU. FT.	———	1/2 BAG	11 GALS.

*U.S. MEASURE

TABLE 4-10 GUIDE CHART: PORTLAND CEMENT FINISH (APPROXIMATE AVERAGE)

PORTLAND CEMENT WHITE OR GRAY	SAND AMOUNT	AUTOCLAVED LIME	WATER*	SILICA SAND
47 LBS. OR 1/2 BAG	21 NO. 2 SHOVELS	2 BAGS	17 GALS.	———
47 LBS. OR 1/2 BAG	———	2 BAGS	17 GALS.	3 BAGS

*U. S. MEASURE

will work more effectively than will sand made up of all fine or all coarse particles.

The coarser the aggregate in the base coats (scratch and brown) the better. For finish coats it may be necessary to use a somewhat finer aggregate, but it should be kept in mind that excessive fineness is one of the principal causes of crazing and cracking. Table 4-9 gives proportions for Portland cement mortar with lightweight aggregate. Table 4-10 gives proportions for Portland cement finish.

Admixtures. Portland cement mortar is a shorter working material than most plastering materials.

To overcome this, materials are used as "lubricators" or "fatteners" to increase workability of the mortar. These are referred to as plasticizing agents. Some of the approved admixtures or plasticizing agents are hydrated lime or lime putty, natural cements, mason's cement, asbestos flour or fiber, and finely divided clays. The volume of these admixtures should not exceed 10 percent of the Portland cement used. The use of asbestos should be held down to 3 percent.

Some Portland cements prepared for use in stucco work contain the proper amounts of plasticizing agents ground in at the mill. If a plasticizing material is added at the job site, take care to use only those materials of highest quality. Use them in limited amounts and in minimum quantities for required workability. Excessive amounts of plasticizing materials added to a Portland cement mixture will reduce strength.

These admixtures increase the plasticity of the mortar and also tend to increase its density, making it more water resistant. New air-entraining admixtures, mixed with the cement at the mill, are coming into use. The entrained air increases plasticity and provides protection against the harmful action of frost.

Ammonium stearate, aluminum stearate, or butyl stearate emulsion may be used, in amounts not to exceed 2.0 percent by weight of the cement. The specific amount that the manufacturer recommends must be followed for a trouble-free job. These materials are added to improve the water repellency or decrease suction.

A retarding admixture can also be added and it should be used only as the manufacturer specifies.

Color may be added to Portland cement but only high-grade mineral pigments should be used. They should be of a fineness exceeding that of the cement. The amount should not exceed 6 lbs. of color per bag of cement or 6 percent of the weight of the cement. Color should be mixed dry with the cement, then mixed again with the sand. This is done to insure even distribution of the color and prevent lumping. Colored aggregates may be used also. These help to produce a stronger color and one that is uniform and long lasting.

The Portland Cement Association recommends the following guide for the selection of coloring materials:

For maximum brightness and clearness of color and for light shades, use white Portland cement. For white, use white Portland cement only. For brown, use burnt umber or brown oxide of iron. Yellow oxide of iron may be added to obtain modification of this color. For buff, use yellow ochre or yellow oxide of iron; add red oxide of iron to obtain modification of this color.

For gray, use small quantities of black iron oxide, manganese black, or Germantown lampblack. For green, use chromium oxide. Yellow oxide of iron may be added. For pink, use a small quantity of red oxide of iron. For rose, use red oxide of iron. For cream, use yellow oxide of iron in small quantities. For blues, use cobalt oxide. For black use regular Portland cement (not white) and black iron oxide, manganese black, or Germantown lampblack.

"White Portland cement" is a specially formulated cement that hardens white. The standard or regular Portland cement hardens to a gray color. To obtain darker colors it is much easier to get the dark shades using regular Portland cement. For example, to obtain black less color additives are needed when using regular cement. Buff or cream requires less coloring additives with white Portland cement. These two cements are the only Portland cement available that have an effect on

145

coloring. High early cement is another cement shown in Chapter 3 and is very seldom used for a finish. Its color would be the same as regular Portland cement.

For scratch coats on metal lath, hair or fiber may be added. This is mixed in with the dry materials. The proper amount depends somewhat on the quality of the hair or fiber. Ordinarily ½ lb. of hair or fiber per bag of cement is used. The material should be pulled apart and spread by hand throughout the dry mix. The Portland Cement Association recommends that no admixtures be added to hasten the time of set.

Mixing. Hand mixing is satisfactory if done thoroughly, but machine mixing usually gives greater uniformity. For hand mixing, dry sand and cement are mixed together first. These materials are mixed until a uniform color is achieved. The mixing must be thorough to insure that each particle of sand will be coated with cement. Admixtures are added to the dry mix in order that they may be completely distributed before water is added.

The dry material is proportioned in one end of the mortar box and is then mixed not less than three times. Chop the material with the hoe a little at a time, then pull it to the opposite end of the box. Now pull it back to its first position and again return it.

When the dry materials have been combined and well mixed, pour the water into the box. With the hoe pull the dry material into the water a little at a time and continue until all of it has been wetted. Mix thoroughly before adding more water. Add water until there is a desired consistency and the mass is uniformly plastic. Proper mixing will bring a seemingly dry mass to a smooth, plastic mortar. Hoe the batch back and forth 10 to 15 minutes after the water is added.

Machine mixing is the best method to use. When you use a plaster mixer, add the water first. Then add about 50 percent of the sand. Next, add the cement and any admixture desired. Last, add the balance of the sand. Mix until the batch is uniform and of the right consistency. Three to four minutes of mixing after the ingredients are in the drum is sufficient. Do not mix the ingredients too long. Overmixing will do more harm than good, for it will tend to reduce the initial time of set.

Setting time is the amount of time necessary for a mortar to become rigid. Close study of the setting time of a mortar will indicate that there are really two stages in the setting process before a mortar reaches its final rigid, strong condition.

The first stage includes a stiffening of the mortar. It has the appearance of being set, but the mortar is still in the plastic state. This first stage is called the *initial set*. If mortar that has reached this stage is to be used again, it will have to be re-tempered. From this point, when the initial set takes place until its final state, the mortar increases in strength until it becomes rocklike.

When mortar has reached the last stage, it is both rigid and strong. As an example, when you mix concrete you perhaps have noticed that concrete *seems* to harden within 24 hours, but that it really takes many days for concrete to "cure" or really achieve its full hardness and strength.

The initial set of Portland cement mortar occurs about two to three hours after the dry materials have been wetted.

Portland cement mortar can be used after the initial set has occurred, but its use depends upon the fact that the mortar must be retempered as mentioned previously.

Within that period of time (two or three hours after the materials have been wetted), remixing and softening of the mortar will do no harm, provided that the mortar is merely mixed and that no water is added to it. The addition of water makes the mortar mixture too plastic. It will, however, make the mortar seem to spread better. But the addition of water after the mortar was first mixed, is not generally recommended, because such a practice tends to reduce the adhesive quality of the Portland cement mortar.

Veneer Plastering

There is no general method or proportion for mixing veneer plastering materials, so in using this type of plaster the manufacturer's directions *must* be carefully followed. As mentioned earlier, this type of plastering can be accomplished by two methods. One method requires two coats, and the other, one coat. When using the two coat system, be careful not to mix more material than can be used in a short period of time, because the strength and quality of these materials is based on their set-ting very quickly. The mixing equipment used should be thoroughly cleaned after each batch. Many plasterers use the paddle drum mixing method for these materials because they are applied very thin, and usually only that amount of material that can be applied in 45 minutes is mixed.

The sand, if needed, is a fine sand with no more than 5 percent by weight being retained on the No. 16 sieve. The limitation on larger particle sizes is necessary because of the thin application. However, the inclusion of limited amounts of sand sized above a 20 mesh is essential in providing adequate thickness and an open, porous surface to receive the finish coat. When added sand is necessary, the proportions are usually one part sand (by weight) to one part gypsum. Some plasterers have found the use of the bagged silica

sand convenient in measuring as well as in workability.

The second coat of the two coat veneers follows the same mixing methods as mentioned in the putty coat or sand finishes.

The one coat veneer system must be mixed only as prescribed by the manufacturer.

Radiant Heating Ceiling Panels

There are two systems for imbedding ceiling radiant heating. One uses special electrical cable secured either to the underside of rocklath, dry wall, or concrete. Another system uses copper hot water pipes varying in sizes from ⅜″ to ¾″ imbedded in the ceiling plaster. It has been found that the ideal construction for hot water radiant heat is obtained by attaching expanded metal lath to the bottom of ceiling joists, securing copper pipe to the metal lath, and finally

applying three base coats of plaster plus a finishing coat.

Electric Radiant Heating Ceiling Panels

There are many specially designed plasters manufactured for this purpose and their individual directions for mixing must be followed exactly. This system can also be covered with ordinary gypsum plaster and finishes, but sand must be used for the aggregate. Vermiculite

TABLE 4-11 GUIDE CHART: PLASTER OVER RADIANT HEAT PIPES (APPROXIMATE AVERAGE)

PORTLAND CEMENT	DRY LIME	SAND †	WATER*	COAT
100 LBS. (1 PART)	50 LBS. (1 PART)	28 SHOVELS (4 PARTS)	13-14 GALS.	SCRATCH AND FILL IN COAT
100 LBS. (1 PART)	50 LBS. (1 PART)	21 SHOVELS (3 PARTS)	13-13 1/2 GALS.	BROWN COAT
GAUGING PLASTER	DRY LIME	SILICA SAND	WATER*	COAT
20 LBS.	50 LBS.	10 SHOVELS	7-8 GALS.	FINISH COAT

*U.S. MEASURE   † NO. 2 SHOVELS

or perlite should *not* be used in the plaster as they stop the proper transmission of heat. In electrical heating these aggregates would cause the elements to overheat and break down.

Hot Water Radiant Heating Panels

Mix to be used for scratch coat on hot water radiant heating panels should be: 1 part finishing lime, 1 part Portland cement, and 4 parts sand to which sufficient fiber has been added. Fill-in coat should be the same as the scratch. Brown coat: 1 part finishing lime, 1 part Portland cement and 3 parts sand to which fiber has been added. Finish coat should be 1 part finishing lime mixed with 3 parts of silica sand and then gauged with 1 part gauging plaster.

Table 4-11 gives proportions for scratch, brown and finishing coats.

Acoustic Materials

These materials come ready-mixed by the manufacturer, under special formula, and in many cases they need only the addition of water.

It is very important that each manufacturer's specifications be followed precisely. The amount of water to be added, the method of mixing, and the length of time the material is mixed should be adhered to.

In mixing, the amount of water per bag should almost always be put in the mixer first, then the dry materials added. Some materials can be mixed by hand, but the best results in all cases are obtained by machine mixing. Rapid and thorough mixing is necessary to combine the various ingredients of the mix, and sufficient air must be entrapped to "fluff" and lighten the batch. Some types of acoustical materials work or spread better if they are allowed to lie in a box for two or three hours after mixing before being applied. This procedure permits the mixed materials to become plastic.

Exposed aggregates

Epoxy Type Binders

Epoxy is a two-component system consisting of a *resin* and a *catalyst* which causes hardening. If these two components are not combined in some way the epoxy resin will remain soft. The mixing of these components is very critical, and each manufacturer's specifications must be followed exactly. The paddle method is used to mix the troweled-type exposed aggregate or stone-seeded exposed aggregate. (As explained in Chapter 3, the troweled-type has the aggregate in an emul-

sion mix and is troweled on and the aggregate becomes exposed after the emulsion dries. The stone-seeded type is composed of a matrix which is applied to a surface and then the aggregates are applied by hand or by machine.)

Generally speaking, epoxy mat-

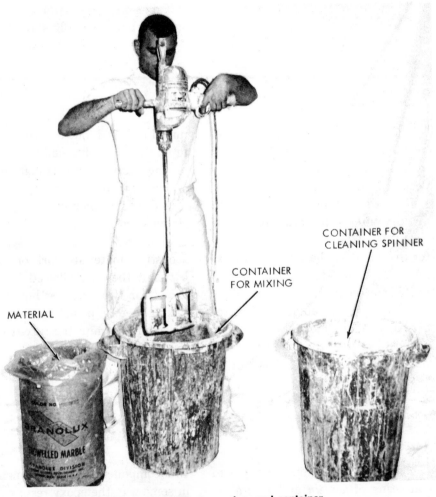

CONTAINER FOR CLEANING SPINNER

CONTAINER FOR MIXING

MATERIAL

Fig. 4-13. Paddle type mixer and container.

rixes are mixed by the following method. Mix the epoxy resin for two or three minutes before adding the catalyst. This is to loosen up the base resin for further mixing; now add the catalyst (hardener) to the resin and mix for about 5 minutes, then add sand, if it is called for, and mix until the sand is thoroughly mixed throughout. To clean tools after mixing, fill a 5 gallon container with water and add detergent and some sand. Heat the water and keep it boiling, run the mixing paddle in the container to remove excess-epoxy and wipe with a cloth. Lacquer thinner can also be used to clean epoxy while it is in a wet state. Once the epoxy has set-up or has become hard it will usually have to be chipped off, and this is very difficult.

Liquid Acrylic, Polymer, Latex, Lastric Type Binders

For mixing the trowel-on exposed aggregate of this type, use the paddle method. Each manufacturer has his own specifications as to how the material is to be mixed, and these specifications should be followed precisely. Generally speaking, these are mixed by the following method. Put the material into a round plastic container and with the paddle type mixer, mix the ingredients for about 3 to 5 minutes to loosen and fluff the material to a desirable texture for workability. See Fig. 4-13.

A small quantity of water may be added, if necessary, to improve workability. Avoid over-mixing, for this may cause the material to run. Mixing should be done in an area with the least amount of direct sun or wind. Heat and moving air will hasten the hardening of the material, and it will be harder to keep tools and container clean if this precaution is neglected. Try to keep all materials covered when not being used.

A handy plastic scoop for getting materials out of the drum after mixing can be made from an empty plastic jug such as a bleach container. The gallon size is the handiest to use. With a sharp knife, cut the bottom off at about $\frac{1}{3}$ the height of the jug; now lay the jug on its side with the handle up and cut away the round side that is up, half way from the bottom at each side to within $\frac{1}{2}''$ of the handle then round off any irregular cuts. See Fig. 4-14.

When mixing the stone-seeded exposed aggregate of this type of binder, follow the specifications of the manufacturer exactly. The paddle type or plaster mixer can be used to mix the materials. This will depend on the quantity needed at one time.

Some manufacturers provide their special liquid binder in a ready-to-use state and others provide the binder in a liquid concentrated form requiring water in specific proportions. Water, when added, should be

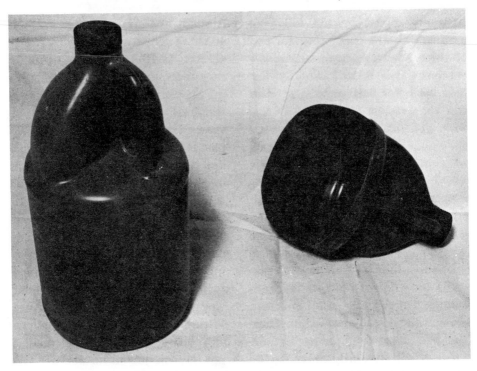

Fig. 4-14. Handy scoop (right) made from bleach container (left).

as cool as possible.

Before removing any binder from the manufacturer's container it should be well stirred. Binders of this type have ingredients that will settle to the bottom in a thick layer during shipping and storage. The area for mixing should be out of the direct sunshine, for heat will decrease the "open time" (the length of time aggregates can be seeded into the applied matrix) and direct sunshine can nearly double surface temperature (as compared to air temperature).

When mixing a large quantity of matrix use the plaster mixer and add the amount of cool, if possible, prepared liquid binder necessary for a bag of Portland cement. This may be 5, 6 or 7 gallons depending on the manufacturer. Now add a cement retarder or a water retainer if recommended by the manufacturer. These are usually recommended only when air and wall temperatures exceed 80°F. The amount to be added is specified by the manufacturer. Note: some manufacturers recommend the addition of these at the end of the mixing. Now add about ½ the required silica sand, then add a bag of Portland cement, either white or regular. There is a color variation in all cements. It is, therefore, essential that succeeding batches used in a combining area be of the same brand and date. If two different brands or dates must be used, plan your mixing to avoid a color differential in panels directly adjacent to each other.

The prepared binder will be used as the wetting agent for the cement and sand. The sand that is usually used for this type of matrix is 120 mesh graded silica sand, sometimes called "bond sand" or "banding sand." However, 60 to 70 mesh silica sand can be used. The finer sand will tend to make a richer matrix which will make the seeding of the aggregates a little easier, because the aggregates will tend to cling more when being applied. Now slowly add the remainder of required silica sand. Let the mixer continue mixing for 5 more minutes, then stop the mixing and let the material remain in the mixer for 2 or 3 minutes. This allows the matrix to "buck up" (stiffen). After the alloted time continue mixing for about one minute, adding, if necessary, a little more liquid binder to bring the material to the right consistency.

When mixing a smaller batch use the paddle type mixer. First, though, determine the manufacturer's specified proportions of binder, sand, and Portland cement. As an example: one bag of Portland cement to 1½ bags of silica sand (120 mesh) using 2½

gallons binder to 2½ gallons of clean cold water. From this we know the ratio of binder concentrate to water is 1 to 1: 1 part water to 1 part binder concentrate. The ratio for cement to silica sand is 1½ to 1: 1½ parts silica sand to 1 part Portland cement. (For practical measuring, it is assumed that a bag of cement is equal to a bag of silica sand by volume or by weight). Now to make the smaller batch of matrix, pour a little prepared binder into a round container and add sand and cement using one measurer. Measure 1 cement and 1½ sand, or for convenience 2 cement and 3 sand until the mix reaches the desired consistency or workability.

Mixed matrixes of these types if dumped on a mortar board will have too much surface exposed to the air and will dry and require retempering. Therefore, keep the mixed matrixes in a pail or container and scoop it out as you use it.

Marblecrete

This is one of the oldest forms of exposed aggregate, and it has been used in western Canada and some parts of the United States long before the modern name of exposed aggregates and their systems were developed. Its appearance is the same as today's patented seeded exposed aggregate; however, it does not have the bonding quality or strength of today's patented exposed aggregates.

The scratch and brown coats are

the same as used for Portland cement stucco. The bed coat (the coat into which the aggregate is seeded) is of different proportions than used for stucco. The materials used to make Marblecrete are readily available to all plasterers. They are Portland ce- ment, autoclave lime, coarse silica sand, and marble dust. There are many formulas that have been used successfully. One formula that has been used is: 1 bag Portland cement (either white or natural), 1 bag auto- clave lime, 1 bag silica sand (fairly

TABLE 4-12 GUIDE CHART: MARBLECRETE (APPROXIMATE AVERAGE)

PORTLAND CEMENT	SAND	AUTOCLAVED LIME	WATER*	NO. 16 MARBLEDUST	NO. 20 MARBLEDUST
100 LBS. (1 PART)	1 BAG	50 LBS. (1 PART)	20 GALS.	100 LBS.	100 LBS.

*U. S. MEASURE

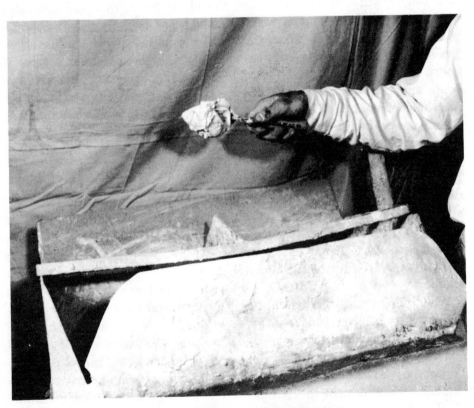

Fig. 4-15. Plasterer testing mix with pointer.

coarse, 20 mesh or "flintshot"), 1 bag of No. 16 marble dust, 1 bag No. 20 marble dust, and about 20 gallons of water. See Table 4-12 for proportions of marblecrete.

To mix marblecrete use a plaster mixer for best results. First add the required amount of water, then add ½ bag "flintshot" silica sand, 1 bag of autoclave lime, and 1 bag of Portland cement. Then add 1 bag No. 16 marble dust, and slowly add 1 bag of No. 20 marble dust. Now add ½ bag of "flintshot" silica sand. Continue mixing for about five minutes, adding water as needed to bring to a good fluffy mix that will be able to carry its own weight when ½" thick. This can be tested by dipping a pointer into the mix and holding it upturned sideways. See Fig. 4-15.

Architectural Aggregates

The architectural aggregates will be mixed as specified by the architect with a definite size, color and combination. Usually a sample is submitted and approved, and from this we can obtain the correct proportions.

After materials of the selected color and sizes have been sent to the job it will be necessary for the plasterer to mix them if more than one color and size are to be used. Colors are classified by name, such as, "polar white," "royal green" and "black obsidian." The sizes are classified as No. 1, No. 2, No. 3, No. 4; or combination "A", "B", and "C". "A" is a combination of sizes No. 1 and No. 2; "B" is a combination of sizes No. 1, 2, 3; and "C" is a combination of Nos. 1, 2, 3 and 4.

Mixing. Have a clean hoe, screen, and mortar box in which to mix the required aggregates. Before mixing, the aggregates should be screened. The reason for screening is not to classify the aggregate as to size, but to eliminate unwanted dust that accumulates in the bags through handling by the suppliers. If this dust is allowed to remain mixed with the aggregate, it could discolor or make unwanted blemishes in the background of the exposed aggregate, and this stain usually cannot be removed. It also protects the plasterer, if he is using the gravity air blower gun to apply the stone, from excessive dust being blown (by the back pressure of the air) into his face. A ⅛" screen can be used for all sizes of stones. It is well to note that the screen is outside the box and the dust that falls through the screen remains outside

Fig. 4-16. Screening of architectural aggregates.

the box as shown in Fig. 4-16. The screen can be arranged so that the aggregates are easily screened and passed on into the box with little effort.

As an example, assume that in a sample submitted there are three different aggregates used: 25% arctic white, size No. 2; 50% dark pearl gray, size No. 2; and 25% Michigan pink, size No. 1.

For this sample, first screen and add 2 bags of dark pearl gray, size No. 2 aggregates; second, screen and add 1 bag of arctic white, size No. 2 aggregates; and third, screen and add 1 bag of Michigan pink, size No. 1. Now take a mortar hoe and mix the stones into an even combination. The mixed stones are now ready to be used for seeding.

Direct-Applied Fireproofing Materials

Most fireproofing materials do not require any mixing and are put into a special machine that breaks them up and feeds them to the plasterer through a hose. There are some however, that need to be mixed with water, and these are the types that usually can be applied with the standard plastering gun. For this type, machine mixing is necessary. A faster than normal mixer speed gives a more uniform mixing operation. Before mixing, check the bag to see the manufacturer's recommendations as to the amount of water to be used, and follow his advice. The average is about 30 to 34 gallons of water for each bag. This should be mixed for a minimum of 4 to 5 minutes to develop fluidity for pumping. When pumping a greater distance more water will have to be used.

Mixing Color in Mortar

Methods used to mix color with the putty vary in different localities. However, to insure complete mixture with the least effort, a dry mix is considered best. Measure out the dry color and add it to the dry lime. Then screen both of these materials through a fine mesh screen. This is necessary in order that any lumps present may be broken up. This screening is also a means by which the two materials may be combined quickly. The combined color and lime are now screened into the water to form a putty.

Be sure to measure each ingredient carefully, even to the water used to soak the lime. Keep a record. If more material should be required later, the same proportions will produce the same color. When the putty is gauged with the plaster, the water and plaster should also be measured. Gauge enough each time to do an entire room or ceiling.

If the putty was soaked without color, the best procedure for adding color is as follows. Measure out a given amount of putty and place it on the finishing board. Form the putty into a small ring, then take the required amount of dry color and mix it with a measured amount of water in a pail or bottle. Stir the mixture well, then pour it through a fine screen or coarse cloth to break up any remaining lumps. Pour the dissolved color into the ring of putty and begin to mix the putty and color using hawk and trowel. See Fig. 4-17.

When the putty and color have been combined it is good practice to screen the mixture again. A lump of color that is barely visible to the eye will spread into quite a large spot when the material is troweled on. Too much care, therefore, cannot be taken to eliminate lumps.

The mixing of colors with matrixes of exposed aggregate must be more precise than the mixing of color for troweled or floated surfaces. This is because the color exposes itself around the colored aggregates and

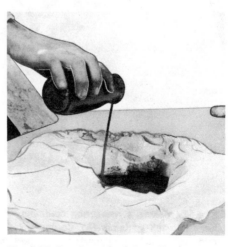

Fig. 4-17. Pouring color into putty ring.

any corrections afterward, such as painting, cannot be done without losing the color effect of the aggregates.

The color on the job will be added in the same amounts and type as was used in the job sample. Color is mixed with the dry cement and sifted through a dry screen to break up any

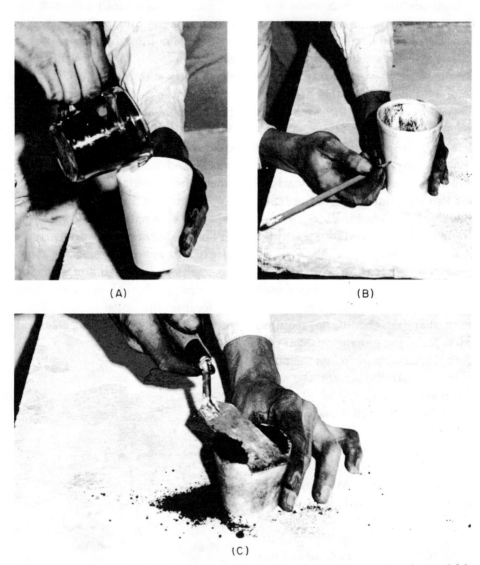

(A)

(B)

(C)

Fig. 4-18. Making a measuring container in three steps. Measured quantity of material is poured into full container (A). Container is marked to level of measured material (B). After cutting off container to marked line, new batch of material is poured into measure and leveled with trowel (C).

lumps of cement or color; then the color cement is used as the cement in the mixing as described earlier.

As an example, assume that a sample reads 32 oz of yellow oxide and 8 oz of black oxide to one bag of white cement.

The ounces are by volume and we can use ordinary kitchen measuring equipment to measure 32 oz of yellow oxide. Now make a container that will equal this amount when sliced flush with the top. We do the same for the 8 oz of black oxide. This will save time when measuring again. Figure 4-18 illustrates how to make these containers.

To mix, place ½ bag of white cement in a clean mortar box and spread it out. Then sift through a screen (an ordinary window screen will do) the yellow oxide and the black oxide coloring measures over the cement. The remaining ½ bag of white cement is spread over this.

Continue this operation until a sufficient amount of cement is prepared for a selected area. Now hoe this material several times back and forth, until it appears to be combined. Then screen the combined material into another box from which the cement for mixing will be taken.

Mixing Drywall Cement

When mixing drywall cement always follow the manufacturer's recommendations regarding the amount of water to be used. This is usually about 2 gallons of water per 25 lbs. of cement powder. Lukewarm water (not hot) will usually mix more easily than cold water. Do not over-thin the cement with water when mixing because this will make the material difficult to apply and will cause excessive shrinkage as well as bond failure. Always add the cement powder to the water and not the water to the cement powder. After mixing, it is best to let the material soak for about 30 minutes before using. When

you are ready to use the material, a final short mix will be necessary to fluff up the material to the right consistency for spreading.

For small amounts, some plasterers use a drywall masher. This is a tool shaped like a potato masher. The required amount of lukewarm water is put in a clean container and the mix is added. The material is then stirred with the masher. As it begins to thicken, start mashing the mix by pushing the masher up and down and around through the mix.

For larger quantities of mix the plasterer should use a round container that is big enough to hold the material and water combined. Follow the same procedure of adding the powder cement into the water, and then use a power driven mud mixer to blend the water and cement.

Checking on Your Knowledge

DO YOU KNOW

1. What are the three things a plasterer should consider before he begins mixing?

2. Name and describe the methods of mixing materials.

3. What is the proportion mix of sand and gypsum for low suction?

4. How does a plasterer prepare lime for finishing?

5. How does a plasterer mix putty coat?

6. What should be added to a putty coat mix which is to be applied over lightweight aggregates?

7. What is sand finish, and how is it prepared?

8. When should Portland cement be used?

9. What are the proportions for cement and sand?

10. How much aluminum stearate can be added to a cement mix?

11. What is the proper material used for hot water, electrical heating panels?

12. Name two types of materials used for exposed aggregates?

13. What is the mix for marblecrete?

14. How does a plasterer mix coloring with this material?

15. How is drywall joint cement mixed?

Lathing and Plaster bases

Plaster is applied over various bases. The type used is, in many cases, dictated by custom in the area, type of work being done, or cost. Over the years there have been many changes and developments in the materials used as the base for plaster.

Thin sticks and reeds were used by primitive man as the base for coats of mud to enclose his hut. The Romans used reeds for much of their plastering work. Gradually this type of lath evolved into the wood lath that remained popular for many centuries. Wood was plentiful everywhere and it was a simple process to split straight grained lengths of logs by hand into strips called *riven lath* (split lath) because the split occurred with the grain of the wood.

The writer can recall his father telling how he and his brothers would split logs into lath in their father's shop during the winter months while work was slack. Later, this method became too slow and sawn lath was developed: first handsawn then later machine sawn. The sawn lath was made uniform in size, but thicker to overcome the warping problem caused by the grain crossing the lath in many cases. As the supply of good wood ran out in many areas, problems began to develop due to buckling and warping.

The plastering industry did much experimenting to develop a better plastering base. After much work and many trials, gypsum lath was developed. The first gypsum lath, or plasterboards, were cast in sections on a bench and then nailed up using large headed nails. Most of these

161

boards were thicker than was desirable and many methods were tried to achieve a strong but thin board.

Shortly after the turn of the century, there appeared a board made of gypsum covered with paper on both sides. A board of this type which gained wide use was called "Sackett Board." This board was constructed of four layers of paper and three alternating layers of plaster.

Owing to their great weight, these boards were used chiefly to form partitions. Later developments reduced the weight and size of the boards. About 1920, marked progress was made in the fabrication of this material. At this time, the board size was fixed at 16" x 48", except in the western states, where the size was set at 16-1/5" x 48". Thickness, at this time, was limited to ⅜" to keep the board as light as possible and to conform to the wood lath thickness.

Improvements in paper and plaster core have been constant. The plaster core is now made up of various admixtures that tend to reduce the weight so that there are now boards of ½" and ⅝" thickness on the market.

Perforated plasterboard came on the market about 1932. This type provides a more nearly fireproof bonding surface than that of the regular board. The perforations allow a more adequate mechanical bond in addition to the adhesive bond caused by the suction that takes place in most plasterboard products. As a result, perforated boards can accept a thicker surface of plaster mortar and are, therefore, more nearly fireproof. A plaster coating ½" thick applied to boards of this type has a fire rating (resistance to fire) of one hour. Additional thickness of lath and plaster will provide added fire rating.

Plasterboard covered on one side with a paper-thin (glued on) sheet of aluminum foil has been in use now for a number of years. This board has a plaster base paper on one side and the aluminum foil on the other; the foil reflects both heat and cold plus preventing moisture penetration so it acts as an insulator and vapor barrier.

Veneer plastering and drywall systems have now changed the sizes and types of boards used for this work. Drywall and some veneer systems use 4' x 8' and up to 16 ft long boards either ⅜" or ½" thick, plus ⅝" thick boards used in special areas. The longer boards are used to eliminate as many joints as possible. This saves time and money and produces a truer overall surface. Some custom jobs call for two layers of board to be applied; the first one nailed to studs and the second one glued to the first.

Fiber boards of various kinds have been in use for about as long as gypsum boards. Insulation and plaster base are combined in one unit. Due to porous construction these boards are lighter in weight and are thicker than the gypsum boards. Various thicknesses produce corresponding degrees of insulating values. Boards are available in both 16″ x 48″ and 4″ x 8″ sizes and the common thicknesses are ½″ and 1″.

Metal lathing was first patented in England about 1797. At that time it consisted of a netting made of wire. The first actual use of a metal plaster base occurred in England about 1841. In 1890, expanded metal lath was produced in Chicago, Illinois by Mr. J. G. Golding. Thereafter, many forms of metal lath appeared. The "Bostick" sheet lath was one of the favored types in use for many years.

Since that time many types and weights of metal lath have been placed on the market. Among the popular metal laths in use today are: diamond mesh, rib lath, woven wire lath, paper-backed wire, and expanded metal lath plus sheet lath.

Wood Lath

Wood Lath is made of white pine, spruce, fir, redwood, and other soft, straight grained woods. The standard size of wood lath is $\frac{5}{16}$″ x 1½″ x 4′. Each lath is nailed to the studs or joists with 3*d* blued lathing nails.

Laths are nailed six in a row, one above the other. The next six rows of lath are set over two stud places; the joints of the lath are staggered

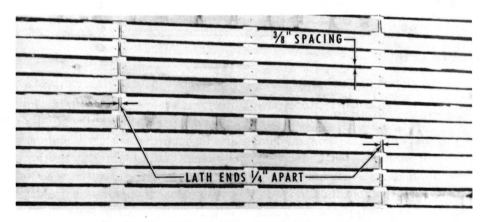

Fig. 5-1. Wood lath with joints staggered every sixth course.

in this way so that cracks will not occur at the joinings. Lath ends should be spaced ¼″ apart to allow movement and prevent buckling. Figure 5-1 shows the proper layout of wood lath. To obtain a good key (space for mortar) the lath should be spaced not less than ⅜″ apart. Fig. 5-2 shows good spacing with strong keys.

Wood laths come 50 or 100 to the bundle and are sold by the thousand. The wood should be straight grained and free from knots and from an excessive amount of pitch. Old lath should not be used, as dry or dirty lath offers a poor surface for bonding of the plaster. Lath must be damp when the mortar is applied or else the dry lath will pull the moisture out of the mortar preventing the proper setting action from occurring. Best method is to wet the lath thoroughly the day before the plastering is to be done. This permits the wood to swell and then reach a stable con-

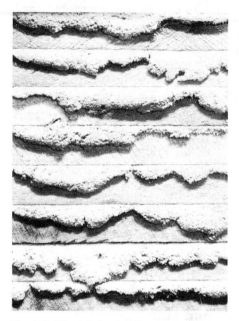

Fig. 5-2. Wood lath, showing proper keys.

dition ideal for plaster application.

Wood lath is seldom used today. It is still used sometimes, however, in remodeling and some patching jobs.

Board Lath

Of the many kinds of lathing materials available, board lath is the most widely used today. Board lath is manufactured from mineral and vegetable products. It is produced in board form, in sizes generally standardized for easy application to studs, joists, and various types of wood and metal furring.

The advantages of board lath are many. It is rigid, strong, stable, and reduces the possibility of dirt filtering through the mortar to stain the surface. It is insulating, strengthens the framework structure, and the gypsum board is fire resistant. Board lath also requires the least amount of mortar to cover the surface.

Board laths are divided into two main groups: gypsum board and insulation board. Each type and its subdivision will be covered separately.

Gypsum Board Lath

Gypsum lath is now made in a number of sizes, thickness, and types. Each type is used for a specific purpose or condition. One basic condition must be kept in mind: only gypsum mortar can be used over gypsum lath. Never apply lime mortar, Portland cement, or any other binding agent to gypsum lath.

The most commonly used size in board lath is the ⅜″ x 16″ x 48″ in either solid or perforated type. This lath will not burn or transmit temperatures much in excess of 212°F until the gypsum is completely calcined. The strength of the bond of plaster to gypsum lath is great, it requires a pull 864 lbs. per square foot to separate gypsum plaster from gypsum lath (based on a 2 to 1 mix of

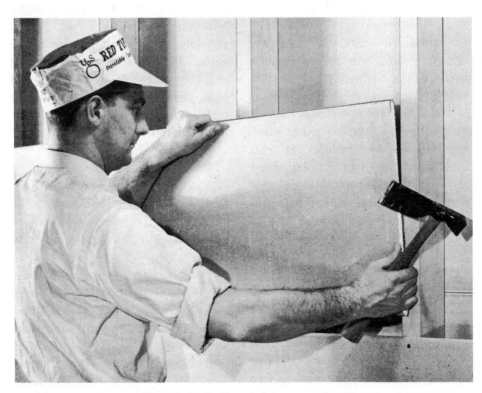

Fig. 5-3. Application of plain gypsum board.

sand and plaster mortar).

There is also on the market special fire retardant gypsum lath called "Type X." This lath has a specially formulated core containing minerals which give it additional fire protection.

The following paragraphs discuss the basic groups of gypsum boards available for numerous patented systems as well as for general use. Each manufacturer has his own brand names and special manufacturing techniques used in producing the boards.

It is wise practice to use only one manufacturer's materials for a specified job or area so that the warranties which back the materials can apply. Always follow the manufacturer's specifications to the letter for materials and conditions of application.

Plain gypsum lath plaster base is for nail or staple application to wood and nailable steel framing, for clip attachment to wood framing, steel studs, and suspended metal grillage; and for screw attachment to metal studs and furring channels. Sizes: 16" x 48", 3/8" or 1/2" thick; 16" x 96", 3/8" thick. It is also made 16-1/5" wide for Pacific Coast area. Fig. 5-3 illustrates this board.

Perforated gypsum lath plaster base is the same as plain gypsum lath except that 3/4" round holes are punched through the lath 4" o.c. (*on center*, meaning between centers) in each direction. This gives one 3/4"

hole for each 16 square inches of lath area. See Fig. 5-4. This provides mechanical keys in addition to the natural plaster bond, and obtains higher fire ratings. Fig. 5-5 illustrates how the keys work. This lath is not recommended for attachment to ceilings, with either wood framing or metal grillage, where the only support provided is by clips at the edges. Perforations weaken this lath, therefore some mid-support is required.

Insulating gypsum lath plaster base is the same as plain gypsum lath, but has bright aluminum foil laminated to the back side. This creates an effective vapor barrier at no additional labor cost. In addition, it provides positive insulation value when installed with the foil facing a 3/4" minimum air space. When used as a ceiling, under winter heating conditions its heat resistance value is approximately the same as for 1/2" insulation board.

Long length insulating gypsum lath is used primarily for furring exterior masonry walls. Sizes: 24" wide, 3/8" thick, lengths mill cut as required up to 12 ft.; formed with square edges.

Veneer plaster gypsum lath base is a special gypsum lath in large sheet form with strength and absorption characteristics designed for use with special veneer plasters. It is available in Plain and Type X core; 4 ft. wide, 1/2" or 5/8" thick, 8 to 12 ft. lengths. The Type X is a fire-rated

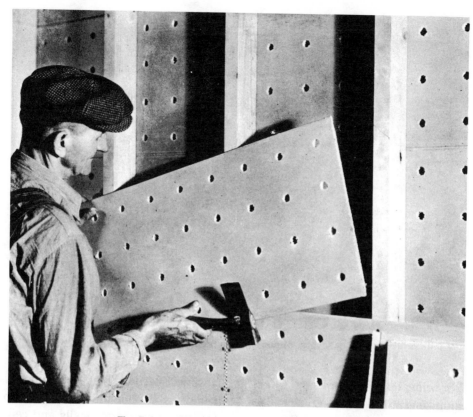

Fig. 5-4. Application of perforated gypsum board.

core which adds fire protection.

Radiant heat gypsum lath plaster base is a specifically fortified large-size gypsum lath, used with special radiant heat plaster in electric cable ceilings. Available 4 ft. wide, regular ½″ or ⅝″ thick; Type X ½″ thick; 8 to 12 ft. lengths.

Gypsum wallboards are made in ½″ and ⅝″ thickness for single layer application, ⅜″ for multi-ply construction; available ¼″, ⅜″, ½″, ⅝″ thick; 4 ft. wide; to 16 ft. long with tapered, beveled or square edges.

Gypsum backer boards are economical base layer boards for multiply wall, wallboard construction, ⅜″ thick; 4 ft. wide; to 12 ft. long. Square edges. Also available fire resistant backer board that meets U. L. requirements for one hour fire rated construction. ⅝″ thick; 2 ft. wide; 8 ft. long. Tongue and groove edges.

Coreboard is used for solid gypsum core partition systems. 1″ thick;

Fig. 5-5. Keys formed with perforated gypsum board.

2 ft. wide; length as requested (maximum length 12 feet). V-groove edges.

Gypsum studs are used for semisolid gypsum core partition systems 3/8" to 1 5/8" thick (in 1/8" increments); 6", 8" wide; length as requested (maximum length 12 feet).

Gypsum sound-deadening board is a special gypsum backer board for sound rated construction. Available in 1/4" thickness, 4 ft. wide; to 16 ft. long. Square edges. Also 1/2" thick, 4 ft. wide, 8 ft. long.

Insulation Board Lath

These boards are used in the same general manner as the gypsum lath. However, because they are primarily insulating, it is common practice to use them on exterior walls and ceilings. There are two basic types in use today: those made of organic fibers, and the inorganic materials such as plastics and glass.

Various thicknesses produce corresponding degrees of insulating values. Boards are available in many sizes and types, each manufacturer producing sizes for specific needs. For plastering purposes the fiber boards are available in 16", 18" and 2 ft. widths; 4 ft. lengths and in 1/2" or 1" thickness. There are also 4 ft. x 8 ft. boards of various thicknesses.

The inorganic types come in numerous sizes because they are used for many purposes other than as plastering bases, such as cold storage rooms, etc. Some of these are: 16" x 9 ft. long; 16" x 4½ ft. long and 2 ft. x 8 ft. long. Thickness: 1", 1½", 2", 3" and 4". All edges are square.

Fiber boards are made with special joints for reinforcing the plaster at the joints. Types available include the following: long edges shiplapped, galvanized wire reinforcing between framing supports; V-lap edge on the long side, beveled on all edges; long edges with tongue and groove; beveled, shiplapped edges; and ship-

lapped on long edges with a 3" diamond-mesh metal lath strip the full length of the long edges.

There are also modifications and variations of these edge treatments but all are intended to perform the same function, namely, to reinforce the plaster at the joints between the individual lath units. Special vapor resisting lath is also available.

Board Lath Attachment Methods

All the small size board are applied with the long edges at right angles to the framing members. Center all joints on framing members except for the clip type attachment

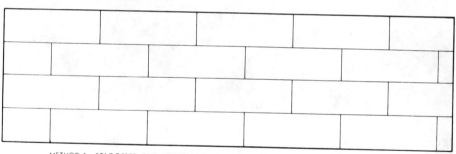

METHOD A--STAGGERED END JOINTS AND LONG SIDE JOINTS IN LINE. (NOT DRAWN TO SCALE)

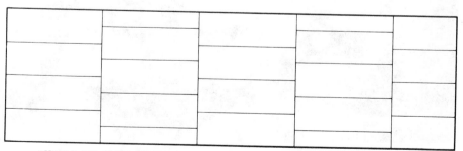

METHOD B--END ALL END JOINTS IN LINE AND LONG SIDE JOINTS STAGGERED. (NOT DRAWN TO SCALE)

Fig. 5-6. Two methods of aligning joints in board lath.

systems which bring the board joints between framing members.

Two methods of joint alignment are in use today, each method has its proponents and opponents. Fig. 5-6 illustrates the two methods. In general, the lath is applied with staggered end joints and long side joints in line. Or, all end joints are butted on a framing member in line and the side joints are staggered a half board width. With this method a three inch strip of metal lath is applied over all the butt joints.

Do not force the boards tightly together, let them butt loosely so that the board is not buckled or under compression before the plaster is ap-

Fig. 5-7. Lather attaching perforated gypsum lath to wood studs, using pneumatic stapling gun. (Bostitch Div. of Textron, Inc.)

plied. Do not use small pieces except where necessary.

The most common method of attachment of the boards has been the lath nail; however, staples are now gaining favor because of the power guns available to drive them. There are also numerous patented clip and rod type systems used to hold the boards in place, each with a particular purpose or reason for its use.

The nails used are of a special

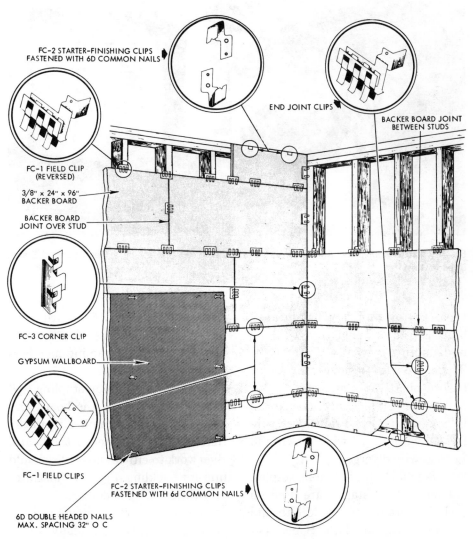

FC-2 STARTER-FINISHING CLIPS
FASTENED WITH 6D COMMON NAILS

END JOINT CLIPS

BACKER BOARD JOINT
BETWEEN STUDS

FC-1 FIELD CLIP
(REVERSED)

3/8" x 24" x 96"
BACKER BOARD

BACKER BOARD
JOINT OVER STUD

FC-3 CORNER CLIP

GYPSUM WALLBOARD

FC-1 FIELD CLIPS

FC-2 STARTER-FINISHING CLIPS
FASTENED WITH 6d COMMON NAILS

6D DOUBLE HEADED NAILS
MAX. SPACING 32" O C

Fig. 5-8. Clips used with backer boards over wood studs. (National Gypsum Co.)

Fig. 5-9. Lather applying perforated gypsum lath, using rod clip lathing system. (United States Gypsum Co.)

type: 1⅛″ x 13 gauge, flat headed, blued gypsum lath nails for ⅜″ thick boards and 1¼″ for ½″ boards. There are also resin coated nails, barbed shaft nails and screw type nails in use. Staples should be No. 16 U. S. gauge flattened galvanized wire formed with a ⁷⁄₁₆″ wide crown and ⅞″ legs with divergent points for ⅜″ lath. For ½″ lath use 1″ long staples.

Four nails or staples are used on each support for 16″ wide lath and five for 2 ft. wide lath. Some special fire ratings, however, require five

nails or staples per 16″ board. Five nails or staples are also recommended when the framing members are spaced 24″ apart.

Start the nailing or stapling ½″ in from the edges of the board and nail first on the framing members that fall on the center of the board, then work to either end—this should prevent buckling. Fig. 5-7 illustrates how the stapling is done.

The clip and rod systems used to hold the boards offer many variations and purposes for their use. They

Fig. 5-10. Lather sticking styrofoam board to masonry exterior wall, providing both insulation and plaster base. (Dow Chemical Co.)

can hold the boards tight to the framing members; they can fur them away from the surface or by the use of resilient clips provide for a reduction in sound transmission; or they can isolate the boards away from the framing members so as to develop a floating surface to avoid transmitting structural movement to the lath and plaster. Figs. 5-8 and 5-9 illustrate a clip and a rod system.

Insulation and gypsum boards can also be glued to the framing members or masonry surfaces. See Fig. 5-10. The use of a special mastic permits the boards to adjust to the surface. This provides further insulation due to the air space while at the same time it isolates the boards from structural and temperature change movements.

Reinforcement and Accessories

Many systems exist for the reinforcement of joints where gypsum and insulating boards meet. Metal lath or woven wire, applied as reinforcement, is nailed or stapled to the boards at crucial points to prevent possible cracking of the plaster finish later. See Fig. 5-11. Initially, the practice was to cover all the joints with a narrow strip of metal lath. Later, however, only the end joints were covered. Some plasterers cover only the angles formed by the tops of doors and windows.

Today, plasterers find that metal or plastic reinforcement is useful at the corners of the room (both where the walls meet and where wall and ceiling join), over angles formed by the tops of doors and windows, or at any other opening in the walls or ceilings.

Fig. 5-12 illustrates the installation of metal stops around a window. This combined reinforcement and plaster stop provides a neat finish at a metal window and prevents damage to the plaster when the metal window sweats in cold weather (condensation). The stop is set about $\frac{1}{8}$" away from the metal window so there will be no direct metal to metal contact. This type of stop is also used wherever the plastering must be terminated without butting into other materials. The stop also acts as a guide to establish the plaster thickness and alignment at this point.

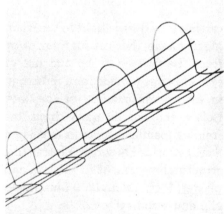

Fig. 5-11. Lather applying "Cornerite", a metal reinforcement, left. Right: "Bentrite", a galvanized welded wire mesh corner reinforcement. (K-Lath Corp.)

Fig. 5-12. Lather installing a metal plaster stop to prevent the plaster from touching the metal window. (Dow Chemical Co.)

The reinforcement materials have specific names that help to identify them. For example, reinforcement used where walls meet or walls and ceilings meet to form interior corners is called "Cornerite". Where walls meet to form external corners various types of metal reinforcement called

175

Fig. 5-13. Lather installing metal corner bead which provides a strong, straight reinforced corner at the external edge of this window jamb. (Dow Chemical Co.)

corner beads (Fig. 5-13) are used. Where boards meet on plane surfaces such as walls and ceilings and the conditions call for reinforcement, the strip of material used is called "Strip-ite". It is used for reinforcing joints and any areas where cracks might develop due to structural movement.

The reinforcements are made in various forms both metal and plastic. Fig. 5-14 shows plastic mesh being used for joint reinforcement.

Fig. 5-14. Lather stapling plastic mesh over joint to reinforce it. (National Gypsum Co.)

Common practice today is to nail or staple the reinforcement into the lath only and not into the structural members. The principle involved is to reinforce the lath and plaster only, permitting the structural members to move independently without cracking the plaster.

Partitions, Veneer Systems, and Wallboard

A great variety of patented partition systems are on the market today. Each manufacturer has developed a full line of systems to suit every need and design condition. This book can cover only some of the basic types to show the difference in construction. The reader should obtain the various manufacturers' catalogs to acquaint himself with the full line each company offers.

The partition systems fall into two broad categories: solid and hollow. These are then broken up into subcategories of stud, studless, solid studs, trussteel stud, gypsum core, gypsum lath, or metal lath.

The most common is the stud type using any one of the various metal studs on the market. To erect this type, the lather places a runner on the floor at the right location, fastens it down; plumbs up to the ceiling and locates the ceiling line. Next a ceiling runner is fastened in place and then the studs are set in between these two runners. In most cases, automatic spacing of the studs is pro-vided in the runners. Various methods are used to tie the studs to the runners: wire ties, clip, screws, punched locking tabs, etc.. The studs are available in various sizes: 1⅝", 2½", 3⅝", 4" and 6" wide.

The lath is then applied to the studs using clips, tie wire, screws or special nails. The lath may be expanded metal or rib lath; gypsum long length lath of various thickness; or laminated gypsum lath, which is used to produce stronger walls. Fig. 5-15 illustrates a typical metal stud construction.

The solid partitions are constructed of either long length gypsum boards or metal lath. These

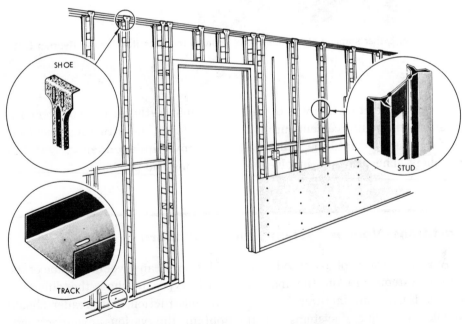

Fig. 5-15. Metal stud construction. (National Gypsum Co.)

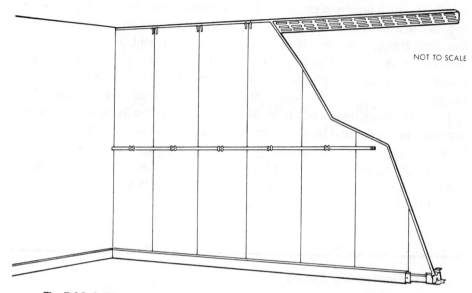

Fig. 5-16. Solid rock lath plaster base partition. (United States Gypsum Co.)

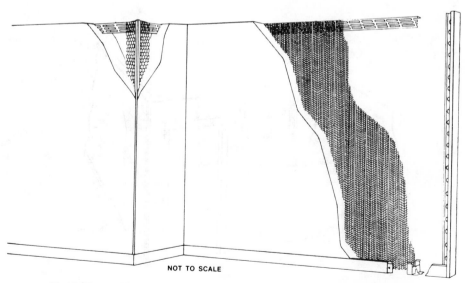

Fig. 5-17. Solid studless ⅜" rib lath partition. (United States Gypsum Co.)

partitions use a floor and ceiling runner on the same general principle as the stud types; however, these are much narrower as most solid plaster partitions are in the 2″ thickness range.

Fig. 5-16 shows how floor and ceiling runners are set and how the long length gypsum lath is installed and temporarily braced with a horizontally fastened channel iron set midway of the partition. This channel is removed after both sides have been scratch coated and the opposite side browned in and set.

Construction of the solid, studless metal lath partition is shown in Fig. 5-17. The metal lath is the ⅜″ rib type set vertically and tied top and bottom. The same system of temporarily bracing the lath during the plastering operations is used as for the gypsum board type.

Veneer Plastering Bases

The term veneer plaster refers to a system of applying an extremely hard, strong plaster with a total thickness of from ³⁄₃₂″ to ¼″ to a specially manufactured gypsum board base. The manufacturer's directions must be followed exactly to insure a good job. Fig. 5-18 shows a single coat veneer plaster system.

The boards used for these systems

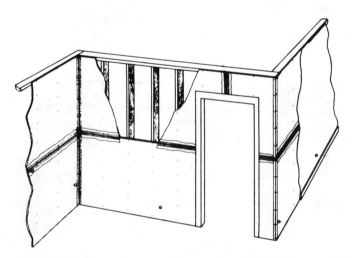

Fig. 5-18. Single coat veneer plaster system. (National Gypsum Co.)

Fig. 5-19. Drywall screw point types. Both types have "bugle" heads. (United States Gypsum Co.)

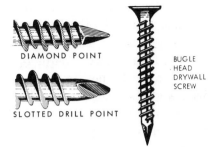

DIAMOND POINT

SLOTTED DRILL POINT

BUGLE HEAD DRYWALL SCREW

Fig. 5-20. Lather using electric screwdriver to screw gypsum boards to steel studs. (National Gypsum Co.)

have a special paper face to create the proper suction for the plaster. All end joints must be set on framing members. Do not force boards tight together and do not butt cut or square edges to tapered edges. Minimize the number of end joints by using the maximum practical length of boards. Stagger joints so that they occur on different framing members and not directly opposite one another on partitions.

For wood studs and furring strips use 1¼″ (3d) or 1½″ (4d) box nails for ⅜″ board, 5d (1¾″) box nails for ½″ board and 6d (2″) box nails for the ⅝″ thick boards. Space the nails 8″ o.c. for studs and 7″ o.c. for

joists. If screws are used they should be 1¼″ type W. For metal studs and furring 1″ type S screws can be used for single layer application. Fig. 5-19 shows various types of screws used for this work.

Special accessories are used with these veneer systems because of the extreme thinness of the plaster coating. The reinforcements must be made to fit tight and straight. A glass fiber tape is used to cover the joints. Staple it on, spacing the staples 24″ apart and keep the tape tight while stapling so it will lay flat against the boards. Fig. 5-20 shows how a lather staples his tape in place.

Wallboards

The plasterer at times is called to tape and cover the joints, nail heads, and corner beads used in the wallboard or drywall systems in use today. The boards used are a specially made product that incorporates special surface papers and has tapered edges to receive the tape and joint cement.

The boards are generally 4 ft. x 8 ft. x ⅜″ or ½″ in thickness. They can be applied either vertically or horizontally depending on the conditions which will produce the least number of joints. It is customary to

apply the ceiling boards first then butt the wall sheet under the ceiling sheets. Fig. 5-21 shows how the boards are applied and how the floating angles are obtained. The proper methods for applying joint cement and tape will be covered later in Chapter 7.

Multi-ply construction calls for a ⅜″ thick board to be applied vertically to the studs and at right angles to the joist. Follow the required nailing pattern. The second or finish board, also ⅜″ thick, is then laminated (glued) to the first layer in a

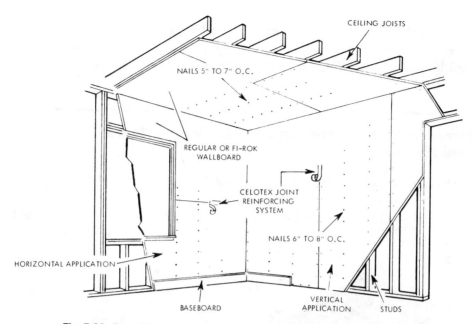

Fig. 5-21. Drywall construction by single layer application. (Celotex Corp.)

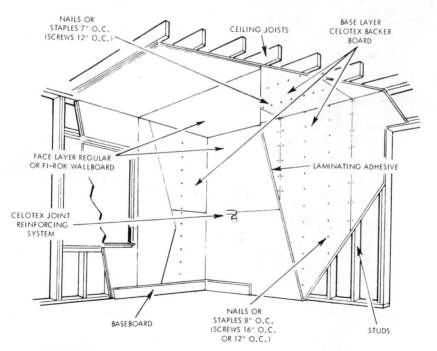

Fig. 5-22. Drywall construction by multi-ply application. (Celotex Corp.)

horizontal position. This system provides fewer joints and only a few nail heads to cover. The first layer can be a backer board as it will be completely covered. Fig. 5-22 illustrates this method.

Fireproofing Structural Steel With Gypsum Board and Plaster

Over the years the industry has developed various systems to protect structural steel and similar framing members from the destructive effects of fire.

One of the most economical and

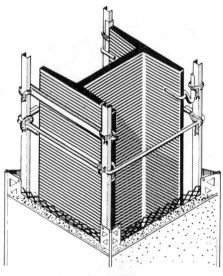

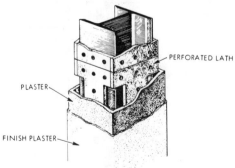

PERFORATED LATH

PLASTER

FINISH PLASTER

Fig. 5-23. Fireproofing structural steel construction. Top, using metal lath and plaster; bottom, using gypsum board and plaster (Bestwall Gypsum Co.).

effective systems employs perforated gypsum lath wire tied in place around the steel and covered with plaster to achieve the required fire rating. There are now systems that will produce 1, 1½, 2, 3 and 4 hour fire ratings. Each rating requires various layers of lath and thicknesses of plaster plus certain types of plaster aggregates.

Fig. 5-23 shows how the gypsum and metal lath and reinforcements are placed around the steel. Specifications must be followed exactly to produce these results. To obtain a four hour fire rating requires the column to be wrapped with 20 gauge galvanized 1″ hexagonal wire mesh. This is done to hold the plaster in place during the intense heat permitting the full thickness of lath and plaster to slowly release its contained water, thereby keeping the heat away from the steel.

Radiant Heat Plaster Base

The type "X", ½″, ⅝″ thick gypsum boards are used as the base for radiant heat ceilings. The boards can be attached direct to wood joist with 5d nails for the ½″ boards and 6d nails for the ⅝″ thick boards. Nails are spaced 6″ o.c. and the joints are covered with glass fiber tape. Electric heat cable is stapled to the boards and covered with a ¼″ thickness of special radiant heat plaster applied in two coats. It is recommended that the perimeter of radiant ceilings be isolated from the intersecting walls by the use of a ¾″ casing bead applied to the joist be-

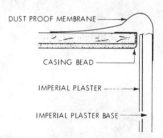

Fig. 5-24. Metal casing bead to isolate radiant heat ceiling from walls to permit expansion and contraction. (United States Gypsum Co.)

fore lathing and spaced ⅛″ away from the walls. Fig. 5-24 illustrates the casing bead and the spacing.

Metal Lathing

Metal lath is perhaps the most versatile of all plaster bases. No other type of lath offers a better key for mortar. Fig. 5-25 illustrates how completely the mortar keys behind the lath to lock the two together.

Metal lath can be divided into four groups: expanded metal lath (diamond mesh), expanded rib lath, wire lath, and sheet lath. Each of these groups has many types and variations which are designed to suit specific needs.

Diamond Mesh

Diamond mesh or expanded metal lath is produced by stamping or perforating a metal sheet and then expanding the sheet by pulling it apart until a meshlike material develops. The standard diamond mesh lath has

Fig. 5-25. Mortar keys forming behind the lath, locking the two materials together. (United States Gypsum Co.)

a mesh size of $\frac{5}{16}''$ x $\frac{9}{16}''$, the mesh is in a diamond pattern. See Fig. 5-25 which shows how the small mesh lath appears in use. The standardized weights for expanded diamond mesh are 2.5 and 3.4 lbs. per sq. yd. Lath is made in sheets of 27" x 96" and are packed 10 sheets to the bundle (20 sq. yds.).

Diamond mesh lath is also made in large diamond mesh which is used for stucco work, reinforcement in concrete work, and as a support for rock wool and similar insulating materials. Sizes and weights are the same as for the small mesh. The small diamond mesh lath is also made into a self-furring lath by forming dimples into the surface which hold the lath approximately $\frac{1}{4}''$

away from the surface. This lath is used for fireproofing columns and beams or on flat wall and ceiling surfaces. Another form is the paper backed lath where the lath has a waterproofed or kraft paper glued to the back of the sheet. The paper acts as a plaster saver and moisture barrier.

Expanded Rib Lath

The expanded rib lath is much like the diamond mesh lath except that various size ribs are formed in the lath to stiffen it. Ribs run lengthwise of the lath and are made, for plastering use, in $\frac{1}{8}''$, $\frac{3}{8}''$ and $\frac{3}{4}''$ rib height. Fig. 5-26 shows a $\frac{3}{4}''$ rib lath. The sheet sizes are 27" x 96" for the $\frac{1}{8}''$ and $\frac{3}{8}''$ rib lath and 29" width,

Fig. 5-26. Expanded rib lath with a $\frac{3}{4}''$ raised rib. (National Gypsum Co.)

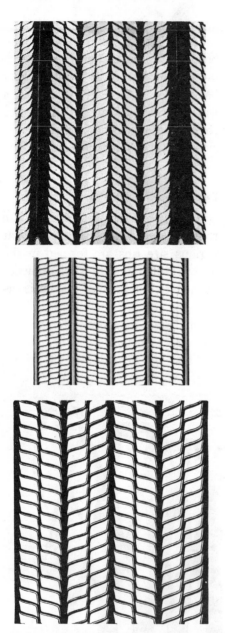

Fig. 5-27. Three types of rib lath. Top, with eight meshes between ribs. Center, with three meshes between ribs. Bottom, with two meshes between ribs. (Metal Lath Manufacturers Association).

and 5, 10, and 12 ft. lengths for the ¾" rib lath. Weights run 2.75, 3.4, 4.0 lbs. per sq. yd., the ¾" rib lath runs .60 and .75 lbs. per sq. ft. Fig. 5-27 shows typical rib lath with varying mesh between ribs.

There is also available a diamond mesh lath with metal rods welded on both sides of the lath back to back. This forms an extremely stiff lath. See Fig. 5-28. All these laths are available in either painted steel or galvanized steel.

Wire Mesh Lath

This type of lath is made in two basic types, woven wire and welded wire. The woven wire lath is made of galvanized wire of various gauges woven or twisted together to form either squares or hexagons. See Fig. 5-29. One popular type called "Keymesh", has 1" hexagonal openings and is woven of 20 gauge wire. It is produced in rolls in various widths and is used over gypsum and fiber insulating lath as a reinforcement; it is also used for cornerite and stripite. It is very popular as a stucco mesh where it is placed over tar paper on open stud construction or over various sheathings.

Another type is made with 2" x 2" mesh, 16 gauge galvanized wire and the wire fabric is interwoven with a fibrous, absorbent paper backing. See Fig. 5-30. The backing is secured by a 17 gauge wire that is corrugated every 4⅜", which provides for an

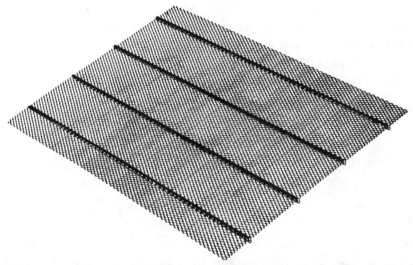

Fig. 5-28. Diamond mesh lath with metal rods welded on both sides to stiffen it. (Metal Lath Manufacturers Association).

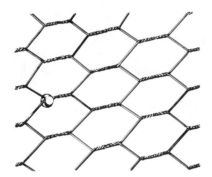

Fig. 5-29. Woven wire lath. (Keystone Steel & Wire Co.)

Fig. 5-30. Paper-backed lath. (John-Manville Co.)

adequate plaster key to be formed. Sheet sizes are 30½″ x 49″. Each shipping carton contains 44 sheets or 50 sq. yds.

This type of lath is made also for stucco use and is popular on open stud construction because of its paper backing.

Welded wire lath has now become very popular because it is very adaptable for machine applied mortar. The K-Lath Corp. produces three types each one designed to suit a specific condition. "Gun-Lath" is, as the name implies, made for machine application of the mortar to the lath. It is made of 16 gauge galvanized wire in a 1½" x 2" mesh with a 13 gauge horizontal stiffener wire every 6". An absorbent, slot perforated,

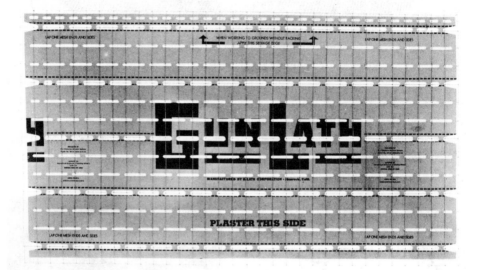

Fig. 5-31. "Gunlath", a paperbacked, welded-wire lath used for either interior or exterior work. (K-Lath Corp.)

Fig. 5-32. "Stucco-Rite", a special waterproof paper-backed welded mesh lath for stucco work. (K-Lath Corp.)

paper separator is fabricated between the face and back wires. See Fig. 5-31.

Sizes are 28″ x 50″ = 1.08 sq. yds., weighing 1.98 lbs. per sheet; and 28″ x 98½″ = 2.12 sq. yds., weighing 3.84 lbs. per sheet. This lath can be used for both interior and exterior work. The lath is also available in a heavy duty type which has an 11 gauge horizontal stiffener wire 6″ o.c. and can be used over supports spaced 24″ o.c.

Aqua-K-Lath is a self-furring lath with a waterproof building paper. Sizes are the same as Gun-Lath, but the weights run 2.07 lbs. per sheet and 4.23 lbs. per sheet. It is used for stucco work and any areas where moisture is encountered. Also available in a heavy duty type for 2 ft. o.c. supports.

Stucco-Rite lath has a special waterproof paper backing and is specifically made for all kinds of stucco work. See Fig. 5-32. It comes only in the 28″ x 98½″ sheet and weighs 4.04 lbs. per sheet. It is also available in a heavy duty type for 24″ o.c. supports. Also available is Pyro-K-Lath, which has a flame spread resistance backing. On special order, sisalkraft, aluminum foil, etc., backings are available.

Sheet Lath

This type of lath was very popular in the early days of metal lathing. It then seemed to fade in popularity,

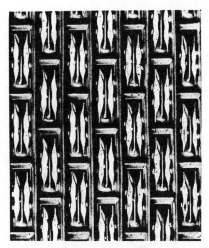

Fig. 5-33. Another example of sheet lath. (Metal Manufacturers Association)

but now with the spread of gun applied plaster and cement mortars this type of lath is gaining favor. The lath is made from sheet steel and has slits punched into it plus ridges and cross bars. The design forms a stiff sheet while the slits provide the keys to hold the mortar. See Fig. 5-33. The design also provides a self-furring condition. Weight is 4.5 lbs. per sq. yd., sheet size 27″ x 96″ also in 24″ x 96″ sheets. Packed in 10 sheets to the bundle (20 sq. yds.) or 9 sheets (16 sq. yds.) per bundle on special order.

Lathing Accessories

Various metal accessories are usually required to complete a lathing job. These accessories serve several basic functions and in some cases serve one or more functions as individual units. See Fig. 5-34 showing some of the items in cross section and their names.

Plaster stops or *casing beads* define the limits of the plastered wall or ceiling and provide a thickness gauge for the plaster at that point. See Fig. 5-34. They also protect the plaster from damage at this terminal edge.

Grounds or *screeds* (sometimes called *base beads*), are available in many types and sizes. See Fig. 5-34.

They are installed primarily to control plaster thickness and alignment but sometimes do double duty in separating the plaster from various other materials, such as a cement base.

Corner beads, as the name implies, are placed at the various external angles to provide protection from damage, to set the thickness of the plaster, and to reinforce the corner. See Fig. 5-34. Historically, in the

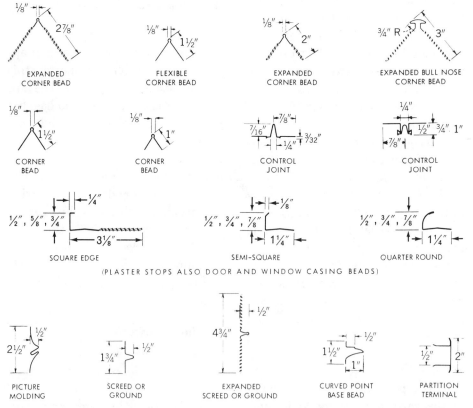

Fig. 5-34. Lathing accessories, showing some of the most commonly used items in cross section.

plastering industry, these items are very modern having come into popular use only during this century. Before this, corners were made by the plasterer, using in some cases special hard plasters or cements. It was also common practice for the carpenter to apply a wooden corner molding over the plastered corner for about four to five feet up from the floor; this protected the corner from damage.

Metal trim consists of base, door and window casing, and various molded shapes providing terminal finishes for plastering. See Fig. 5-34.

Concealed picture molding (Fig. 5-34) provides a flush groove in the finished plaster into which picture hooks can be inserted anywhere along the wall.

Control joints or expansion and contraction joints are available in many types and sizes for both interior and exterior work. See Fig. 5-34. These items help to provide room for expansion and contraction that takes place in any material due to temperature changes or structural movement.

All of these items are made in galvanized or zinc finish to prevent rusting. There are now on the market various accessories made from plastic and these are used for both exterior and interior work. Because they will not rust or stain they are used for some of the exposed aggregate and veneer coat plastering. This is only a partial listing as there are hundreds of variations available, each designed for a specific purpose or need.

Installation of Lath

All metal lath is installed with the sides and ends lapped over each other. The laps between supports should be securely tied using 18 gauge tie wire. In general, metal lath is applied with the long length at right angles to the supports. Rib lath is placed with the ribs against the supports and the ribs nested where the lath overlaps. Generally, metal lath and wire lath is lapped at least 1″ at the ends and ½″ at the sides. Some of the wire lath manufacturers

specify up to 4½″ end lapping and 2″ side laps. This is done to mesh the wires and the paper backing.

Lath is either nailed, stapled, or hog-tied, (heavy wire ring installed with a special gun) to the supports at 6″ intervals. Use 1½″ barbed roofing nails with $\frac{7}{16}$″ heads or 1″, 4 gauge staples for the flat lath on wood supports. For the ribbed lath, heavy wire lath, and sheet lath the nail or staples must penetrate into the wood 1⅜″ for horizontal application and at least ¾" for vertical application. When common nails are used they must be bent across at least three lath strands.

On channel iron supports the lath is tied with number 18 gauge tie wire at 4" intervals using lathers' nippers, for wire lath the hog tie gun can be used. Lath must be stretched tight as it is applied so no sags or buckles will occur. Start tying or nailing at the center of the sheet and work towards the ends. Rib lath should have ties looped around each rib at all supports as the main supporting power for rib lath is the rib.

When installing metal lath at both internal and external corners the lath is bent to form a corner and is carried at least 4" in or around the corner. This provides the proper reinforcement for the angle or corner. (In some special cases the specifications may call for a floating or isolated ceiling or wall and then this rule does not apply.)

Suspended Ceilings

A suspended ceiling is one that is composed of a metal channel iron framework that is suspended from the main structural supports and to which the metal lath or other types of laths are fastened. The common hanger used is the No. 8 galvanized wire; however, various other size wires, rods, and flat iron are used for specific purposes. The range of hangers used is as follows: ³⁄₁₆" pencil

rod; 1" x ³⁄₁₆" flat iron; and 12, 10, 9, 8, 7, and 6 gauge wire. The heavier the metal or wire the more load it will carry.

The common No. 8 wire is rated to carry 16 sq. ft. of complete ceiling per hanger. Hangers are attached to concrete slabs by looping the end and embedding it in the concrete, or securing the hanger to inserts cast in the concrete or tying it to the reinforcing rods set in the concrete. For structural steel and steel joists the hangers are wrapped around the steel, bolted on, or attached to clips that are made for this purpose. For wood joists the hanger can be inserted in holes drilled 3" or more above the bottom of the joist or attached to large spikes driven through the joist. See Fig. 5-35 which shows various hanger attachment methods.

The lower end of the hanger is wrapped around the carrier channel and twisted at least three turns around the hanger wire. Flat iron hangers are usually bolted to the carriers.

The basic framework (See Fig. 5-36) that is supported by the hangers is composed of carrier or runner channels which vary from the standard 1½" hot or cold rolled channel irons to 2" channel irons (some special ceilings may use channels as large as 3" and 4"). The carrier channels are spaced from 2 ft. o.c. to 4 ft. o.c. The common spacing is 4 ft. for the 1½" channels and the hang-

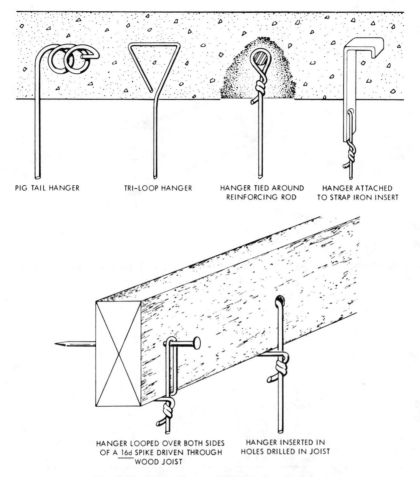

PIG TAIL HANGER   TRI-LOOP HANGER   HANGER TIED AROUND   HANGER ATTACHED
                                        REINFORCING ROD   TO STRAP IRON INSERT

HANGER LOOPED OVER BOTH SIDES   HANGER INSERTED IN
OF A 16d SPIKE DRIVEN THROUGH   HOLES DRILLED IN JOIST
        WOOD JOIST

Fig. 5-35. Several types of hanger installations.

ers spaced 4 ft. along this channel.

The cross furring channels are usually ¾″ hot or cold rolled channels with some 1″ channels used for certain conditions. The cross furring channels are spaced from 12″ to 24″ apart depending on the type of lath to be attached. For 2.5 lb. diamond mesh the 12″ spacing is required;

3.4 lb. lath can span 16″ while the rib and sheet lath can span 24″.

The cross furring channels are tied to the carrier channels with a saddle tie (See Fig. 5-37) made with two strands of No. 18 gauge galvanized wire. Each crossing of channels is tied tightly together to form a strong, firm grillage. To this frame-

195

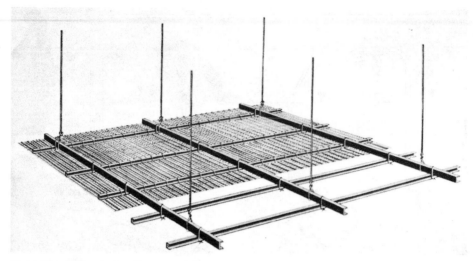

Fig. 5-36. Suspended ceiling, using hangers, carriers, and furring channels. (Inland Steel Products Co.)

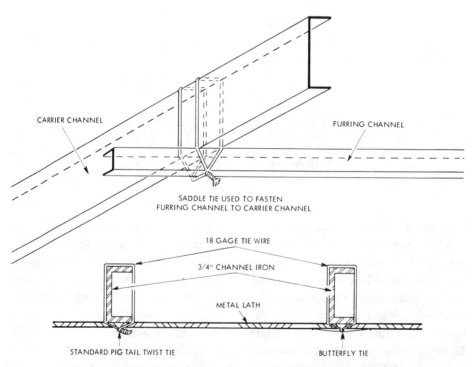

CARRIER CHANNEL

FURRING CHANNEL

SADDLE TIE USED TO FASTEN
FURRING CHANNEL TO CARRIER CHANNEL

18 GAGE TIE WIRE

3/4" CHANNEL IRON

METAL LATH

STANDARD PIG TAIL TWIST TIE

BUTTERFLY TIE

Fig. 5-37. Cross furring channel tied to carrier channels with a saddle tie.

work the lath is tied at 4″ intervals along the furring channels. Use single strand No. 18 gauge wire looped over the channel and twisted together under the lath. Leave at least three twists of wire before cutting and flatten the twist (See Fig. 5-37) or use the butterfly tie which leaves two opposing wings of wire flat against the lath.

Metal Wall Systems

Various methods are used to attach metal lath to supports to form partitions and furred walls. The use of all-metal framing or support for the plaster offers freedom from warping, buckling, shrinking, and swelling. These systems also offer high fire resistance as there is no wood involved.

There are two basic types in use: the stud partition and the studless. Each type has endless variations and patented systems. Some of the things these systems offer are sound control, insulation, pipe space, space saving, and low cost.

Stud Partition. The stud partition in all its variations is the most common because it fills the need for pipe space, built-in fixtures, and sound control. There are solid metal studs, studs with holes punched at intervals, truss type studs, channel iron studs, nailable studs, and so on. The studs come in various widths and lengths to suit almost any need. It would be impossible to list them

all in this book. For more details check manufacturers for their free catalogs which list their systems in detail and give full installation instructions.

Curtain Wall Systems

There is a growing field for the lather and plasterer in the development of the exterior curtain wall (combined exterior and interior wall built as one unit). These systems offer great savings in time, cost, and weight. The lather and plasterer as a team can finish both the interior and exterior of a building in one operation. There are many systems in use now and each manufacturer has his own patented design. The specification on materials and installation must be followed exactly to insure the desired results.

There are many variations but the basic principle is the same for this type of curtain wall. Prime consideration must be given to flashings, drips, expansion joints, and the full thickness of the exterior stucco.

Almost unlimited types of finishes can be applied to the basic stucco coat. Exposed aggregate finishes are ideal for this type of wall. The interior treatment can also include many variations depending on conditions and architectural design.

Wall and Ceiling Furring

The wall and ceiling furring systems are variations of the partition

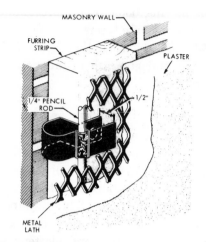

Fig. 5-38. Clip system for wood furring on masonry.

systems already covered. They can be classified as one half of a given partition system. There are now many forms of metal furring strips, clips, and fastening devices in use.

Other methods use either wood or metal furring nailed or screwed to the exterior walls. Clips are fastened to the wall or furring to hold the lath or channel irons fastened to the floor and ceiling but independent of the wall. See Fig. 5-38. The furred wall ends up securely fastened at the floor and ceiling (except in very special sound control cases) but the in-be-

Fig. 5-39. Ceiling system using channel irons attached to wood joists.

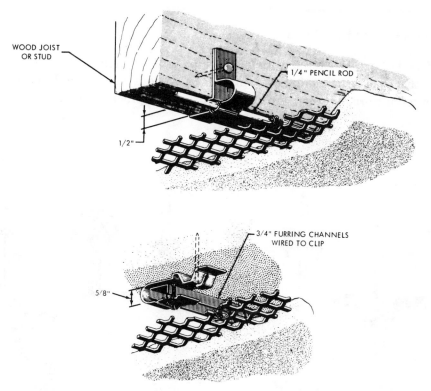

WOOD JOIST
OR STUD

1/4" PENCIL ROD

1/2"

3/4" FURRING CHANNELS
WIRED TO CLIP

5/8"

Fig. 5-40. Two different methods of furring a ceiling, using clips. (United States Gypsum Co.)

tween supports are spring clips, as shown, which prevent actual contact of the lath and plaster to the supporting wall. This type of furring helps sound control and also helps prevent cracking due to structural movement.

Furred ceilings are constructed using various channels, rods, and clips. They are designed to hold the lath to or close to the structural members. Very often the lath is held away from the structural members

for sound control purposes. In ceiling furring the prime reason is, usually, to provide the properly spaced members to which the lath may be tied. See Fig. 5-39. Furring provides a lower cost method of constructing a metal lath ceiling than the suspended type while still providing many of the advantages of the suspended over direct attachment of the lath to structural framing. Fig. 5-40 shows two different methods of furring a ceiling.

Exterior Metal Lathing

Metal lathing for exterior work follows the basic pattern followed in the interior work, but there are some important differences. One, all surfaces to be lathed should be covered with waterproof building paper or

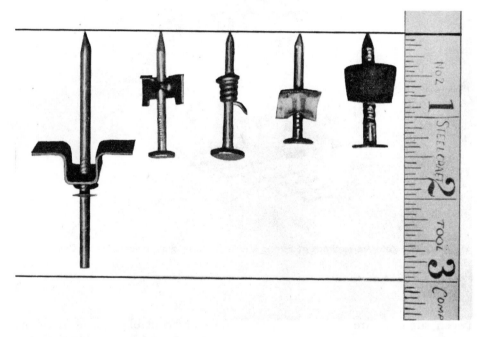

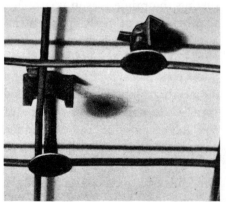

Fig. 5-41. Furring nails used to hold out lath from framing for exterior stucco work. Top, several types of furring nails. Bottom, nails used to space out welded wire lathing. (Portland Cement Association)

plastic. Two, the lath should be of a heavier gauge and with large openings or meshes. Three, the lath must be applied in such a manner that it is not in direct contact with the background or base. Recommended practice is to use furring nails which hold the lath out ¼″ to ⅜″ from the framing. See Fig. 5-41.

One of the most important factors in good exterior plastering is the use of the correct flashings, drips, expansion joints, and stops. Unless the exterior material is protected from water seepage behind it, there will be trouble in a short time. Not only is there a need to provide proper protection against the entrance of water behind the finish materials but also

from the effects of freezing and thawing cycles.

Flashings

The exterior work that is done by the plasterer is, in many instances, ruined in time because of the neglect of proper flashing. Most materials that are used are water resistant; that is, they are capable of shedding water from the exposed surface. But what happens when there is water entering behind its exposed surface? This is where the failure develops, first, by breaking down the bond of various coats to one another, and then finally deteriorating the material itself. Some special materials are designed so they may breath, but

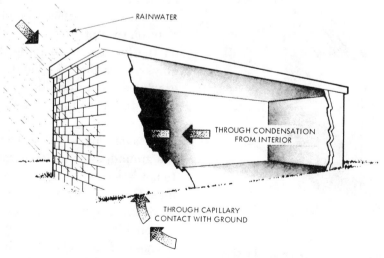

Fig. 5-42. Three ways water penetrates a structure.

these are only effective to the degree that allows moisture to escape as a vapor. They cannot handle heavy amounts of water. If good flashing conditions do not exist, these materials will also fail. In the areas where

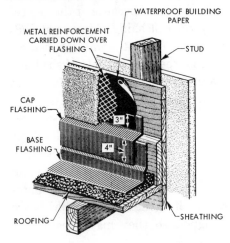

Fig. 5-43. **Flashing where porch roof and dormer meet. (Portland Cement Association.)**

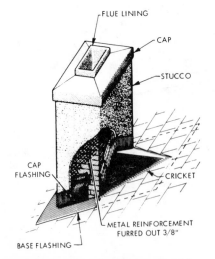

Fig. 5-44. **Flashing placed around a chimney. (Portland Cement Association)**

there is a thaw and freeze cycle, the failure will occur more extensively. See Fig. 5-42.

All openings where moisture can seep behind the material applied should be protected. Moisture may cause rusting of the metal reinforcement and attachments that may be used to hold this material to a building.

Flashing consists of pieces of metal worked in between various members of a building or any place where there is a danger of leakage from rain or snow. Adequate flashing with rust and corrosion resisting metal is of utmost importance. Place flashing at the following places: tops and sides of wall and roof (Fig. 5-43 and Fig. 5-44) and at other points where moisture might gain entrance. Also, prevent any collection of water flow wherever possible by using overhanging roofs. All members that project from the wall's surface should have a slope on the upper surface and a drip on the lower surface to shed water quickly. See Figs. 5-45 and 5-46 for examples of projections that use slopes and drips to shed water effectively.

To avoid moisture seepage from the ground, apply exterior materials to not less than 4″ to 6″ above grade line. Running materials below the grade line is used in some areas with reasonable success because of the dryness of the climate. Generally, however, running exterior materials

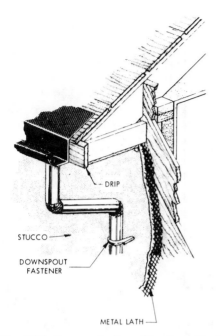

Fig. 5-45. Eave projection. (Keystone Steel & Wire Co.)

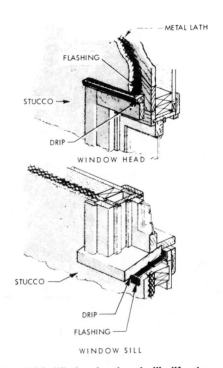

Fig. 5-46. Window head and sill. (Keystone Steel & Wire Co.)

below the grade line is undesirable since mud and dirt are likely to splash upon the finished surface. This is even more so because moisture in the ground (through capillary action) will be drawn up behind the surface. Also, if it is extended to concrete surfaces, such as sidewalks, water may accumulate on them, or they may become damaged in certain areas by the action of frost in the ground that often can move these slabs 2″ or 3″.

Do not apply material on flat horizontal surfaces when they are exposed to weather. If these areas must be covered, there should be a good pitch made so that the water will run off. The water flow then may not affect the material, but it may cause undesirable stains to occur. Also, with rough surfaces in cold climates having a thaw-freeze cycle, damage may occur because sometimes moisture in snow will lay on this surface heavily even if it is pitched. Under certain conditions some of the snow melts next to the surface of the material during the day and then freezes again at night. In most cases, this will weaken the material and eventually cause more and and more water to enter this surface.

203

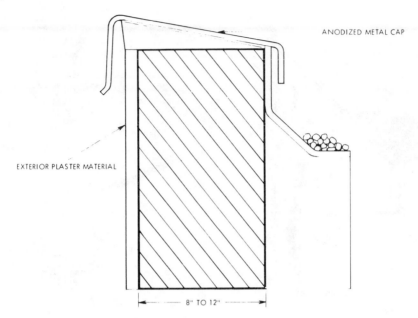

Fig. 5-47. Roof detail, showing use of metal coping.

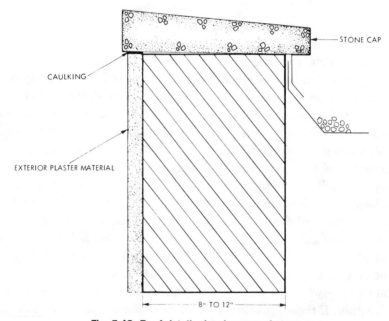

Fig. 5-48. Roof detail, showing use of stone cap.

The following suggestions can help in preventing failures in exposed materials as a result of moisture.

1. Surfacings used on exterior walls should be terminated at the roof under a cap flashing or coping. See Figs. 5-47 and 5-48. Surfacing should never be applied across the top of a parapet (a low wall along the roof edge) wall or other horizontal sections where water, snow or ice can be retained for extended periods.

2. Surfacing used on exterior wall should terminate at a line at least 4″ above the ground level. This can be done with special beads or a marble or stone base. See Figs. 5-49 and 5-50, showing installation.

3. Wherever possible, avoid the use of reglets (special molding used to separate different materials) to create the joint between the surfacing and roof flashing. Experience has shown that reglets often can permit water problems as caulking deteriorates or movement occurs, thus allowing water seepage behind the material.

4. The flashing that is used should be made of rust proof materials to prevent oxidation of the material which would cause staining of the finished surface.

5. Make use of drips at the underside of window sills or drip caps on the top side of doors or windows or

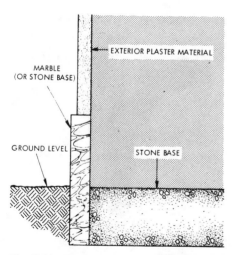

Fig. 5-49. Lower wall detail, showing use of marble or stone base above and below ground level.

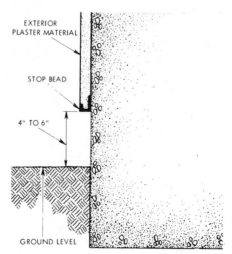

Fig. 5-50. Lower wall detail, showing use of stop bead.

drip moldings at soffits or canopies. See Figs. 5-45 and 5-46. Fig. 5-51 shows a stop bead used for drip-off.

6. Always apply black roofing paper or water vapor barrier between exterior materials and any wood structure it may be connected to, such as plywood or open frame construction using wood studding. This will prevent moisture entering the wood during application or after by condensation. Moisture would cause the wood to expand and possibly warp and then upon drying would shrink, and in most all cases cracks would appear in the material which would then enable water to enter freely and further damage the material. See Figs. 5-52, 5-53 and 5-54.

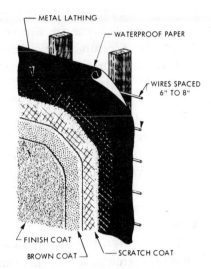

Fig. 5-52. **Preparing stucco base on open frame construction.**

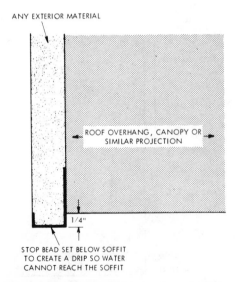

Fig. 5-51. **Cross section through roof projection.**

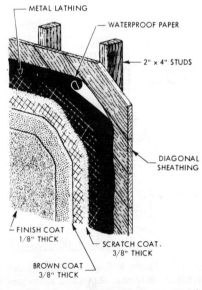

Fig. 5-53. **Stucco on sheathed construction.**

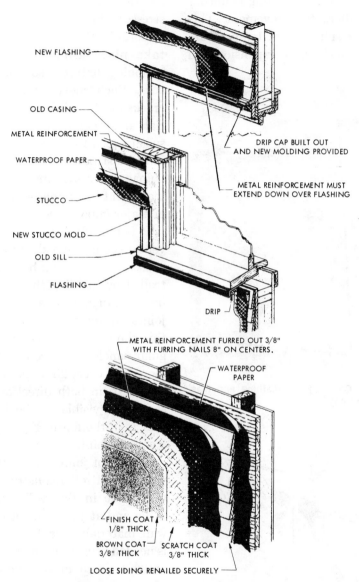

NEW FLASHING

OLD CASING

METAL REINFORCEMENT

WATERPROOF PAPER

STUCCO

NEW STUCCO MOLD

OLD SILL

FLASHING

DRIP CAP BUILT OUT
AND NEW MOLDING PROVIDED

METAL REINFORCEMENT MUST
EXTEND DOWN OVER FLASHING

DRIP

METAL REINFORCEMENT FURRED OUT 3/8"
WITH FURRING NAILS 8" ON CENTERS.

WATERPROOF
PAPER

FINISH COAT
1/8" THICK

BROWN COAT
3/8" THICK

SCRATCH COAT
3/8" THICK

LOOSE SIDING RENAILED SECURELY

Fig. 5-54. Details of waterproof paper, heavy gage lath, flashings, drips, stops, and coats of Portland cement used in overcoating old wood construction.

Expansion or Control Joints

Another cause for failure in exterior materials is the lack of adequate expansion or control joints to allow for growth and shrinkage with changing temperatures.

an external wall or ceiling is uncontrolled differential movement. This usually results from temperature changes which cause expansion and

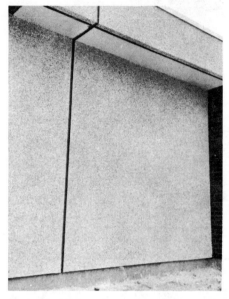

Fig. 5-55. Control joint. (Lathing and Plastering Institute)

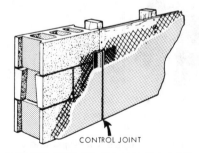

CONTROL JOINT

AT CONTROL OR CONSTRUCTION JOINTS IN WALLS
AND AT JUNCTURES OF DISSIMILAR WALL MATERIALS
(STUCCO DIRECTLY APPLIED OR ON METAL REINFORCEMENT)

Fig. 5-56. Junction of dissimilar wall materials. (Portland Cement Association)

contraction of all materials or load stresses which cause settling of the structure. Also, contraction cannot take place around lighting and plumbing fixtures, so cracking occurs. Such movements can cause cracks or other failures in the materials and break the continuity of any surfacing. Water can then penetrate into the structure and further cause damage, so provisions for controlled movement in any design is therefore of major importance.

It is difficult to anticipate or prevent cracking from all these possible causes, but they can be largely controlled by dividing the large panels into rectangular areas using control joints. If metal lath base is used, this should be divided into rectangular panels with a control joint not more than every 20 ft., preferably every 10 ft. in both directions. These should be visible in the lathing as well as in the finish. Fig. 5-55 shows a control joint.

A control joint should be placed where a control or construction joint was made in the wall by another trade and at junctures of dissimilar wall materials. See Fig. 5-56. This is needed because each part will generally tend to move independently of the other.

In the case of direct applied materials that adhere to surfaces, like cement asbestos boards, each board panel should be finished as a panel. Any attempt to obtain a monolithic

Fig. 5-57. Material applied to asbestos board. (Cement Enamel Development, Inc.)

surfacing effect across these discontinuous sections will surely result in cracking at the joints wherever differential movement occurs. See Fig. 5-57. Each board of this type must be completed as an individual unit.

Crossing of joints with the plastering material to create a larger monolithic surface will result in cracking at the joints regardless of the reinforcing installed across them.

Masonry Bases

Masonry walls are about the oldest form of plaster base known. Rough brick or stone constituted the materials of which these walls were formed. Therefore, thick coats of lime mortar were required for filling

209

in the open spaces and producing a smooth wall.

Today, masonry units are made and laid to a more uniform surface, so that a thick overlay of mortar is no longer required. Modern mortars applied in the proper thickness produce hard and strong surfaces.

Masonry bases to which plaster is applied can be divided into three classes: (1) materials of medium or average suction, (2) materials with a high suction, and (3) materials of low or practically no suction.

There are also in use today various plaster bases which at one time were impossible to plaster over successfully. Some of these are asbestos cement boards, various metals both sheet and structural, plywood, and smooth concrete. New materials and techniques plus machine application make possible the plastering of these bases.

Medium Suction Group

Included in the medium group are: concrete block and cinder block, face or medium hard brick, medium hard clay partition tile, and some of the better grades of common brick. Also, many forms of soft stone belong to this group.

This type of base is the one most commonly encountered. It is the best of all bases over which to apply plaster. Proportions for mixing mortars for these bases are 3 to 1 (three parts aggregate to 1 of cementing material).

Its suction ability is such that there is ample time to apply a fair amount of mortar before the work must be rodded and darbied. Because of the lower suction ability of materials in the medium suction group, mortar will spread better and take less effort to apply. In machine application this medium suction is great enough to hold the mortar in place and permit the gun-man to stay far enough ahead of the rod men so as not to interfere with each other.

Suction is the friend of and prime aid to the plasterer. It can be defined as the ability of a plaster base to absorb wet plaster mortar into its minute pores. If the suction is sufficient, the bond or holding power will be adequate.

Though suction is a needed condition, it must be controlled. While water is often applied to a masonry wall before the application of the plaster, it should be applied with reason, as too much water will fill up the pores temporarily, thus preventing or killing the suction.

The reason for applying water to a masonry base is to keep the base from acting as a sponge and drawing out the water from the mortar before it has a chance to set or harden. A bone dry base will require some water to kill the extreme suction. This water should be sprayed on ahead of time to permit it to be drawn into the masonry before the mortar is applied. Most of the masonry units listed in this group would not need water except in extremely hot, dry weather. Only experience and the testing of the suction will tell the need for water.

High Suction Group

The high suction group consists of soft common brick, soft clay partition tile, gypsum partition tile and some forms of tile made from highly porous materials.

The great suction ability of bases of this category dictates that the mortar used must be "made poor," that is, more aggregate must be used per bag of plaster. Common practice is to make the mix for this type of work in the proportions of 3½ to 1 (3½ parts of aggregate to 1 part of cementing material).

Coarse sand is best for bases of this classification. Sand has the ability to retain a large amount of water, therefore, it takes more water to mix it. In mortar of high sand content, the water retained will exceed the amount normally desired. When applied over a high suction base, however, the excess amount of water retained by the sand is drawn out by the base, leaving only enough to set the plaster. The mortar will be hard and strong because a minimum amount of water is left. The same rule applies here as in the case of Portland cement; that is, the less water used, the stronger the mortar.

Bases of this type may be wet in advance of the plastering. However, little good is accomplished by this as the water is absorbed almost as fast as it is applied and overwetting would fill the pores and spoil the bond.

Mortar should be applied in two coats, one right after the other. Keep the area of operation small. Rod and darby the work before the suction prevents proper straightening of the work.

Gypsum partition tile are precast gypsum tile used for building non-load bearing, fireproof partitions. They are made in sizes from 2"x12"x 30" (solid) through 6"x12"x30" (hollow). The 6" size plastered both sides has a 5 hour fire rating. These tiles have an extremely high suction and require a very poor mix of mortar using a sharp (angular shaped grain) sand.

Low Suction Group

In the third or low suction group are glazed tile, hard burnt brick such as road brick, and the hard stones such as granite.

For bases of this type a different method must be used. First, the mortar must be rich; that is, the aggregate content per bag of plaster must be comparatively low. Common practice is to use a 2 to 1 mix (2 parts of aggregate to 1 part of cementing material).

These proportions are the same as

those used for the scratch coat over metal lath, or the brown coat on gypsum or insulation lath. The mortar is scratched on the low suction base, and after the mortar has been scored with a scarifier, it is allowed to set.

The mortar should be disturbed as little as possible while it is being applied and scored. When the suction is low it is easy to loosen or break the bond. Machine application of the scratch coat over these low suction bases is ideal as the mortar can be applied uniformly and is not disturbed in the process.

Many new bases are now on the market which have little or no suction at all. Using special materials, and in some cases machine application, it is now possible to successfully plaster over such bases as cement asbestos board, masonite, plywood, metal, and smooth concrete. Suction plays no role in bonding these materials to the base. The holding power is developed by special chemical action, and perhaps to some degree by the vacuum developed between the two materials.

Concrete Bases

Plastering directly over monolithic concrete has long been a problem for the plasterer. When it must be done, a mechanical key of some kind

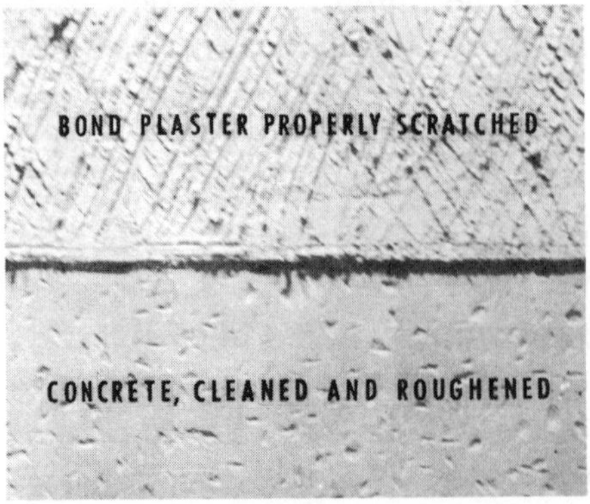

Fig. 5-58. Preparation of concrete to receive plaster.

should be provided. One way is to score or pit the concrete. See Fig. 5-58. Metal or wire lath secured to the concrete will provide an ideal key for the permanent support of the plaster.

Various plaster manufacturers make a special plaster material for plastering directly onto concrete called "bond plaster." This material when applied to a clean concrete surface will provide a bonding surface to which the regular plaster coats will adhere. Fig. 5-59 shows plasterers applying bond plaster to a concrete ceiling.

Modern concrete poured over plastic coated plywood and metal forms is now so dense and smooth that it

Fig. 5-59. Plasterer applying bond plaster to a concrete ceiling.

is hard to insure a long lasting bond of the plaster to this type of concrete. There are now available a number of latex or plastic bonding materials which have shown great promise in developing good holding power and reliability.

These new bonding materials are brushed or sprayed on and can be used for both new and old work. There are two types, one for general plastering use and the other for bonding Portland cement to concrete or for mixing in the concrete to give it greater bonding power.

Many plasterers use these products in doing repair work by coating the surface before applying the patching materials. In plaster patching the application of the bond material seals off the moisture of the new plaster from the old and will prevent the loosening of additional old plaster around the patch work.

Checking on Your Knowledge

The following questions give you the opportunity to check up on yourself. If you have read the chapter carefully, you should be able to answer the questions. If you have any difficulty, read the chapter over once more so that you have the information well in mind before you go on with your reading.

DO YOU KNOW

1. Name three basic types of plaster bases in common use today.
2. What are bonding materials? Name two types.
3. Explain the differences between stud partition, solid plaster partition and curtain wall.
4. List the five basic parts or items that make up a suspended ceiling.
5. What are the differences between expanded metal lath, wire mesh lath and sheet lath?
6. What is a resilient clip lathing system? Why is it used?
7. List five lathing accessories in common use today.
8. Name the three classes (for plastering purposes) of masonry bases.
9. Is monolithic concrete an ideal plastering base, explain?
10. What is the difference between wallboard and veneer plastering?
11. What is the difference between a suspended and a furred ceiling?
12. What plastering base replaced wood lath in the residential field?
13. What are three ways water can penetrate a structure?
14. What are some ways to prevent water damage to an exterior surface?
15. What value is there in using expansion or controlled joints for exterior surfaces?

Applying Plaster

Chapter

6

The plasterer cannot apply plaster and related materials properly without a knowledge of the tools, ingredients of the mortar used, and lathing materials that make up a good plastering job. These subjects were taken up in chapters just covered.

Applying plaster and other mortars, however, is the main function of the plasterer. Before the application of the scratch and brown coats (Chapter 7) and the application of the final coat (Chapter 8) a few points need to be discussed. The methods of using the hawk, trowel, rod, darby, slicker or shingle should be explained. How to produce walls and ceilings that are plumb and level should be covered. To understand this the use of the level, plumb bob, gauge dots, straightedge and featheredge, plus how to cut and trim the

angles in the browning operation, must be studied.

How do you plaster a beam to be sure it is straight? How do you make certain that the walls of a room being plastered are square to each other? To answer these questions, it is necessary to discuss lining, dotting and stripping beams. It is necessary to explain the various ways to square a room so that each wall in a given room is perpendicular to the other.

Before the base and finish coats are applied it is important to have these considerations well in mind. Also, all the mechanical means used to apply plasters, cements, acoustical materials, insulating and fireproofing materials and exposed aggregates must be understood. What was formerly a purely hand manipulative operation has now become a combined hand and machine process.

Applying Mortar With Hawk and Trowel

The two principal tools used by the plasterer to apply mortar are the hawk and trowel. The hawk holds a supply of mortar sufficient for immediate needs. The trowel is used to pick up mortar from the hawk and to apply the mortar to the plaster base.

In Fig. 6-1 the hawk and trowel are shown on the mortar board, together with a pointing trowel and some mortar. Fig. 6-2 shows the plasterer using the trowel to cut about a hawkful of mortar from the pile. The separated mortar is now pushed onto the hawk with the trowel, as shown in Fig. 6-3. To do this successfully, the hawk and trowel are moved at the same time. The tools should be moved towards each other and as the trowel meets the hawk, both tools are lifted from the mortar board. Then the trowel is used to center the mortar on the hawk, and any mortar excess is trimmed off as shown in Fig. 6-3. The mortar is picked up from the hawk by means of the trowel.

Fig. 6-4 shows the motion of the trowel as it cuts into the mortar just before lifting it. As shown in the illustration, the hawk is tipped upward when the mortar is lifted. Notice that the mortar is cut from the hawk at a point farthest from the plasterer. This is done in order to permit just a trowelful of mortar to be lifted from the hawk cleanly.

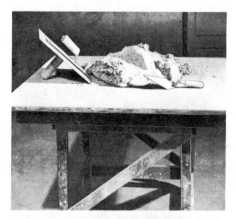

Fig. 6-1. Hawk, trowel and mortar on mortar board.

Fig. 6-2. Cutting hawkful of mortar from pile with trowel.

Fig. 6-3. Pushing separated mortar onto hawk.

Fig. 6-4. Cutting trowelful of mortar from hawk.

The mortar on the trowel is then applied to the plastering base. The thickness of the coat the trowel applies will depend upon the angle at which the trowel is held as it is moved across the base. Held very flat, it produces a thick coating. Held at a sharp angle, it will lay on a thin coat and the mortar will spread further.

As each trowelful is lifted, the hawk is turned a one-quarter turn. That is, a new side is brought into operating position. This is done to keep the mortar in the center of the hawk. The hawk can be handled more easily if the mortar is always kept centered. This operation must be practiced over and over again to develop it into a smooth, rhythmic motion. Keep picking up mortar from the board with the hawk and trowel until this is an easy, flowing movement. Then practice picking up mortar from the hawk until this, too, is effortless.

The next step is to start applying mortar to a wall surface; that is the simplest application procedure. Hold the hawkful of mortar quite close to the wall, waist high, and cut off a trowelful of mortar; with a continuing motion place the trowel against the wall at a slight angle with the bottom part of the trowel touching the wall. Now move the trowel upward with a steady pressure applied on the trowel so as to force the mortar against the wall in a thin coat as shown in Fig. 6-5.

217

Fig. 6-5. Applying first coat on gypsum lathing.

Notice that the mortar is applied from right to left so that the heel of the trowel rides over the previously applied mortar and ties the mortar flowing off the trowel into the existing mortar.

Left handed plasterers would work from left to right. When working on ceilings, left handed plasterers should face the wall while applying the plaster, his right handed partner normally works facing towards the center of the room. This method permits each man to start at opposite ends of the room and work out of each other's way. Teamwork will permit each man to do his job without endangering his partner.

Move the trowel upward in a full stroke until it is empty of mortar. Repeat the operation until the area is covered. Apply a second layer of mortar over the first if a thicker coat is necessary. With the trowel empty and held flat, smooth down the ridges left by the end of the trowel using sweeping strokes.

The area to be covered at one time will depend upon the type of base the plaster is being applied over, the type of mortar used, and the job conditions. These items will be covered thoroughly in the next chapter.

Using the Darby and Slicker

When applying mortar over gypsum or insulating lath in residential work, it is common practice to use only the darby or a slicker to straighten and smooth the work. Area practices vary as to which of the two tools the plasterer will use to do this work.

The darby was the popular smoothing tool for many years when the mortar was applied over wood lath and masonry bases. With the development of gypsum lath and its wide acceptance in the residential field, the slicker has became the most widely used tool.

When darbying or slicking, first go over the work, holding the tool at a sharp angle so the tool acts as a straightedge moving the soft mortar into a level or straight plane. Fill in, with additional mortar, hollow spots that show up in the first operation. Fill in as neatly as possible without excessive mortar buildup.

Now splash or spray water over the surface, using a browning brush or spray gun. Use the water sparingly, as an excessive amount will leave a film or scum of *killed* (unable to harden) plaster and sand on the surface. This film offers a poor surface upon which to apply the finish coat. Used sparingly, however, the water serves as a lubricant and permits the mortar to flow under the darby or slicker without dragging or sliding loose from the lath.

The second darbying or slicking is now performed at right angles to the first. Hold the tool flat and at a slight slant. This helps to smooth the work. The slanting of the tool permits the excess water and soft plaster to flow

Fig. 6-6. Plasterer using slicker to smooth brown coat.

to one end of the tool where it can run off without the annoyance of dropping on the plasterer's hands or other parts of his body. Fig. 6-6 shows a plasterer using a slicker to smooth brown coat.

During the first and second operations of straightening and smoothing the plaster, the darby or slicker is placed into the angles (corners) and pulled out in repeated moves until all the angles have been straightened out. Work both sides of the angles carefully, placing the tool so as to clean out any excess of mortar in the angle. Many plasterers use a feather-edge in the last operation in the angles to make sure they are straight.

Trimming Angles

After all the work has been darbied or slickered, the angles must be cut or trimmed to remove the surplus mortar that the other tools could not clean out sharp enough. Again, area practices vary and each plasterer has his own version of how to do this job best.

One method is to use a clean trowel with the toe of the trowel doing the cutting and the heel held away from the wall or ceiling slightly. Use just enough pressure to make the trowel cut off the surplus left in the angle, but slight enough pressure to keep the trowel from digging into the soft surface.

Do this by cutting on both sides of the angle to form a clean, sharp corner. Fig. 6-7 shows the trowel cutting the angle. Some plasterers prefer to use an angle float or an angle plow to form a clean, smooth angle. The float and angle plow do not cut into the mortar like the trowel and may, therefore, produce a stronger angle.

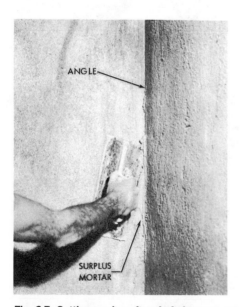

Fig. 6-7. Cutting angles after darbying.

Establishing Dots and Screeds

To make walls plumb, to make ceilings level, and to develop the proper plaster thickness, the plasterer on commercial, industrial and some custom house work puts up screeds (narrow bands of mortar). The thickness of the coat of plaster is controlled by the use of screeds and grounds. Since the thickness of the plaster on many plastering bases is an important element in good plastering, the correct use of grounds and screeds becomes important. While grounds and screeds are both used to establish plaster thickness and straightness, they are two entirely different things.

Grounds are either wood or metal and are placed along the base line, around window and door openings, or near the top of the walls for nailing purposes to hold picture moldings or similar moldings. Basically the grounds, although they establish the wall lines and plaster thickness, are actually installed by the carpenter for the nailing of baseboards, door and window trim, and picture moldings.

At times wood grounds are not installed because the architect has specified metal trim such as metal base or base bead, metal picture molding, metal plaster stop, etc. In these cases, the metal trim serves a dual purpose:

these items are both a finish trim and a guide for the plasterer to produce a straight and true surface. The metal trim is installed by the lather.

Screeds are narrow bands of mortar which are plumbed, leveled or built up between dots (small spots of plaster set plumb with the grounds or level on a ceiling) to function as guide lines upon which the straightedge may ride when rodding the mortar on walls or ceilings.

A first class plastering job will usually use both grounds and screeds. The ground establishes the wall line at the floor and wall openings while the screeds are developed from the grounds at positions laid out so that the rod used to straighten the plaster surface can bear on the grounds at one end and the screed at the other. When the area or field between the grounds and the screeds is filled in with mortar, the rod is used to straighten the surface — cutting off high spots and showing up the slack or hollow places. Rodding, then, means to move the rod over the area while it bears on both the ground and screed or on screeds alone, as the case may be.

Wall Dots and Screeds

Two methods are used to establish screeds. The first method is to place dots or small spots of plaster at about straightedge length apart, and to a height measured by the reach of the arm above the floor or scaffold. These dots are set out to the face of the finished wall by plumbing with a level.

Fig. 6-8, left, shows the plasterer pressing a dot to a plumb line with the wood grounds below, using a straightedge and a level. Notice that he has placed a piece of paper over the dot so that the straightedge, when it is removed, will not stick to the mortar and drag off the surface of the dot.

Some plasterers prefer to use small, thin strips of wood placed on the face of the soft mortar and pressed into the dot with the straightedge until it is plumb. Practice will establish that the rod must be pushed in a little more than plumb because as the pressure on the rod is released, the dot will swell out again just a little.

Dots can also be pressed without paper or wood strips by keeping the rod clean and as the plumb point is reached in pressing in the rod, it is moved upward in a sliding motion, thus freeing the rod from the surface

Fig. 6-8. Establishing wall dots, using rod and level, left; and plumbing wall dot, using plumb bob and gages, right.

of the dot without damaging it.

When all the required dots are pressed, trim off any excess mortar and recheck each dot with the rod and level to make sure it is plumb.

Dots can also be plumbed by using a plumb bob and line with two gages. The gage is a device used by the trade to transfer a measurement, mark, or line from one plane to another. Also, it is used as a means of repeating a given measurement a number of times. In the case in discussion, the gages permit the plumb bob to hang free without touching the wall while transferring the ground line to the dot. See Fig. 6-8, right.

The gages are made by cutting two pieces of wood such as 1″x2″ wood grounds, 1″x4″ beam strip, or similar material into identical gages as pictured in Fig. 6-9. The measurements are only suggested sizes, any size gage can be used.

To plumb the dot with this meth-od, place the dot on the wall as before, then hang the plumb bob line in the cut out section of the top gage and hold this gage against the soft mortar dot. Now the other plasterer holds his gage against the base ground and the upper gage is pressed slowly and carefully into the dot until the lower gage's cut out section clears the line exactly. Hold both gages out level or otherwise the measurement taken will not be accurate.

The mortar of the dot is then trimmed off to the impression line the gage made. The dot is then checked by repeating the operation, and this insures that the dot is plumb with the base ground.

When all the required dots are in place the screeds are built up between them keeping the mortar in a narrow band about 4″ wide. Do not make the screeds too wide as they can develop twists and out-of-plumb areas. Build the mortar up slightly thicker than the dots and then using the rod resting at each end on the dots work it back and forth in an upward motion cutting off the excess mortar. Fill in the hollow spots and rod again.

In Fig. 6-10 the plasterer is shown rodding a horizontal screed to the dots just plumbed.

The second method used to form screeds is to plumb up a vertical screed on each side of the angles in the corners of the room and to plumb

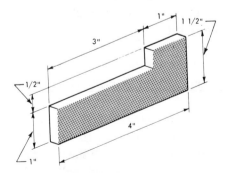

Fig. 6-9. Gage used with plumb bob.

Fig. 6-10. Forming horizontal screed.

mortar spots at needed intermediate points between the corners. The mortar is applied screed width, the full height and the rod is used to cut off the excess mortar moving the rod up and down in a see-sawing motion while drawing the rod across the width of the screed.

Now check the screed for plumb by using the cleaned rod and level; cut off or add on as needed to bring the screed to true plumb. This operation is repeated around the room. The plasterer in Fig. 6-11 is shown checking the screed to make sure it is plumb.

When all the vertical screeds and intermediate spots are plumbed, form horizontal screeds between them to complete the screeds needed to guide the rod when filling in and straightening the balance of the wall. Lightly darby all the screeds to bring them to a smooth surface. Trim off any extra wide parts of the screeds if they will stand overnight or longer before the walls are filled in.

The field or balance of the wall area between the screeds is now filled in and rodded straight as shown in Fig. 6-12. Notice that the rod is held slightly forward at the top as it is moved up and down to rod off the mortar. This is done so the mortar, as it is cut off by the rod, will fall free and not hit the hands.

Fig. 6-11. Plasterer checks vertical screed to make sure it is in plumb.

After the work has been rodded straight and the hollow spots filled in and rerodded, the surface is darbied. The slicker is not used on mortar applied over masonry or similar bases as it is too flexible for this type of work. The first darbying is done by holding the darby horizontally and drawing it upward from the bottom of the wall as shown in Fig. 6-13. This smooths the ridges left by the rodding. Finally, holding the darby vertically, as shown in Fig. 6-14, and pressing it flat against the surface, smooth the mortar into a true flat plane.

Some further spotting up of hollow spots with mortar may be necessary after the initial darbying. Apply the mortar carefully now in order not to add more than the required amount.

Fig. 6-12. Rodding work to the screeds.

Fig. 6-13. Darbying—first up-stroke.

Fig. 6-14. Final darbying—holding darby vertically.

ameter. Its length may vary to suit the need or preference of the plasterer; the average length is about 50 feet. In each end of the hose a glass tube about 12″ long is placed. Boiler gage glasses are about the size for this purpose. Fig. 6-15 shows an enlarged cross section of a water level glass set in the end of the hose. Notice how the water, when it is at rest or still, is inverted or depressed in

A little water sprinkled over the surface will help to smooth the work.

After darbying, the angles are cleaned and the mortar at the grounds and corner beads is trimmed back in order to provide room for the finish coat. This permits the last coat to finish flush with all beads and grounds.

Ceiling Dots and Screeds

To establish a level ceiling, the plasterer uses a water level and a gage. The water level sets level marks around the walls of a room; from these marks, using a gage, dots are pressed on the ceiling to develop a level and straight ceiling. The water level is a light (rubber or plastic) hose, usually ⅜″ or ½″ inside di-

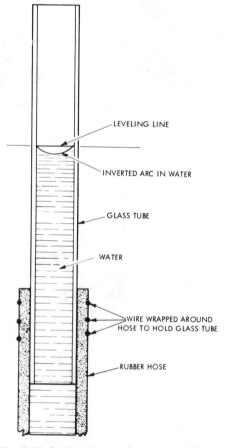

Fig. 6-15. Sectional view of water level glass.

the center. This is caused by the water adhering to the sides of the glass while the center of the water sags or is depressed.

To use the water level it must be filled with water. Water seeks its own level at all times; therefore, no matter where the gage glasses are held, the water will come to a level position. To fill the level properly, put one end in a pail of clean water that is placed on a stand or bench. Let down the other end into a pail which rests on the floor. Start a siphoning action by sucking on the lower end, or pour some water in the upper end and plunge it into the water quickly.

Let the water siphon through the hose until all the air has been removed. Place a finger over each gage glass opening at the same time so as not to lose any of the water in the hose. Bring the two glasses together and shake them up and down gently with the ends open. If no air is trapped in the hose, the water will be brought to a level line across both glasses. When moving the level into position, be sure to hold a finger over the end of each glass, otherwise the water will run out.

To use the level, place the middle of one glass over the mark that is to be leveled around the room. Hold the other glass at about the same height, and at the place where the next level mark is needed. As shown in Fig. 6-16, the first plasterer (the one on the left) has adjusted his glass up or down until the water is exactly even with the mark; the second man does not move his glass but holds it still. The water contained in it rises or falls until the first man has brought his end to the mark. Now the second man places a pencil mark on the wall at the top of the water or ring formed inside the glass.

Note that the water sags in an inverted arc (Fig. 6-15), therefore, the plasterer could take either the highest or lowest point of the waters' surface. To be accurate, both men must use the same point of water in each glass. The outer ring is best because it is the point most easily observed. Do not make the mark until the water is at rest (still).

The hose must remain free of kinks and loops. When the water is being brought to the proper point, the ends of the glasses are left open. To sight the marks, and to make the marks at the other end, the plasterer's eye must be about level with the water; otherwise, the angle of sight may distort the apparent position of the line of the water and the mark. Always check the last mark made back to the original mark. If the hose is not long enough to permit all of the required marks to be placed from the first one, the first plasterer moves his glass to the last mark made by his partner and the operation is repeated until all the marks needed are made. Always cross check back to a few previous marks to make sure the

Fig. 6-16. Water leveling marks on a screed.

marks are correct.

Both glasses in the water level must be of the same diameter; otherwise, the level will be off as a result of the surface tension of the water. The level will be consistently lower in the larger glass because, the greater diameter provides more area for the water to spread out. If a level with different glass sizes has to be moved a number of times because the room is large, the accumulation of error could be great enough to render the readings valueless. The water level is a fine, accurate and sensitive instrument, and care in its use is necessary to give dependable service.

The water level marks are usually established at an easy working height. Sometimes, however, a specific height or place is required because of later needs, such as a cornice or other work requiring a level line.

With a series of level marks made around the room (or area) at points where needed, the plasterer now needs a gage to transfer these level wall marks to level dots on the ceiling. A gage is made to equal the distance from the wall mark to the required ceiling height.

The gage is made of light wood strips, such as 1″x2″ grounds, and is made into the form of a square with the long leg on the bottom, and the short leg held upright to the ceiling. The short leg has a cross piece of wood nailed to its end which will form the actual dot. It is therefore equal to the length of the dot required. A brace is nailed across both legs of the gage at a 45° angle to stiffen it and to hold the upright leg in place and plumb. See Fig. 6-17.

To the bottom leg of the gage tie a short level using two pieces of tie wire to hold the gage and level tightly together. Fig. 6-17 shows the plasterer pressing a dot on the ceiling to the required height using the gage just described. Over-press slightly as the dot will sag back a little when the pressure on the gage is removed. Always recheck the dots after they have been trimmed.

Press as many dots as are required to guide the straightedge to form the

Fig. 6-17. Pressing ceiling dot using gage and level.

ceiling screeds around the room or area. The ceiling field is filled in after the screeds have been completed. In some large ceilings it is often necessary to form intermediate screeds across the ceiling at regular intervals so the straightedge can rest on two screeds at all times. Such screeds are formed by laying on narrow bands of mortar across the ceiling from screed to screed and then rodding this down until the screed is straight and true for its full length. Repeated checking for bumps or hollows by placing the longest rod available at various points on the screed will develop a true screed.

If the room is to have a cornice run at the ceiling angle, the water level line is then set at the line where the cornice mold's slipper (bottom board of the mold) will run on the cornice strip. The water level line then serves a twofold purpose in that it levels the ceiling and later will be used to set the cornice strips. In many cases when a cornice is to be run, the wall screed is finished to a true straight surface using high gauged putty and plaster. Then the cornice strip line is water leveled and the cornice strips nailed in place. The gage to press the ceiling dots can then rest on the cornice strip as it is pressed up to level the dot.

The lining of wall dots for long walls where base grounds are not available will be taken up in Chapter 9, *Ornamental Plaster.*

Machine Application of Plaster and other Materials

The application of all kinds of plastering materials by mechanical means is now a common practice. The operation of the machines has been covered in Chapter 2. The method of applying the mortar or other materials will be covered here, while the actual techniques involved in the application of specific materials will be covered in detail in Chapters 7 and 8.

Plastering machines are appliers of mortars and finishing materials. They are basically pumps which force the mortar or other materials through hoses to the point of application; the mortar at this point is then blown onto the desired surface in various patterns and at controllable rates of flow. The machines do not screed, rod or darby. They must be knowledgeably directed and controlled to properly perform their function in the plastering operation.

The same general steps followed in the hand application of plastering materials are followed for machine application. Screeds are established, if called for, and the same number of coats of mortar are applied depending on the plastering base involved. The same rodding, darbying and trimming of the angles is required.

There are some important differences, however. One is that the work must be laid out so that the nozzleman and his follow-up men can proceed without too much delay from room to room or from area to area. This means scaffolds, if needed, should be built ahead or set so they can be placed quickly as needed. Windows, doors and other surfaces that need protection should be covered ahead of the work. The spraying should be planned so the proper sequence of coats of material can be applied with the fewest possible moves of the spraying equipment.

Mortar boards should be set up in each area that the nozzleman will operate in, so he can catch the mortar that will flow out of the nozzle after the machine is turned off. This overflow is useful in spotting-up by the follow-up plasterers. Fig. 6-18 shows a nozzle-man spraying brown mortar on a wall.

When spraying, the nozzle is pointed towards the surface to be covered and the machine turned on. The nozzle should be held about 18″ to 24″ away from the work. It must be moved from side to side in a continuous sweeping pattern in order to place a uniform amount of mortar over the entire area. The rate of movement of the nozzle will deter-

Fig. 6-18. Nozzle of mechanical pump is held 18″ to 24″ from surface. (United States Gypsum Co.)

mine the amount of mortar that is placed in one particular spot.

The output of the machine can be controlled by adjusting the speed of the pump. Speed control is particularly important in spraying on various acoustical plaster and finishing materials. (See details on control of the machines for specific materials in Chapters 7 and 8.) Each manufacturer of materials has detailed information on how their material should be handled under varying conditions. It pays to follow these instructions to the letter.

The nozzle spray can be controlled by the plasterer in such a way that the area not to be covered with plas-

ter can be protected. The nozzle can be held at an angle; in this way, the line of spray can be controlled so that the mortar will not reach those places not intended to be plastered. Bringing the nozzle closer to the work also limits the area being covered at one time. If the operator is properly trained, the mechanical pump will not produce any more waste than the average plasterer with the hawk and trowel. The machines will pump and spray more mortar in a given period of time than can be rodded and darbied by one or two men. Therefore, the work must be planned in this respect also. The number of angles and other slow work areas in a room, as well as the type of mortar being sprayed, will determine the number of follow-up men needed behind the nozzle-man.

For normal gypsum lath plaster base and brown mortar in residential work, a four man crew can handle this work quite well. One man sprays and three men straighten up the work. The nozzle-man is usually alternated around with the other men because this is the heaviest part of the work, even though some plasterers would rather spray than straighten.

On this type of work the nozzle-man must complete his operation early enough to permit the other men to complete the straightening operation before quitting time. This usually means stopping at least one hour earlier. The nozzle-man then helps the other men complete the straightening and cleaning-up operation.

For commercial work involving metal lath, the lath must, in most cases, be *fogged in* (covered with a fine mist of mortar) with the nozzle held at a sharp angle to the ceiling or wall so the mortar will not be blown through the lath. When this light coating is partially set the full scratch coat can be applied with ease. Paper-backed lath and sheet lath can be given a full scratch coat without any preliminary coating.

On masonry bases and other high suction bases the brown coat can be applied to any required thickness (3″ thickness can be built-up by the machine with ease because the mortar is compacted by the force of its blown application). In most cases, however, it is best to spray on a thin scratch coat to cover a couple of walls in a room and then turn right back and fill into the full required thickness. This method will provide a better working mortar for the rodding and darbying crew that follows.

Plastering large, open areas of ceilings may require twice as many follow-up men, since machines used on this type of work can apply mortar at their full capacity. Usually, in these cases, two men operate the nozzle, alternating between spraying and moving the hose so the nozzle-man does not have to drag the hose as he moves ahead.

Many types of finishing materials can be applied to all kinds of surfaces using various types of guns. Some of these materials are: sand finishes, texture finishes, acoustical materials, glitter materials and exposed aggregates. Each material grouping requires a special machine designed to handle that type of material. Basically, the differences are in the weight of the materials to be pumped and sprayed. Heavy materials require heavy-duty pumps and large air compressors.

The special techniques required to apply each of these finishing materials will be covered in detail in Chapter 8.

Applying Portland Cement Mortars

Portland cement mortar is applied in the same general manner as the other mortars, but there are some important differences when succeeding coats are applied. There are also some differences depending upon the surfaces that are to be covered.

Portland cement mortar can be applied over most masonry bases and metal reinforcements. One rule to remember is that any masonry base that is weaker than the mortar being applied must have metal reinforcement applied over it to make sure the mortar will not break off.

Never apply Portland cement mortar over any base that is painted, dirty, dusty or soot-covered unless it has been covered with metal reinforcement to hold the mortar in place. All exterior areas to be covered with cement mortar must have adequate flashings of rust and corrosion-resisting metal to prevent water from getting behind the mortar.

Overhanging roofs and projections that have built-in drip slots or edges provide good protection for this purpose. Flashings and drips must be applied over all door and window heads, also under all sills and similar openings. Chimney and parapet wall caps require flashings also. Where the chimney and parapet walls meet the roof, flashings and counter flashings are required. A metal stop bead should be provided at approximately 6″ above any ground surface to prevent moisture seepage from the ground. These stops also serve to prevent frost damage in cold climates. (See flashing, drip and stop details in Chapter 5 for more information.)

Cleaning Surface

When applying Portland cement over monolithic (cast in place) concrete, make sure the surface is clean and rough enough to provide a good bond for the mortar. Various methods can be used to insure a good clean surface. One method is to wash concrete surface with 1 part muriatic acid to 6 parts water. See Fig. 6-19. Wet surface with water first to prevent acid from being sucked deeply into the pores of the concrete. Use rubber gloves and a fiber scrub brush to make one or more applications of the acid solution to the concrete sur-

Fig. 6-19. Washing concrete surface with an acid solution. (Portland Cement Association)

Fig. 6-20. Roughing cast-in-place concrete by brushing. (Portland Cement Association)

Fig. 6-21. Dashing first coat on cast-in-place concrete. (Portland Cement Association)

face. Wash down the surface with clean water to remove the acid. Use wooden or plastic bucket to hold the acid, as a metal bucket would be eaten away by the acid.

Another method is to roughen the surface of new cast-in-place concrete with a heavy-duty brush or scoring tool. This can be done if the forms are removed early. Fig. 6-20 shows the rough surface obtained if the wire brushing can be done early enough.

A third method is shown in Fig. 6-21. Dash on a thin coat of Portland cement mortar to the clean concrete using a fiber brush. Let this coat cure until hard — keeping it moist (wet with a fine mist spray) all the time. Usually seven days is required to insure initial strength. This coat can also be machine applied and then cured. There are now various pat-

ented liquid bonding materials that can be brushed or sprayed on which will provide a good bond.

Application

Portland cement mortar should be applied in three coats to insure the proper thickness and strength. The scratch coat is applied, scored and let set and cure not less than two days and preferably seven days. Figs. 6-22, 6-23 and 6-24 illustrate the application of the scratch coat. Do not disturb this scratch coat by excessive troweling or movement when it is being applied. Excessive troweling or movement will break the bond created between the mortar and the surface, either masonry or metal. Fig. 6-25 shows machine-applied scratch coat.

The brown coat is applied next, and this coat must be built up to

Fig. 6-22. Scratch coat applied direct to masonry wall having coarse texture. (Portland Cement Association)

Fig. 6-24. Scratching the first stucco coat. (Dahlhouser; Keystone Steel & Wire Co.)

Fig. 6-23. Applying the scratch coat. (Dahlhauser; Keystone Steel & Wire Co.)

Fig. 6-25. Machine application of scratch coat to paper backed lath.

straighten the surface and bring it to the proper plane. On exterior work always try to work on the shady side of the building so as to keep the mortar from drying out too rapidly.

Fig. 6-26. Applying the brown coat. (Dahlhauser; Keystone Steel & Wire Co.)

Dampen the scratch coat ahead of the browning if it is *hot* (that is, if the surface is extremely dry and therefore has excessive suction). Some plasterers add a small amount of waterproofing to the scratch coat. This prevents the loss of water and develops a working surface which is *cool* (low suction surface). This practice permits the plasterer to rod and straighten the brown coat without too much drag or pull.

In Fig. 6-26 the plasterers are shown applying the brown coat and in Fig. 6-27 they are stripping an *arris* (external corner). To bring the arris out to its proper position for both sides of the corner first apply a screed of mortar, of the proper thickness, at the corner on one side. Then press a straight strip of wood into the soft mortar, setting it plumb and out to the required face of the adjoining wall.

The soft mortar will usually hold the strip firmly for a short time. Otherwise one man holds it while the other fills mortar against the edge of the wood strip and rods it off. After a few minutes, if the base coat suc-

Fig. 6-27. Stripping an arris (external corner) using a wood strip to act as a guide. (Dahl-hauser; Keystone Steel & Wire Co.)

tion is normal, the strip may be removed. Use a sliding motion working backwards and upwards, allowing the strip to free itself without damaging the soft corner. If the strip seems to stick, tap it lightly to loosen it.

After the brown coat has been rodded and floated straight (Portland cement brown coat is floated after rodding, using a large wood float to compact the mortar and to bring it to a good straight surface) it is scratched to form a good key for the finish coat. Cross scratch uniformly and lightly. Deep scoring weakens the brown coat. Some plasterers feel the scratching of the brown coat is not needed, as the floating produces a rough enough surface.

Thickness is very important; the brown coat should never be less than ⅜″ thick and preferably not less than ½″. Moist cure for at least 48 hours, then allow it to dry for about seven days before the finish coat is applied.

237

Fig. 6-28. Proper construction for stucco on metal reinforcement. (Portland Cement Association)

Fig. 6-28 shows a typical stucco job done over large mesh metal lath. Note how the mesh is embedded in the scratch coat because it is held away from the waterproof paper by spacing nails. Total thickness of the three coats should be approximately 1".

Before applying the finish coat dampen the brown coat evenly with a fine spray and let this soak in so the surface is damp, but has suction. The thickness of the finish coat should be sufficient to obtain a good surface texture, but in no case less than ⅛".

If possible, follow the sun around the building in hot weather, staying in the shade to keep the sun from drying out the mortar too fast. In very hot weather the work should be shaded with canvas hung outside of the scaffolding.

Start from the top of the wall and work down. Enough plasterers must be working on the area to be finished so that the whole wall surface may be covered without joinings (laps or interruptions). On jobs of ordinary size, one plasterer applies the finish coat and a follow-up man floats or textures the surfaces. When the area is large more men are needed.

All arrises should be stripped, if possible, ahead of the finish application to insure well-defined, sharp corners. For irregular surface textures, however, stripping is not necessary; corners are formed with the trowel or float as the work progresses.

Many types of finishes and textures can be obtained using either hand or machine application. With the use of various machines the plasterer can now create many new finishes and cover large areas without joinings, thereby solving a problem that has caused much trouble for hand-applied finishes.

There is a new field of thought on the application of Portland cement mortar. With the development of the plastering machines and the ability to build up thicker coats of mortar through power application, some plasterers advocate building up each coat of mortar one on top of the other as quickly as possible so as to make the total thickness one homogeneous mass. This is thought to make a stronger wall or ceiling, having a more uniform mass with equal stress throughout. Over masonry surfaces the scratch, brown and finish coats are applied one right after the other. When the finish coat requires extensive manipulation to produce a heavy texture, the application must be delayed until the next day to permit the scratch and brown coats to set hard.

Special Cements and Admixes

Using "High-Early" cement reduces the curing time required for the cement to reach its initial strength (three days instead of seven). "Air-Entraining" cement is also used on some jobs to resist the problems caused by freezing action in the wintertime.

Various admixes can be used to increase workability, prevent freezing and to waterproof the mortar. All of these must be but a small percentage of the volume or weight of the portland cement used. Follow the Portland Cement Association's directions when using these admixes.

Plastering Beams

An important part of the plasterer's work is the plastering of beams—usually structural members which project below the ceiling or false beams formed to conceal pipes, etc. There are three architectural terms used in describing the parts or surfaces of a beam. These are: *soffit*, the underside or bottom of a beam; *face*, the sides or cheeks of the beam; *arris*, (external corner) the sharp edge or angle formed by the meeting of the soffit and face of the beam. Fig. 6-29 shows a cross section of a typical beam with the correct terminology.

Lining and *stripping* of beams are terms used to describe the operations involved in producing straight beams. *Lining* means to set dots to a line on the sides of a beam to produce a finished product that is true in width, depth and straightness. *Stripping* means applying wood strip to the sides of a beam to develop the soffit and arrises.

Basic to forming a plastered beam are the *beam strips*. These are long pieces of wood usually 1" x 4" x 12' long with lengths of 14' and 16' also available. White pine, fir or redwood are the common woods used. Some plasterers use cornice strips for this purpose. Beam strips must be

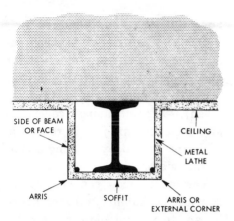

SIDE OF BEAM OR FACE

CEILING

METAL LATHE

ARRIS

SOFFIT

ARRIS OR EXTERNAL CORNER

Fig. 6-29. Soffit, sides (also known as faces or cheeks) and arrises or external corners of a beam.

straight-grained lumber, usually without knots.

Beam strips are attached to both sides of the beam and these two strips will, in a sense, act as forms. By plastering the soffit between these strips, a level soffit is developed and the beam's straight edges or arrises are formed. After the soffit is finished the beam strips are removed, the sides or faces of the beam are filled in and finished. These basic steps form a completed beam; however, there are a number of ways to do these steps, and these will now be outlined.

Stripping: First Method

The plasterer uses two methods to strip a beam. The first is used for very accurate work. The second is accomplished by eye and is used where workmanship is less exacting.

In the first method, the rough scratched-in beam is lined and dotted; then a water-leveled line is established on each side of the beam and strips are set to these lines.

To line up the sides of the beam thin sticks of wood about the thickness and length of a lead pencil are stuck to the soffit using high-gauged putty and plaster. These sticks are spaced so that there is one near each end of the beam and the rest are set so the average length straightedge will ride over them. Fig. 6-30 shows a typical setup with the sticks stuck in place and a chalk line set at end of the beam to the proper point. The remaining sticks are now marked at the line, thus establishing a straight line at the required thickness for that side of the beam.

The chalk line is now removed and, as shown in Fig. 6-31, the sticks are cut off at the marks, using a fine bladed coping saw. Dots are now pressed on the face of the beam over each stick, using a hand level to plumb them. Next, cut a stick equal in length to the desired width of the beam.

In Fig. 6-32 the plasterer is shown plumbing the dots on the opposite side of the beam, using the stick just cut to gauge the width of the beam. Hold one end flush with the dot just pressed on the other side and then plumb the dot placed on the side shown. Do not forget to place a piece of paper over the soft plaster dot be-

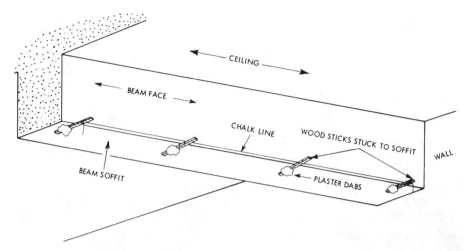

Fig. 6-30. Line set to face of beam. Strips marked, ready to be cut.

Fig. 6-31. Establishing straight side of beam by cutting off sticks to a line.

Fig. 6-32. Plumbing dot on face of beam, using level and gage.

fore pressing the level against it.

With all the dots plumbed on the sides of the beam, the next step is to fill in the sides of the beam be-

tween the dots. Two methods are used to do this. In one, brown mortar is used to fill in to the dots and the mortar is then rodded straight. The

241

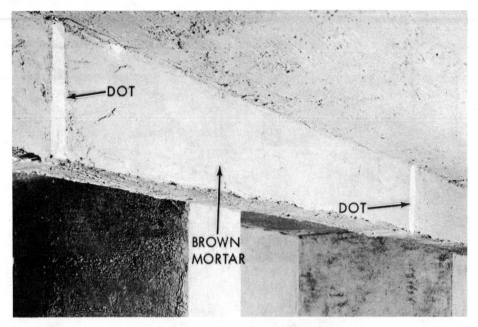

Fig. 6-33. Face of beam browned in to dots.

other method uses high-gauged putty and plaster. The method selected depends upon job conditions, specifications and construction.

A beam formed of channel iron and metal lath will be stronger if the sides (faces) and soffit are browned in to the dots. A solid concrete or other solid beam can be coated with high gauged putty and plaster over the bonding coat because it will not require the stiffening and strength afforded by the brown mortar. Fig. 6-33 shows the beam browned in.

When the sides (faces) are filled in and straight the beam is ready to be water-leveled. Establish a mark at one end of the beam that will equal the top of the beam strip to be used. Water level this mark to the other end of the beam and to the other side of the beam at both ends. Fig. 6-34 shows the water leveling being carried out. The chalk line has been struck to show its position on the side of the beam.

Now nail the beam strips to the chalk lines on both sides of the beam. Use 6d or 8d nails for this purpose and use only enough nails to hold the strips to the line and tight to the beam. Slanting the nails in alternating directions as they are driven in will help to hold the strip

WATER LEVEL GLASSES

WATER
LEVEL
HOSE

Fig. 6-34. Establishing water-level line for beam strips.

tight to the beam side. Check the alignment of the strips by sighting along the strip from the bottom side. If there is any bulging out use additional nails; any inward curving can be wedged out with slivers of wood. When the strips check out for straightness, stick then solidly to the beam sides with daubs of high gauged putty and plaster spaced about 2 ft. apart.

The beam soffit is now ready to be filled in. Common practice is to fill in the soffit with high gauged putty and plaster so that the strips can be removed and the job completed without the need to wait for the brown coat to set and dry. Fig. 6-35 shows the plasterer filling in the beam soffit. The soffit is filled in and rodded off as shown; spot up any hollow spots and rod again until the soffit is filled in straight and true. Finish off the soffit with a new mix of the same finishing material and trowel it down tight to the strips.

The strips are now ready to be removed; tap them lightly to release them from the plaster soffit; chop off the dabs of plaster that were holding the strips in place and pull the strip straight out and away from the beam. Do not let the beam strip drop down while removing it because this

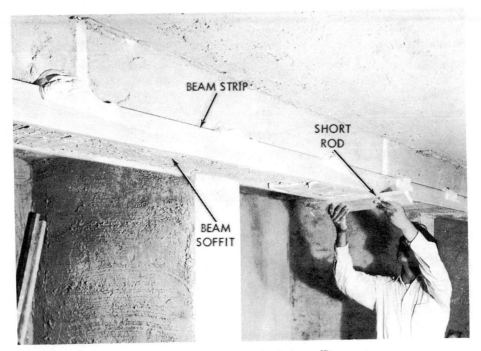

Fig. 6-35. Browning the beam soffit.

Fig. 6-36. Finished beam with beam strips removed.

Fig. 6-37. Beam sides and soffit completed.

may cause the strip to damage the arrises.

Fig. 6-36 shows the beam with the strips removed and the soffit finished. The sides are now to be finished. Fill in any hollow spots along the arrises with high gauged putty and plaster before finishing the sides with regular putty coat finish.

Fig. 6-37 shows the completed beam. Note the sharp, straight arrises. In some cases the sides are not finished separately, but are left to be finished with the ceiling as a complete unit. The procedure just described produces a straight, plumb and level beam with parallel sides.

Stripping: Second Method

The second method is used on jobs where the standard of workmanship is less exacting. A beam of fair appearance will be produced, but the accuracy of the work must rely on the eye alone. No lines are set, the water level is not used and no measurements are made.

By this method, the plasterer prepares a gauging of high gauged putty and plaster. He should prepare enough to place about the same number of dabs of plaster on the beam sides as were used to secure the strips by the method first described. The dabs are applied to the beam sides at intervals of approximately 2 feet. The dabs must be quite stiff in order that a dab a good size may be formed. Now take the beam strip and press it into the soft plaster dabs. In doing this let the top of the strip tilt in towards the beam slightly. This will make it easier to finish the sides later.

Make sure the beam strips have been cut to fit the beam length. If more than one strip is required to complete the side of the beam, place an extra-wide dab where the two strips will join.

Now tap and press the strip until it is securely in place, then sight along it to see that it is straight and at the proper height. Repeat this operation for the other side; then check with a hand level to make sure that both strips are level with each other. Also measure their width apart at each end and at various points.

These several operations must be performed while the plaster dab is soft. Yet the plaster must start to set as soon as the strips are applied if the strips are to remain in place and not fall off. Most plasterers do their work effortlessly and achieve a straight looking beam; but speed and ready manipulation are essential to the task.

After the strips have been set in place, more plaster is applied over the strips to hold them in place securely. The soffit is now ready to be filled in with high-gauged putty and plaster. Repeat all the operations used in the first method to complete the beam.

Squaring a Room

Most rooms are of square or rectangular shape, they have square 90° corners and parallel walls. There are, of course, rooms and areas that have curved or slanting walls, but these are the exception rather than the rule.

A number of methods may be employed to square a room. Three of the methods commonly used by the trade are as follows: right-angle triangle method, the square method, and the bisected center-line method.

Right-Angle Triangle Method

In Fig. 6-38 the plasterers are shown establishing a line along a wall that is square to a joining wall that was chosen as the reference or base line. In other words, one wall was picked as one thought to be straight. All further work is based upon this wall. As seen in the picture (Fig. 6-38) the 3 ft. line was chosen as the base line. The method used in this illustration is called the right angle triangle method.

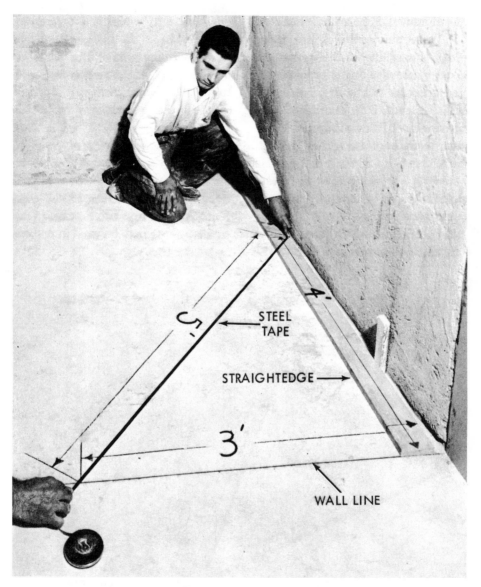

Fig. 6-38. Squaring a room—right-angle triangle method.

A right-angle triangle is one having one angle of 90°; that is, it has one of its three sides perpendicular to another side. Because one of the sides is perpendicular to another, a perfectly square corner is the result.

Along the wall you have chosen as

the reference or base line, mark off 3 ft. from the corner of the room. Along the other wall forming the corner mark off a line approximately perpendicular to the first line measured. Measure 4 ft. this time; then measure out 5 ft. on a diagonal line in order to connect the end of the 3 ft. line and the end of the 4 ft. line. Use a steel tape to do this, as the measurements must be carefully taken to insure accuracy.

Hold one end of the tape on the 4 ft. mark then measure out on the diagonal 5 ft. where the 5 ft. mark crosses the 3 ft. mark establishes the exact square line. The base line point is fixed, the other line is moved to make the intersection.

These dimensions may be multiplied by any given number, so that the formula will be applicable to areas of any size or condition. As an example, multiply them by 6 to produce a formula reading 18'-24'-30'. Note that the first two numbers (the smaller ones) may be used in reverse order. In this example, 24'-18'-30'.

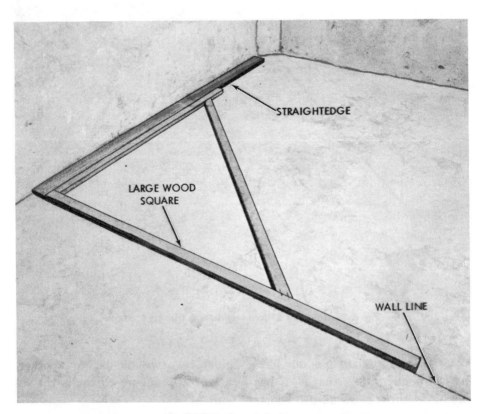

STRAIGHTEDGE

LARGE WOOD SQUARE

WALL LINE

Fig. 6-39. Large wooden square.

Square Method

For the square method, a large wooden square is used to lay out a wall line that is at right angles to the starting wall. As can be seen in Fig. 6-39, the square is laid against a straightedge. This is done to obtain the true line of the starting wall. The square may be used to rod a screed for the length of the square. The screed is then extended from this section on, in a straight line as needed. The wooden square is made up and checked by the formula given for the first method.

Both of the methods previously described required a starting line or wall. These methods do not insure a perfectly square room, but they do form square corners and reasonably square floor or ceiling areas. To form a truly square area for either floor or ceiling, the best method to employ is the bisected centerline method.

Bisected Centerline Method

In Fig. 6-40, you will find a plan of a room illustrating the centerline method of squaring a room. To use this method, measure the width of the room shown at A or Fig. 6-40; then find the center point of this line. The center will be one half the distance A. See B of Fig. 6-40

Go to the opposite side of the room and repeat the operation, establishing center point D. Now snap a chalk-line between these two points B-D.

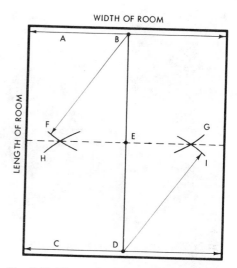

Fig. 6-40. Plan illustrating centerline method.

Measure this line which runs the length of the room and find its center at E. In Fig. 6-41 two plasterers have measured the center line and are determining point E as seen in Fig. 6-40.

The next step requires the use of a steel tape, wire or stick (longer than distance B-E). Start at B of Fig. 6-40; using B as a base, rotate the opposite end of the tape, or whatever is used, to form arc F. See Fig. 6-42 where the two plasterers are forming arc F as represented in Fig. 6-40. Do the same thing for arc G. Remember, the same length of tape must be used to form all arcs. If care is not used and there is a variation in the length of the tape used, or the pencil marks are not carefully made, the crossing of the arcs will be inaccurate.

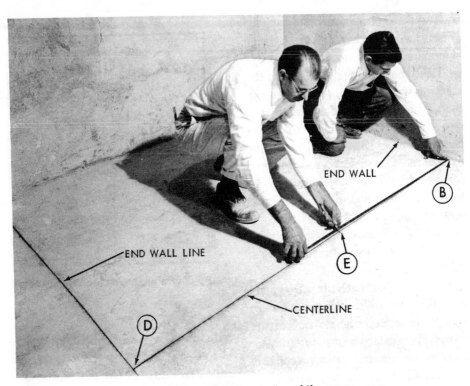

Fig. 6-41. Finding the centerline of the room.

Fig. 6-42. Bisecting the centerline.

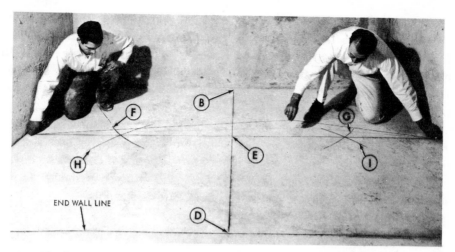

Fig. 6-43. Snapping a chalkline across the intersections of the bisected area.

Fig. 6-44. Using centerline as guide, mechanic uses gage to form wall dots.

Cross over the center line and repeat the operation at point *D*, forming arcs *H* and *I*. Arcs *F* and *H* cross to form a point, so do arcs *G* and *I*. If a line is drawn through the points formed by these two sets of arcs, you will find that it also goes through point *E*, as shown in Fig. 6-40. In

251

Fig. 6-43 the plasterers are snapping a chalk line across the intersection of the bisecting arcs. If the line snapped through these two sets of arcs does not cross the center line at point *E* exactly, some error has been made in forming the arcs or measuring center lines. Recheck to discover the error.

The method just detailed is the most practical and accurate method to square a room or ceiling area. It is the only method to use in laying out center lines for ornamental plastering or accoustical tile work. No other method produces the accuracy and the required center lines as part of the squaring operation.

From the established square center lines, the plasterer uses a gauge to form a series of dots on each wall. The gage is used as shown in Fig. 6-44 to form the dots.

Splays

A splay is the angle that one surface or line makes with another when the angle is greater than 90°. Fig. 6-45, left, shows how two splay angles occur when a wall is built across a corner. Any angle may develop depending on how the wall is set. When a straight wall is interrupted by a change in its direction of more than 90° a splay angle angle is formed.

For the plasterer the splay angle can present problems. If both walls (each side of the splay angle) are not perfectly plumb the resulting angle will be out of plumb, and this will be very noticeable.

In developing a true splay angle there are various methods that can be used. For the very best class of work the splay is run using a splay mold. (See Chapter 9 for details on molds.) In this method screeds are formed plumb on each side of the angle; a splay mold with the required angle is made to fit the splay. A cornice strip is set plumb on one screed and fastened. The splay mold is then run up and down the angle to form a perfectly straight and plumb angle. See Fig. 6-45, right, which shows a splay mold set in place ready to run the angle.

Another method is to plumb both walls carefully in the brown coat. See Fig. 6-46. The angle is then featheredged in the finish coat (as done in normal angle work) and a straight cut is made in the angle, using the featheredge as a guide, and a long joint rod (miter rod) is used to make the cut. The angle is then

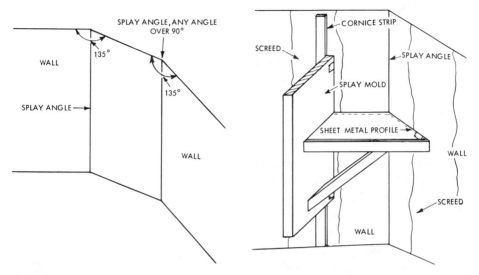

Fig. 6-45. Splay angle (more than 90°), left; and splay mold set to run a splay angle, right.

Fig. 6-46. Splay angles may also be run by carefully plumbing both walls in the brown coat, as shown, then featheredging the angle in the final coat.

carefully troweled out working out on each side from this plumb straight cut. (See chapter 7 for more detailed coverage on forming straight angles.)

For most residential and light commercial work the requirements are not as rigid, so the plasterer works out the splay by eye. Some plasterers brown out the main wall into the splay angle. A straightedge is then set flat against the wall and plumbed, and a trowel is run along the edge of the straightedge to form a mark in the brown mortar. Finally, the other side of the wall is worked to this mark so as to achieve a straight angle. The trick is to always make sure both sides of the wall at the angle are straight in the brown coat.

Curves and Irregular Shapes

To produce various curved and irregular surfaces the plasterer either works to grounds set by other trades or produces his own screeds to develop the required shapes. The principal equipment used in this type of work is the template.

There are numerous types of templates, each one designed to suit a specific condition or situation. One of the most commonly used templates is the arch template. This type can be made in two forms. One is a positive or pressed screed type, where a wood board is cut to the required arc and placed in position in the rough opening as shown in Fig. 6-47. High gauged putty and plaster is pressed between the template and the rough arch to form a plaster screed.

The other is the reverse template. Here the wood board is cut out to the required arc and, using two templates in pairs, the arch is formed and

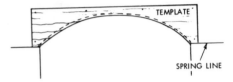

Fig. 6-48. Solid wood reverse type arch templates (used in pairs).

finished in one operation. As shown in Fig. 6-48, the templates are nailed on each side of the opening and the soffit filled in to complete the arch. Greasing the edge and sides of the wood will help to release it from the plaster.

For large arches and many other curved surfaces, either single or compound curves (concave and convex combination), the templates can be made from boards nailed together and cut to shape or made up of light wood strips bent over struts to the required shape.

Fig. 6-49 shows the development of a simple arch template constructed with a lattice or cornice strip top and 1″ x 2″ grounds for the struts. The 1″ x 4″ beam strip is used as the base of the template. For larger templates this base can be enlarged to a 1″ x 6″, 2″ x 4″ or even a 2″ x 6″ plank to provide a ridged backbone.

To lay out a template of this type

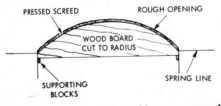

Fig. 6-47. Solid wood positive or pressed screed arch template.

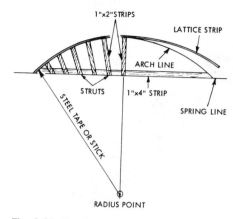

Fig. 6-49. Wood strut type arch template.

the plasterer usually works on a clean section of floor. Using a steel tape with the end ring hooked over the radius point, strike the required radius on the floor using a pencil or crayon to make a good clear mark. Snap a chalk line across the arc at the spring line (level line from which the arc springs or rises.) See Fig. 6-49. Now lay out the 1″ x 4″ wood strip, or heavier board if the arch is large, on the spring line. Place the bottom of the board on the spring line, cut each end to fit the curve of the arch, but cut back the thickness of the cornice strip which will form the top of the template. Nail this strip down with a couple of nails so it will not move from the line. If the floor is concrete place a couple of cement blocks on the strip.

Next, lay out a series of 1″ x 2″ strips from the spring line strip up

to the arc drawn on the floor. Cut each strip the thickness of the top strip short of the arc and nail it in position to the 1″ x 4″ base strip. Space the strips so they will support the top cornice strip without deflecting.

When all the struts are nailed in place the template is ready for the final step. Starting at one end, nail the cornice strip securely in place to the 1″ x 4″ base strip. Now bend the strip over the struts, nailing it to each strut as you go. Make sure the cornice is long enough before you begin nailing

The three basic types of templates (solid wood, positive or reverse, and wood strut) can be adapted to suit almost any job conditions. Each template is a separate type but they all can be used to form the same arch. The first template (Fig. 6-47) sets inside the arch opening and is used when there is no room to use the second or reverse type (Fig. 6-48) which is nailed on the face of the walls to form the arch. Job conditions set the type of template to use. The third or strut type (Fig. 6-49) is normally used when the arch opening or similar shape is very large. It is very strong but light in weight and can be made to almost any required size.

Each template needed will be some variation of the three types described (Figs. 6-47, 6-48 and 6-49). You can also combine convex and concave

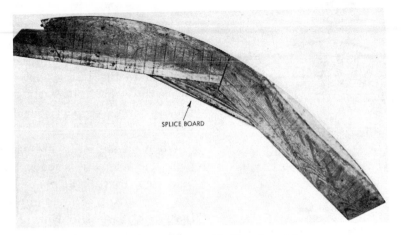

SPLICE BOARD

Fig. 6-50. Solid wood built-up template.

Fig. 6-51. Strut type template set in position and the plaster screed formed.

shapes (equal to an "S" shaped figure) to produce templates that can form the screeds for some of the modern ceilings and walls found in today's buildings.

Fig. 6-50 shows a large solid wood template built up using two 1"x12" boards nailed to a splice board to hold the two boards firmly together. Fig. 6-51 shows a strut type template in position, and the high gauged plaster screed formed by it. The plasterer here is using only a part of the template; on the next move he will use it all to form the required screed. Also shown in the background (to the left) is a lunette template in position. (The construction details and use of the lunette template will be covered in Chapter 9.)

Checking On Your Knowledge

The following questions give you the opportunity to check up on yourself. If you have read the chapter carefully, you should be able to answer the questions. If you have any difficulty, read the chapter over once more so that you have the information well in mind before you go on with your reading.

DO YOU KNOW

1. What are the two principal methods in use today to apply plaster?
2. What is the difference between a darby and a slicker?
3. What is meant by the term "trimming the angles?"
4. What are dots, screeds and why are they important?
5. Why are gages used with a plumb bob?
6. Why is the water level used by the plasterer?
7. How is a level wall line transferred to produce a level ceiling?
8. Why are screeds required on walls and ceilings?
9. What is a plaster pump?
10. What are the two methods used in stripping beams?
11. What three methods are in common use to square a room?
12. What are the three parts of a plastered beam?

Chapter 7

Base and Finish Coats

When to apply two coats and when to apply three coats of plaster? This is a difficult question to answer, since the use of two or three coats of plaster depends upon many different considerations. The various surfaces upon which the plaster is to be applied sometimes dictate the number of coats required. For example, plaster applied over masonry seldom requires more than two coats. Metal lath generally requires three - coat treatment. Resilient lathing systems also require three coats of plaster.

It is generally believed in the trade that three coats of plaster (the scratch coat, brown coat and finish coat) make a better plastering job.

City fire laws and city building codes sometimes alter this. One city may require just two coats of plaster over gypsum lath, while another city, not too distant, may require three coats of plaster over gypsum lath.

The number of coats used, then, varies with the job location and the type of base being used. We may generalize, however, to say that it is usual to apply two-coat work over wood laths, gypsum laths, insulation laths, and masonry work. Three-coat work is generally applied over metal lath but it is sometimes applied over gypsum lath as well. Sometimes only two coats are applied. It depends upon the locality.

Base Coat Plastering

Base coat plastering, as noted in Chapter 1, consists of the *scratch* coat and the *brown coat*. The third coat, which is applied over the other two is called the *finish coat*. The result is three coats of plaster called three-coat work. Two-coat work consists of one base coat and, of course, the finish coat. The finish coat is always the last coat.

The reason for using a scratch coat in three-coat work is that it stiffens the lath. It forms either a mechanical key or an adhesive bond, depending upon the lathing used. In other words, the scratch coat (together with the lathing) is the foundation for the structure of the plaster wall. The scratch coat also provides uniform suction for the brown coat that follows.

After the scratch coat is applied, and before it sets, the surface is raked or scratched. The surface is usually scratched in two directions so that there is a cross-hatched effect of roughened surface. This allows for an adequate mechanical bond with the brown or second coat.

In two-coat work the base coat (just one in this case) is applied by either the single coat (lay-on) method or by the scratch double-back method. In the single coat (lay-on) method, the mortar is applied to slightly less than the thickness of the grounds and screeds and then it is rodded and darbied to a true, even surface. It is left rough enough to receive the final coat. In the scratch double-back method, mortar is applied in a thin coat, and then the plasterer returns before the mortar just applied has time to set. He then fills in to the grounds and screeds, rodding and darbying in preparation for the finish coat.

When the plaster is being applied during cold weather a minimum temperature of 40°F should be kept in the building until the plaster is dry. If this is not done the plaster might freeze before it has a chance to set. After the plaster has set ventilation should be provided to eliminate too much moisture in the building. In hot, dry weather all openings should be closed while the plaster is being applied. Plaster should not be allowed to dry before setting.

Gypsum Scratch Coat on Various Bases

Of the basic cementing materials that are used as binders in mixtures of base coat mortars, gypsum and its by-products are perhaps the most widely used.

The scratch coat is always the first coat of mortar to be applied. This first coat is applied for two reasons: to provide a base which offers sufficient suction power to hold the coats laid over it and to insure a good bond with the base upon which it is applied. Only by using a thin coat laid under good pressure can a good bond with the base be secured.

When a scratch coat is put over metal lathing, or is used as a suction-developing coat over hard brick, concrete or other dense materials, the coat is scored or scratched over its entire surface with a wire scratcher. This is done to provide a mechanical key that will lock the next coat applied.

Always score or scratch plaster that will be set (allowed to harden) before the next coat is applied. Cross scratching is preferred. When coats of plaster are to be laid in quick succession, scoring is not necessary but the base must have suction of sufficient ability to bind the mortar securely.

Three-coat work provides the strongest and straightest job. The first coat is a scratch coat put on and permitted to set hard without being disturbed. This coat does double duty: (1) it contributes good suction, and (2) it stiffens the lath, preventing it from buckling or sagging. In many of the clip systems, it is necessary first to scratch the surface of the lathing with a fast-setting mortar to keep the lathing from sagging. Gypsum or insulation board laths, if not supported firmly throughout the center, will sag and buckle because of the softening of the laths beneath the wet mortar.

Diamond-Mesh Metal Lathing

In Fig. 7-1, the plasterer is applying scratch mortar to metal lathing,

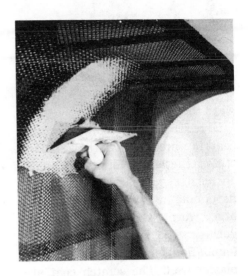

Fig. 7-1. Applying scratch coat to metal lath.

both interior and exterior. Light pressure is used and the trowel is held at an angle sharp enough to prevent the mortar from screening through the holes while the plaster is being applied to the metal lath. To screen more mortar than is necessary is to waste material and add unnecessary weight to the unit treated.

The mortar must be stiff and rich. Fibered plaster is usually used because the fibers help hold the mortar more firmly together than the non-fibered. A point to remember is that when mortar is applied over metal laths, overlapping of strokes must be avoided. Overlapping of the stroke is a duplication of effort in work already accomplished.

Gypsum and Insulation Board Lathing

When doing three-coat work on gypsum laths or insulation laths, or styrofoam, the lathing is first plastered or scratched with a thin coat of rich mortar. See Fig. 7-2. When scratched and allowed to set, the coat will provide the suction needed for a good, straight job of browning. In Fig. 7-3 the plasterer is seen scratching the soft mortar with a scratcher. The entire coat is cross-scratched in order to provide the best possible key for the coat next laid.

To strengthen and reinforce the plaster, especially on ceilings, a light-gage wire mesh can be nailed over the gypsum or insulation lath. The mesh provides an all-over reinforcement of such light body that little weight is added to the job. Fig. 7-4 shows the plasterer applying the first coat of

Fig. 7-2. Applying scratch coat to gypsum lath.

Fig. 7-3. Scratching the first coat applied over styrofoam. (Dow Chemical Co.)

261

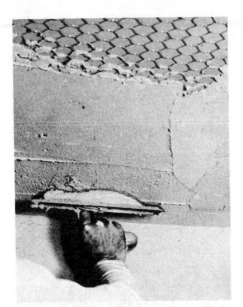

Fig. 7-4. Gypsum lathing and wire mesh reinforce plaster against earthquakes. (Dahlhauser; Keystone Steel & Wire Co.)

Fig. 7-5. Plasterer applying scratch coat of mortar over diamond mesh metal lath. (United State Gypsum Co.)

plaster over wire mesh attached to gypsum lathing. Now, both mechanical and adhesive bonds can operate. Another way to reinforce gypsum and insulation boarding is to use strip lath across ceiling joints and cornerite where ceiling and wall join.

Wire Lathing

Wire-cloth or woven-wire lathing requires a stiffer, more fibered mortar than regular diamond-mesh metal lathing. This is especially true if wire mesh is used without gypsum or similar backing. Fibered motar is necessary because the thin, round wires provide too little support to

hold the mortar securely while it is soft. Once the mortar is set, however, this type of lath provides excellent reinforcement for plaster, particularly in earthquake areas.

Excessive troweling or lapping on this type of lathing may cause the mortar to drop. Fig. 7-5 shows the keys formed as the mortar squeezes through the lathing. Dropping can be avoided by using light pressure with the trowel when applying mortar to metal lath.

Paper-backed Lathing

Paper-backed lathing and similar constructions were designed to save plaster. A stiff, heavy paper or fiber is fastened to the back of the lathing.

Various means of attachment are employed. One method is to press the lathing onto strips of asphaltum laid over heavy paper. Another is to clip the lathing to the paper. A third method is to weave a wire through stiff paper in such a manner as to produce an overlay of woven wire lathing upon a solid backing of paper.

In Fig. 7-6 the scratch coat is seen applied to paper-backed lathing. Use sufficient pressure on the trowel when scratching to insure flow of the mortar around all exposed portions of the lathing. Push the paper away from the lathing enough that the mortar can get behind the metal and form good keys.

Tarred Surfaces

For a number of years, many plasterers and architects thought that if the inside surfaces of masonry walls were covered with a coat of tar, or some other bituminous material, dampness could be prevented from penetrating and damaging the plaster.

It was also believed that these materials would insure a good plaster bond. It was customary to scratch this type of surface with a rich scratch mortar and then let it set. With suction provided, the work could be browned. Fig. 7-7.

On many of these jobs, it was dis-

Fig. 7-6. Applying scratch coat to paper-backed lathing. (Pittsburgh Steel Products Co.)

Fig. 7-7. Applying scratch coat over a tarred surface.

covered that after ten to fifteen years the plaster had loosened. When it was removed, or it fell, the tar or bituminous material was observed to be in a dried-out condition. The oil in the tar seemed to be absorbed by the plaster.

Only when some form of mechanical key is produced should this type of damp-proofing be used. The scratch coat of mortar must be very rich. It must not be disturbed after it has been applied and scratched, but allowed to set and harden. Ordinary brown mortar should never be used directly over a low-suction base of the kind described here.

Concrete Surfaces

Plastering over concrete beams and slabs is not recommended. If it must be done, the concrete should be hacked or roughened in order to produce a mechanical bond. Then the surface of the concrete should be washed with a 10 percent solution of muriatic acid to remove any grease or oil deposited by the forms. Remove the acid by washing, using plenty of clean water.

The concrete may now be scratched with bond plaster. Bond plaster is a special plaster made so that, in setting, it will neither expand nor shrink. Regular plaster, when setting, first shrinks slightly, then swells. This action would loosen or break the slight mechanical key produced by the plaster on the surface of the concrete. Apply this material neat, that is, as it comes in the bag without any additions except water. Use a thin coat and do not disturb after the mortar has been applied.

Concrete should be damp when the bond plaster is laid over it. To keep the plaster from drying out too fast in warm, dry weather some plasterers spread a thin coat of regular scratch mortar over the soft bond plaster. This must be done with care in order not to break the adhesion built up during application of the first coat. In either case, the surface is scratched to form a good key. Fig. 7-8 shows a concrete surface during the various stages of preparation to receive coats of a smooth-surfaced plaster.

Because of the weak bond that concrete affords, even under the most favorable conditions, the coats of plaster applied should not exceed $\frac{1}{2}''$ in total thickness.

The safest and surest method to use when plastering over concrete is to provide some form of strong mechanical bond, such as metal lathing fastened with concrete nails, and furred out away from the beam or slab.

Liquid bonding agents which are applied by brush, roller or sprayer to any sound surface are being used today for concrete surfaces. The surfaces to be coated must be clean. Remove all loose and scaling paint,

After the surface has been coated the liquid bond should be allowed to dry for about 45 minutes, depending on the drying conditions. When it is dry there will be a glossy finish. If dull spots are evident recoat these spots so the entire surface to be plastered has a completely uniform glossy appearance. The tools used for applying the liquid bond may be cleaned with warm soapy water. The application of plaster now proceeds in the usual manner with consideration given to the fact that there will be very little suction. If a heavy coat of plaster is to be applied, a scratch coat will be necessary to prevent sagging or sliding of material down the wall or dropping if the work is overhead.

Plastering over Foam Plastics

The foam plastics will give very little suction and a scratch coat in most all cases will be necessary. Apply a normal ¼″ thick scratch coat with firm pressure to insure a good key to surface cells of the foam, and then scratch surface and allow material to dry firm and hard before applying brown coat.

When using a plaster gun for scratch coating metal lath, the air volume control at the nozzle should be reduced and the consistency of the mortar should be a little stiffer. To apply, the nozzle should be held at an angle to the metal lath so that as little as possible will be blown directly through.

PUTTY-COAT FINISH

BOND PLASTER-SCRATCH COAT

HACKED SECTION

SMOOTH CONCRETE

Fig. 7-8. Preparing concrete to receive mortar.

grease, dirt and efflorescence by scraping or wire brushing, followed if necessary by washing. Before using, stir the liquid bond thoroughly. Most manufacturers will not recommend thinning. Usually this material is colored to help show that all the surface is completely covered.

265

Gypsum Brown Coat on Various Bases

Browning is the term applied by the plasterer to a coat of mortar applied for the purpose of building up and straightening the surface. Its thickness is subject to existing conditions. Usually, however, the brown coat is from ½″ to ¾″ thick.

A brown coat is made poorer than a scratch coat. One reason for this is that a rich mortar is hard to rod because of its extreme stickiness. Another reason is that the brown coat is most often applied over surfaces that have fairly high suction power. When the brown coat is applied as a first coat, it is applied over bases with some suction power in them. When the brown coat is applied as a second coat, it is applied over the scratch coat that has previously been applied and that provides the necessary suction.

Three-Coat Work on Gypsum, Insulation and Metal Lath

The brown coat in three-coat work is the second application of mortar. It is applied after the first or scratch coat has set firm and hard. The brown coat is then straightened to a true surface with rod and darby, and left rough, ready to receive the finish (third) coat.

The brown coat to be applied in three-coat work is usually mixed in proportions of 2½ to 1. Because gypsum, insulation and metal laths are more flexible than masonry units, which are used as a backing, a stronger mortar must be used. Fig. 7-9 shows the plasterer applying the brown mortar over previously scratched foam plastic. The scratch coat is dry; therefore, he is applying the brown coat to a small area right after troweling on a thin coat to insure bond. This procedure permits him to rod and darby the work before suction has caused the mortar to stiffen too much.

Fig. 7-9. Plasterer applying the brown mortar over previously scratched foam plastic. (Dow Chemical Co.)

Two-Coat Work on Various Bases

The brown coat in two-coat work is the first and only base coat for this type of application. The base (first) coat should be applied with sufficient pressure and enough mortar to form a good bond on gypsum, fiber insulation lathing or masonry, as the case may be.

The base coat in two-coat plastering should cover well. Then it should be doubled back to bring the plaster out to grounds, straightened to true surface with rod and darby, and left rough, ready to receive the finish (second) coat.

Fig. 7-10. Applying first coat on brick wall.

Two-Coat Work on Masonry Bases

Two-coat plastering can be applied in either of two ways: the single coat (lay-on) method, or the scratch (double-back) method. The latter is the method most often preferred on masonry.

When browning over brick, tile, and other masonry bases the mortar is usually made in proportions of 3 to 1. This proportion gives a mortar that spreads easily and that can be rodded and darbied without dragging or sticking to the tools.

After the dots and screeds are in place the brown mortar (first coat) is spread in two applications. The first application is spread over the masonry surface and then the mortar is scratched. This procedure creates a good bond with the base and equalizes the suction. Fig. 7-10. The sec-

ond application is doubled-up; that is, it is put on right after the first application. The purpose of the second application is to fill the wall or ceiling out to the screeds.

Depending upon the suction ability of the base material, apply the mortar in small enough areas so that it can be rodded and darbied easily. On high-suction bases, use a poorer mortar and work smaller areas at any one time. In Fig. 7-11 the plasterer is working on a common brick wall of exceptional suction ability. He is doubling up the small area scratched in order to have a plastic surface to rod and darby. This operation is repeated section by section until the wall is finished.

On low-suction bases, greater areas are scratched in before the doubling-up coat is applied. Practice

Fig. 7-11. Applying second or double-up coat on brick wall.

Fig. 7-12. Applying first coat on gypsum lathing.

and experience are needed to determine the area to be worked in a doubling-up operation.

Two-Coat Work on Gypsum and Insulation Lathing

When browning over either of these types of board lathing, the mortar must be mixed in proportions of 2 to 1. A rich mortar is needed to insure a good bond to the lath. As these lathing materials do not offer much suction, large areas may be applied ahead of the darbying. In most cases, an average room ceiling and the upper part of the walls may be applied before proceeding with the darbying.

Again, the mortar is laid on in two operations. See Figs. 7-12 and 7-13. To provide adequate strength, ½″

Fig. 7-13. Applying second or double-up coat on gypsum lathing.

of mortar should be applied. Plaster cracking and failure on these types of bases can, in most cases, be attributed to excessive thinness of the base coat applied.

Lime Mortar: Scratch and Brown Coats

Lime mortar may be applied either as three-coat work or as two-coat work. Three-coat work can be applied over metal and wood lathing. Two-coat work can be applied over brick, tile, gypsum block, and other masonry surfaces. Lime mortars must not, however, be applied to gypsum or insulation laths, because lime mortar does not supply the bonding action that is so vital for good results on these bases.

Lime mortar must be well-aged in order to work properly, and it must not be too rich. Many plasterers will make this type of mortar rich in the mistaken belief that a mortar of rich body is necessarily better than one that is poor. The truth of the matter is that a rich lime mortar will shrink while setting and, in doing so, will crack. It commonly happens that a rich mortar will shrink so much that the bond breaks and the mortar falls from the base. Since it is slow to set, lime mortar must be protected from dryouts caused by exposure to dry winds.

The scratch coat, whether used for three-coat work or for two-coat work, is made of 3 parts of aggregate to 1 part of stiff putty (3 to 1). If the scratch coat is to be put on wood or metal lathing, the addition of fiber is necessary. The scratch coat should be applied with sufficient pressure to ensure a good key or bond, as the case may be.

In three-coat work the scratch coat is allowed to set. The first coat is scratched, however, before the coat dries. It should be scratched to insure a good bond with the second coat. When the scratch coat has become dry, the second or brown coat can be applied. By using a mix of 3 to 1 (3 parts sand to 1 part lime) and aging the mortar for about a week, the plasterer may safely proceed to brown-in the work. Suction can be controlled by dampening the base before applying the mortar. The day after the mortar is put on, it should be floated by means of a large, cross-grained float.

It is a good practice to drive a nail in the forepart of the float to a depth that permits the nail to protrude only far enough to form a scratching device. See Fig. 7-14. By this means, the surface of the mortar can be "roughed" sufficiently to insure firm adherence of the finish coat. Floating is done to compact the mortar, thus keeping shrinkage cracks from forming.

In the case of two-coat work over masonry bases, a coat of this mortar (3 to 1) is applied to insure a good bond. Then the surface is doubled

269

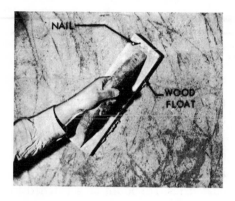

Fig. 7-14. Floating lime mortar, using wood float with nail driven in it.

back with the same mortar. This doubled-back coat is then brought out to grounds and is rodded and darbied to a true surface. When the base coat is thoroughly dry, it is then ready for the final coat of finishing plaster.

Admixtures of Portland cement or Keenes cement help to prevent the formation of shrinkage cracks, and they produce a stronger mortar as well. All operations, applying the mortar, rodding and darbying, are the same as when gypsum mortar is used, except for the final floating. Final floating is an extra operation.

Veneer Plaster

Veneer plaster is a classification used throughout the industry to describe any thin application of a monolithic troweled finish to walls and ceilings. There are two systems: One is the use of two coats, the other is the one coat. There are several manufacturers and each has developed its own particular system. Two of the many manufacturers' systems of two-coat veneer plaster will be discussed here.

One of the two-coat systems of veneer plaster uses the standard gypsum lath (48″ x 16″) as a base. A catalyst is sprayed on the lath before plastering to accelerate the setting time of the plaster. This should be applied just before the application of the plaster to achieve a quick setting time and a good bond.

This catalyst can be a manufactured bagged material identified as gypsum catalyst, or a solution of 2 or 3 pounds of powdered aluminum sulphate to one gallon of water. An ordinary garden sprayer can be used to successfully spray the catalyst onto the lath, and it is a good practice to screen the solution into the

Fig. 7-15. Plasterer applying fiberglass mesh covering. (United States Gypsum Co.)

can to collect any undissolved particles.

Should the catalyst dry out prior to application of the plaster, the surface should be re-wet with the same solution. The base coat plaster should be applied with sufficient material and pressure to form a good bond on the gypsum lath to a thickness of not less than $\frac{1}{16}''$. This can be applied by hand or by machine. Before the plaster sets, the surface should be left in the same manner as conventional plaster to aid in the bonding of the finish coat. The surface can be straightened true and level by using a slicker or a long trowel. Angles can be cleaned up with an angle plow. Remember the surface of this plaster will become too hard to scrape down after it has hardened and set. Therefore extra care should be used not to leave too many slobbers or unwanted roughness on the surface when plastering

with this type of plaster.

Another veneer plaster, also using two coats, is one that is applied over a special base which is designed with highly absorbent paper and is usually in lengths and widths resembling drywall. There is a fiberglas mesh covering over the joints of the base and in the inside corners. Fig. 7-15 shows the plasterer applying the fiberglas mesh. The mesh is covered first by squeezing on a tight coat of the plaster with the trowel, and then the whole wall or ceiling is covered. Whether the method of application is by hand or machine, the mesh is first covered by hand troweling, and then re-covered when the remainder of the surface is plastered. The plaster used for this type calls for a catalyst to be added before using, and if it is to be applied by machine the special catalyst gun is used. This gun adds the catalyst to the material at the nozzle.

Lightweight Aggregates: Scratch and Brown Coats

Lightweight aggregates are used in mortar mixtures that are applied in three-coat work and in two-coat work. In three-coat work lightweight aggregate mortar is applied first as a scratch coat and then as a second brown coat. In two-coat work only one base coat is applied. It consists of the initial application and the double-up application.

Lightweight aggregate mortar can be applied on all bases. Wood and metal laths generally take three coats. Masonry bases generally take only two. Gypsum and insulation laths take two- or three-coat work depending upon the locality and its specifications.

These materials are a pleasure to work with because they are light. On the other hand, the mortar made of lightweight aggregate is a little more gummy than when sand is used. As a result, lightweight aggregate is slightly more difficult to spread. When mortar containing lightweight aggregates is applied over bases having exceptional suction power (gypsum partition tile for example), lightweight aggregate mortar will be found hard to darby. To overcome this difficulty, rodding and darbying must follow application of the mortar at a shorter interval than when sand is used. Most plasterers add a few shovels of sand to the batch when working on high suction bases. The added sand reduces the gumminess and to some extent prevents shrinkage cracks from forming as the result of the extremely high suction. Wetting the base just before the mortar is applied will also prove helpful.

There have been many fire tests conducted on identical construction which show that when sand aggregate is used with gypsum or Portland cement plaster, it does not give the same fire resistance as an equivalent amount of lightweight aggregate. When exposed to fire, both gypsum and Portland cement plasters release their chemically combined water in the form of vapor which maintains the plaster temperature at about 212°F until all the water has been driven off as steam. The insulating action of lightweight aggregate protects the gypsum, delays the release of steam and retards the transmission of heat, thus improving the fireproofing quality of the plastered construction.

Most all gypsum manufacturers have a listing of fire ratings using combinations of plaster bases, aggregates, and various thicknesses. (Fire ratings were discussed in Chapter 5.) The plasterer should consult these listings when a definite fire rating is

TABLE 7-1   FIRE RATING WITH LIGHTWEIGHT AGGREGATE PLASTER

RATING	DESCRIPTION
Structural Steel Columns	
2 hours	Steel Column protected by lightweight aggregate gypsum plaster 1" thick on the face of metal lath spaced 1 1/4 "from the flanges. No plaster fill behind metal lath.
2 hours	3/8" perforated gypsum lath applied vertically and tied with double strands 18 ga. tie wire spaced 2" from ends of lath and approximately 15" at intermediate points. Expanded metal corner beads. Plastered with 1" gypsum lightweight (LW) plaster.
Floors and Ceilings	
3 hours	Steel column protected by LW plaster 1" thick on the face of metal lath spaced 1 1/4" from the flanges. Space between lath and outer faces of flanges filled with plaster.
3 hours	1/2" long-length gypsum lath. Two 3/4" coats LW plaster applied separately with 1" mesh fabric wrapped around columm over first coat of plaster. Expanded metal corner beads.
3 hours	Two layers 1/2" long-length gypsum lath, wrapped with 1" hexagonal mesh Expanded metal corner beads. Plastered with 1" LW plaster.
4 hours	Steel column protected by LW plaster 1 1/2" thick on the face of metal lath spaced 1 1/4" from the flanges. Space between lath and outer faces of flanges filled with plaster.
4 hours	Two layers 1/2" long-length gypsum lath wrapped with 1" wire mesh. Expanded metal corner beads. Tie wires surrounding columns. Two 3/4" coats LW Plaster.
4 hours	One layer 1/2" long-length gypsum lath temporarily wrapped with wire between coats. Total thickness of plaster 2 5/8". Expanded metal corner beads.
4 hours	Floor and beam construction consisting of steel floor units mounted on steel members with 2" LW concrete top flooring. Metal lath suspended ceiling spaced at least 3" from the steel members and covered with LW plaster 1" thick on face.
4 hours	Suspended ceiling consisting of metal lath and LW plaster 1" thick on face of lath spaced at least 3" from non-combustible structural members without a finished floor above. (Note: This rating applies to suspended ceilings beneath poured or precast concrete roofs, gypsum roofs, auditorium ceilings, etc.
1 hour Combustible	Wood joists supporting wood floors protected on underside by 3/8" perforated gypsum lath nailed on and covered with 1/2" LW plaster.
1 3/4 hours Combustible	Wood joists supporting wood floor protected on underside by 3/8" plain gypsum lath and 1" wire mesh nailed to joists and covered with 5/8" LW plaster measured from back of lath.
1 1/2 hours Combustible	Wood joists supporting wood floor protected on underside by ex- panded metal lath nailed to joists covered with 1/2" LW plaster.
Walls and Partitions	
2 hours	Non-bearing hollow partition of 4" trussed steel studs protected on both sides with LW plaster 1" thick on metal lath furred 1 1/8" from studs.
1 hour Combustible	Hollow load bearing partitions of 2 x 4 wood studs 16" on centers protected on both sides with 1/2" thick LW plaster on 3/8" perforated gypsum lath.
1 hour	Solid partition consisting of metal lath on 3/4" steel channels with LW plaster. Total thickness 1 1/2".

273

called for. Table 7-1, using light-weight aggregate, shows how the various bases and thicknesses of plaster will change the time of fire rating.

Radiant Heating Panels

When plastering over hot water radiant heating was first introduced, many failures occurred. Upon studying these failures, it was found that gypsum mortar would not hold up to the terrific heat carried through the pipes. It started the recalcination of the gypsum which in essence weakened the plaster, creating cracks and causing the falling out of pieces of plaster. We will discuss a method that has been used successfully.

Hot Water Radiant Heating Panels

For plastering hot water radiant heating panels using tubing of a small diameter, the use of the normal three-coat system can be used: scratch, brown, and finish coats. When covering tubing more than $3/8''$ in diameter, four-coat work is usually necessary to completely imbed the tubing within the plaster. A scratch coat, fill-in coat and the brown coat of at least $3/8''$ is applied over the tubing, and then the finish coat. Allow time between coats for drying to obtain the suction necessary for the application of this thickness of plaster. Fig. 7-16 illustrates the four-coat method.

Note that before any plaster is applied, the pipes must be checked for leaks, and they must be O.K.'d by the heating contractor.

Remember that cement is the selected material for this work and it should be treated in the same way as regular Portland cement work. The scratch coat, fill-in coat, and brown coat should all be double scratched with a scarifier before they become hard.

The finish, however, will be applied in a somewhat different manner. The finish or white coat shall consist of 1 part finishing lime mixed with 3 parts of silica sand; then it is gauged with one part of gauging plaster. The finish coat should be applied with a thickness of $3/16''$ to $1/4''$ laid on evenly and then, before setting, burlap (running 7 meshes to the inch) should be laid on dry and troweled into the finish coat until it is invisible and the surface finish troweled in the usual manner. If a sand finish is desired, additional sand should be added at the time of mixing. It will sometimes be necessary to trowel on another thin coat with a second plasterer floating the surface to an even texture; this will be governed by the amount of suction. To spray texture, protect all adjacent areas. It would be wise to wear eye

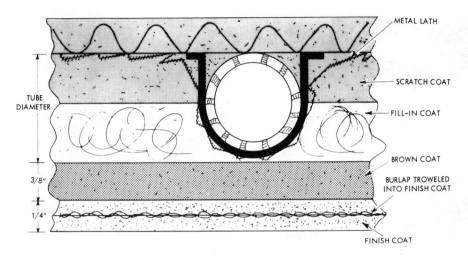

Fig. 7-16. Cross section of hot water radiant heating panel.

protection. Spraying can be done immediately after the burlap has been troweled in.

After the ceiling has been plastered these precautions should be followed. The plaster should be thoroughly dry (about 3 weeks) before heat is introduced into the system. Heat is first applied at low temperature and gradually increased to full operating temperatures.

Electrical Radiant Heating Panels

There are two controversial opinions concerning the methods of lathing and plastering electrical radiant heating panels. One opinion is that a special high-heat-resistant plaster must be used in order to resist the tendency of gypsum plaster to calcine at prolonged high temperatures. With this method it is felt that there will be a greater flexibility in the expansion and contraction in heating or cooling a room when *thin* high-strength coverings are used.

The other opinion is that this is not so and that electrical radiant heating panels can be covered by almost the same system as for ordinary plastering provided sand is used for the aggregate. This is thought to be true because the electrical cables seldom carry temperatures high enough to cause damage. Also a thickness of plaster will provide more storage area for the heat.

Application. The plaster is applied with particular care to thoroughly imbed and cover the cable without disturbing its attachment. Some manufacturers feel it is a good practice to put the system in operation after the first coat is applied to

Fig. 7-17. Applying brown and finish coats to electrical radiant heating panels: top left, applying the brown coat; top right, straightening the brown coat, using the cables as grounds; bottom left, applying finish coat; bottom right, spraying on texture. (Huron Radex Co.)

force expansion pockets which will relieve the strain under adverse operating conditions. The system should, of course, cool down for finishing operations and be left off until the plaster is thoroughly dry before again gradually introducing operating heat.

Special Radiant Heat Products. There are special radiant heat products for plastering a thin coat over electrical heating cables. These must be applied in accordance with the manufacturer's directions. It can be applied to gypsum drywall backing that has been prepared with all joints and angles conventionally covered with joint compound and tape. It is not necessary to fill and finish joints or nailheads. It can also be applied to any sound, clean, dry concrete surface which is free of grease, dust or foreign matter.

The radiant heat cables should be installed with a rust-proof fastener and according to manufacturer's recommendations. One such type uses two separately mixed materials. One is for the brown or base coat; the other the finish coat.

Each of these coats can be mixed by the paddle and drill or plaster mixing machine. Fig. 7-17, top left,

shows the plasterer applying the brown coat. Fig. 7-17, top right, shows the plasterer straightening the brown coat using the cables as grounds.

After the base coat has dried sufficiently (usually overnight), the plasterer applies the finish material, troweling it to as smooth a surface as possible. Fig. 7-17, bottom left, shows the plasterer applying the finish with hawk and trowel. After he has troweled on the finish, it is then textured sprayed using the finishing material as shown in Fig. 7-17, bottom right.

Portland Cement: Scratch and Brown Coats

Portland cement plaster is one of the best materials to use wherever surfaces will be subjected to hard use or where a condition of dampness is normal to the area. No other material applied with hawk and trowel is as strong or as resistant to moisture as this is.

A ratio of 3 to 1 (3 parts sand to 1 part cement) is used for all coats of Portland cement plaster, whether for interior application or for exterior application (sometimes called "stucco"). Mix the materials as detailed in Chapter 4. Note that in cement work, suction in the base is controlled by dampening the base

instead of changing the mix, as when lime or gypsum is used. Suction is controlled this way because Portland cement plaster must be as poor a mix as possible to prevent the formation of shrinkage cracks. Moreover, any mix made in proportions more than 3 to 1 would be too hard to rod properly.

Portland cement plaster may be applied over many types of bases. A good rule to remember is that the base material must, in all cases, be harder and stronger than the plaster mortar applied over it. Proper bases are cast-in-place concrete, concrete masonry, hard brick and tile, metal lath, woven wire cloth and the exterior of curtain wall construction. (A curtain wall is a thin wall supported by the structural steel or concrete frame of the building, independent of wall below.) Never apply Portland cement plaster to gypsum lath or gypsum tile because when Portland cement plaster comes in direct contact with gypsum products a chemical reaction takes place between the two materials and this action destroys the bond. Also, do not apply directly to tarred or painted surfaces. Metal lath should be fastened over tarred or painted surfaces to provide the key necessary to hold the plaster securely.

Portland cement may be applied with trowel or a plastering machine. The use of the plastering machine has many advantages such as speed of application and the many attractive surfaces and textures that can be obtained for the finish coat. A mixture of high air content (that is, air trapped in the mix) can be used with the plastering machine with the excess air being removed at the nozzle. The impact of the plaster on the surface improves the bond.

Portland cement plaster is applied in the same manner as most other plastering materials. The scratch coat may be eliminated on masonry bases. Fig. 7-18 shows the application of cement plaster to a masonry base.

On *cast-in-place concrete* or on

Fig. 7-18. Applying cement plaster to masonry base. (Note that this is a left-handed plasterer.)

any base with little ability to form a mechanical key, it is best to apply a thin coat of mortar. The mortar must be thin in body (soupy) and may be mixed 1 part Portland cement to 1½ parts sand. Splash this mixture on, using a whisk-broom or stiff-fibered brush. The mixture can be put in a pail, the brush dipped in it and the material picked up, and then the brush may be whipped toward the wall or ceiling with force. Cover the surface uniformly and do not disturb the applied material. Let mortar set hard, then keep it wet for at least two days to cure. Good suction will occur when the mortar is finally allowed to dry. Applying the mortar in this way insures a good bond by excluding air, which sometimes becomes trapped between the mortar and the base to which it is applied.

Liquid bonding agents for Portland cement may also be used. Follow manufacturers recommendations.

Portland cement plaster should not be applied to surfaces that contain frost. Provisions should be made to keep the plaster above 50°F during application and for not less than 48 hours after the plaster has been applied.

Generally, the *first coat* (scratch) should be applied to a thickness of approximately ⅜″. Before the first coat has hardened it should be evenly scored in two directions for ceilings. Walls should be scored horizontally. Scoring provides a good mechanical bond for the second coat and helps to hold water when dampening the surface. The first coat should be kept continuously moist until the application of the second coat.

To obtain a good monolithic bond the *second coat* (brown) should be applied to a thickness of approximately ⅜″ about 24 hours after the application of the first coat. It may be applied earlier if the first coat has hardened enough to receive it, except on wood frame construction where 24 hours are necessary. (The 24 hours is required on this type of construction to assure the added strength of cement and also to allow more time for the wood to move. Because of weight and moisture the wood will expand and contract.) The surface of the first coat should be dampened evenly before the second coat is applied. The surface of the second coat should be brought to a true and even surface by rodding. It should then be floated with a wooden float to further flatten the surface. This will also compact the cement which will compensate for shrinkage cracks that may occur. The surface should be left rough (no slick or smooth spots) to receive the finish coat. Water should not be applied during the rodding or floating operation. The second coat should be moist cured until the application of the finish.

Brown Coat for Exterior Cement

Fig. 7-19 shows protective paper, metal lath, cross-scratch, brown coat and finish coat for exterior cement. (Follow manufacturer's instructions on adding waterproofing.) The brown coat is ready for application when the scratch coat has been properly cured. Apply the brown coat in the same manner as you would apply the brown coat for interior plastering. Work on the shady side of the building when possible.

Remember to dampen the scratch coat if it is very dry. A good way to

Fig. 7-19. Proper construction for stucco on metal reinforcement. (Portland Cement Association)

Fig. 7-20. Two methods of applying brown coat: top, hand application; bottom, machine application. (Portland Cement Association)

retain normal suction is to add some waterproofing to the scratch coat, following manufacturers instructions. This prevents the loss of water and develops a working surface known as a "cooler." A cooler is a low suction base and one over which it is easy to work.

Fig. 7-20 shows the plasterer applying the brown coat both by hand and by machine. Thickness is very important. The brown coat should never be less than ⅜" thick and preferably not less than ½". Moist cure the brown coat for at least 48 hours. Then allow it to dry before the final coat is applied.

In exterior work, corners are sometimes formed by stripping; that is, a

Fig. 7-21. Two methods of producing sharp exterior corners: top, arris developed by hand stripping, bottom, machine application. (Portland Cement Association)

Fig. 7-22. Rodding close to the applied brown coat. (Portland Cement Association)

straight piece of wood is held or stuck to one side of the corner and the cement filled into it on the other side. To bring the arris out to its proper position for both walls and sides, first apply a screed of mortar of the proper thickness at the corner on one side. Then press the straight-edge strip of wood into the soft mortar, setting it plumb and out to the re-quired face of the adjoining wall. The soft mortar will usually hold the strip firmly for a short time. The plasterer now fills in to the edge of the strip and rods the mortar off to it. Fig. 7-21 illustrates the procedure. After a few minutes, if the base coat suction is normal, the strip may be removed. Use a sliding motion, working backward and upward. This allows the strip to free itself without damaging the soft corner. If the strip sticks, a light tap will loosen it.

Do not work too large an area before rodding. Fig. 7-22 shows a plaster rodding immediately after the application of brown coat. Cement mortar is harder to rod than gypsum or lime mortar.

After the brown coat has been rodded and floated straight it is scratched lightly to form a good key for the finish coat. Do not score too deeply as deep scoring weakens the brown coat.

Finish Coat

The finish coat (both interior and exterior) should be applied to a thickness varying from ⅛" to ¼" depending upon the size of the aggregate granules used. The finish coat goes on about 24 hours after the application of the second coat. The surface of the second coat should be evenly dampened when the finish coat is applied. The finish coat should be *moist cured* for at least 24 hours after it has been applied. The total thickness of Portland cement plaster must never be less than ¾". A coat of less thickness would almost certainly crack.

Job Precautions

When doing exterior cement work follow the sun around the building if possible. Try to work in the shade. Working on the shady side will help to keep the mortar from drying too fast and, in very hot weather, will be more comfortable for the workmen. If the sun is very hot, the work should be shaded with canvas hung on the outside of the scaffold. Fig. 7-23 shows a protective enclosure used to prevent rapid evaporation and to provide proper curing of exterior Portland cement.

Start from the top of the wall and work down. Enough plasterers must be working on the job that the whole wall area may be covered without joinings. Dampen the brown mortar well ahead of the application of the finish coat. By doing so, the water will have time to soak in and help suction to develop.

On jobs of ordinary size one plasterer applies the finish coat and a follow-up man floats or textures the surface. When the area is larger, more men are needed.

All arrises should be stripped, if possible, to insure well-defined, sharp corners. For the irregular surface textures stripping is not necessary; the corners are formed with the trowel or float.

It is essential that Portland cement plaster be adequately moist cured. This helps develop strength, density, and water resistance. It reduces shrinkage and helps prevent cracking and crazing. For interior work, curing can usually be accomplished by keeping the rooms well enclosed (free of drafts) and maintaining a temperature of 50°F or

Fig. 7-23. A protective enclosure used to prevent rapid evaporation and to provide proper curing of exterior Portland Cement. (Portland Cement Association)

higher. In large rooms the use of a fog spray, or other means for providing moisture, may be required to obtain adequate moist curing. For exterior work, the surface should be kept damp by applying a fog spray uniformly over the entire surface as soon as the cement has hardened enough so that it will not be damaged. This should be repeated as often as necessary to keep the cement damp during the curing period.

Finish Coat

All finish materials are applied over suitable and properly prepared bases like those just described. Finishing materials are used to form a pleasing surface for walls and ceilings. The plastered walls and ceilings of a room take up about three-fourths of the surface area of a room. Consequently, the beauty of the room depends upon the fine workmanship of the finish coat.

Under the final coat, the base coat or coats are applied and should be brought to a true plane. A finish coat should in no case be applied to a base that has not been given a coat of brown mortar, with the exception of special materials such as one-coat veneer plaster. The brown mortar supplies a solid, rocklike, coarse-textured surface. It is this base which provides the strength and bond to which the finishing materials will adhere securely and thereafter remain intact and crack free. Most finishing materials are hard and brittle. These properties contribute to the smooth, dense surface required of a finish coat of plaster.

Finish coats are often applied in two operations or applications. First, a very thin application of finishing material is forced into the slightly roughened brown coat. Then, when the first application has bonded with the base coat, a second application of the finish coat is applied smooth and troweled until hard.

Putty Coat (or Smooth Coat)

Putty coat is the most popular finishing material in use today. It is mixed as described in Chapter 4. To apply this finish, follow the step-by-step procedure outlined below.

Step One: After the required amount of material (see Chapter 4) is mixed or gauged on the finishing board, the first man of the two-man work team applies a coat of finishing material to each side of each angle (corner). The second man uses a featheredge in order to straighten the angles. See Fig. 7-24.

Straight angles are important to the appearance of the finished plastering job. Consequently, the straightening of the angle with a featheredge requires some explanation. As mentioned above, both sides of the corner or angle are covered with putty coat. Then the angle itself is wiped out to define the angle and to make it as straight as possible.

To wipe out means to rub the

Fig. 7-24. Straightening the angle by rodding it.

featheredge along one side of the angle until the joint or intersection of the two wall surfaces is reached. Then the plasterer wipes or slides the featheredge from the joint across the second side of the angle. In this operation, the featheredge is pushed into the angle. At the same time, it is moved up and down slightly. When it reaches the actual intersection of the two planes, it is then turned so that the side of the tool that has the soft plaster on it is pressing against the opposite side of the angle. The

featheredge, then, by pulling or wiping out the angle, forces the soft plaster to form the second of the two sides of the angle.

If the browning material that was applied as the second coat of plaster is not straight, it may be necessary to go over the angles a second time to bring them out straight and true. When all of the angles in the room have been featheredged, the walls between the angles are scratched in with the first of the two applications. Apply this coat thickly enough to cover the brown mortar and use sufficient pressure on the trowel to insure a good bond.

Step Two: The doubling-up or second coat of finishing material is laid over the first coat immediately. Never double up over set or dried putty coat unless it is first well roughened by scratching with a wire scratcher. The second coat is laid on as smoothly and evenly as possible with a trowel. This coat performs the same function as the brown coat in the base coats, that is, it builds up a straight surface of sufficient thickness to cover all irregularities in the browning. In Fig. 7-25 the plasterer is shown applying the doubling-up coat.

Note that in these finishing operations, the plasterer begins to apply the material at the side of the wall to his left. He then works to the right. This procedure has been found to be the most efficient way of apply-

Fig. 7-25. Applying the doubling-up coat of finishing mortar over the first coat.

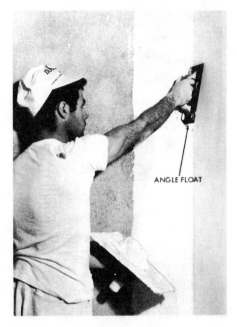

Fig. 7-26. Floating the angle, using an aluminum angle float.

ing the finishing coat. All finish coats are worked from the toe of the trowel. This is done to enable the plasterer to lay the material down smoothly and to flatten out any ridges and trowel marks he finds as he goes along.

Step Three: After the application of the doubling-up coat, the angles are floated. This step consists of filling-in and squaring-up the angle, using an aluminum, plastic, or wooden float and some finish material worked into a square, straight angle. The float is drawn up and down both sides of the angle until it is smooth and straight. This operation is illustrated in Fig. 7-26. Note that the bottom of the float

is held slightly away from the wall. This is to prevent the tool from digging into the soft plaster.

Step Four: After the angles have been floated, the work is gone over once more. Use only enough material to cover and fill in the *cat faces.* Cat faces are blemishes or small rough depressions resulting from uneven thicknesses of brown coat beneath. This step is called *drawing up* or *laying down* (both terms signify that the surface is being compressed and all imperfections filled in). Some plasterers eliminate this final operation, but to do so is to run the risk of spoiling the job.

287

Fig. 7-27. Water troweling the plaster surface.

Fig. 7-28. Sharpening the angle, using the angle paddle.

Step Five: When the work has been drawn up, or "tightened up," as some plasterers express it, it is water-troweled. Here the plasterer wets the surface with a large brush. The trowel is used immediately after the wetting of an area. The trowel is held in one hand, the brush in the other, and the two are manipulated at the same time. This troweling and brushing (wetting) action is shown in Fig. 7-27.

The plasterer trowels the surface in long up and down strokes, moving across the room or area. He holds the trowel at a sharp angle in order to polish the surface.

Step Six: The angles are gone over again either by troweling or by paddling. The latter operation (see Fig. 7-28) is performed with a small paddle. The paddle is, itself, a time saver, because paddled surfaces on either side of the angle do not necessarily require troweling.

Only by troweling out the angle, however, can a perfect surface be produced. The angle may be troweled and finished before or after the final troweling of the entire surface, depending upon the degree of suction present.

Step Seven: The plasterer now

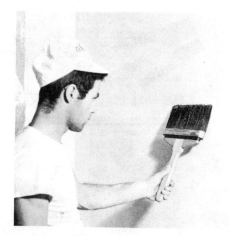

Fig. 7-29. Brushing water over surface to finish work.

duces dribbles fat. The final brushing washes off such remainders and leaves a smooth, clean surface.

These seven steps are the necessary ones for producing a good putty coat finish. One other factor must be considered, namely the condition of the base coat. It must be dry enough to provide suction, which is necessary to insure a good bond and create a hard, dense surface. However, if the base ground is too wet when the finish coat is applied, blisters will appear, and, because of the moisture present, the work cannot be troweled to a hard surface.

On the other hand, extreme dryness of the brown mortar should be avoided, as this condition causes chip cracking. This means that fine cracks show up in a honeycomb pattern covering the entire surface. A poor bond will result and there will be a bad surface upon which to apply paint. Too much retarder will cause chip cracking, but only on a very dry base.

The precise degree of dryness is not easy to achieve. Only practice and experience will enable the plasterer to judge the correct time for the application of putty coat. Slight variations of dryness can be overcome by increasing or decreasing the amount of gauging plaster used; or some of the high-strength gaugings that are designed for low suction may be used.

When putty coating over a base coat containing lightweight aggre-

brushes water over the surface with the large plasterer's brush, or felt brush. See Fig. 7-29. This is the last operation, and it follows the water troweling. The surface should be troweled at least twice to insure a smooth, highly polished surface. Suction, plus the amount of retarder used, will determine the number of times the work must be troweled to achieve a good surface. The final brushing with water insures a face free of all dribbles of fat left by the polishing operation.

When the trowel is rubbed over the plastered surface in the troweling or polishing operation, a certain amount of plaster is ground off. The excess amount of mortar produced in such a way, together with the water applied with the brush, progate, it is necessary to add at least 50 lbs. of fine silica sand or ½ cu. ft. of vermiculite or perlite fines per 100 lbs. of gauging plaster. This improves the compatibility of this type finish with the lightweight brown coat. This is because the tensile strength of the two materials will become more alike at the point of contact or bond. There is usually extreme suction to the lightweight brown coats, and this will avoid shrinkage cracks produced by the suction.

Putty coat finish or lime putty and gauging mixture will require a longer time to set in cold temperatures; therefore, the retarder used can be lessened accordingly.

Fig. 7-30. Sand finish: applying first coat.

Sand Finish

Sand finish is the oldest form of finishing used by the trade. It can be either hand (float finish) or machine textured. Its composition and mixing are detailed in Chapter 4.

Sand finish has one advantage over putty coat—it can be applied over wet, brown mortar. In fact, it is good practice to apply it the day after the browning. At that stage the brown mortar is damp and provides a slight but uniform suction. Uniform suction is important in sand finish.

Irregular suction causes variation in the wall texture. Wet spots produce coarse sections. Dry spots will be noticeably smoother than surrounding areas. These variations in

texture make a poor appearing finish. Dry walls should be sprayed with water to kill or reduce suction.

Sand finish (float finish) is applied by the same procedure as putty finish; that is, featheredge the angles, scratch in, double-up, float the angles. Sand finish is a coarse material. Consequently, greater effort is required in applying this coat than is needed for putty coat. In Fig. 7-30 the plasterer is seen applying the scratch coat or the first of the two applications of finished coat.

After the angles have been floated and the work tightened up (troweling to compact the material), the next step is to float the surface, using a wood float. See Fig. 7-31. Floating performs three functions in sand finishing: first, it compacts the material; second, it straightens the

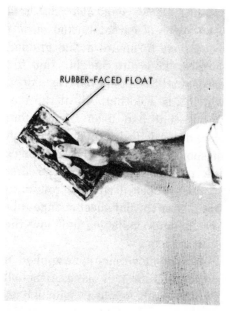

Fig. 7-32. Finish floating, using a rubber-faced float.

Fig. 7-31. Rough floating, using a wood float.

surface; and third, it creates the texture.

To ensure an even surface use a large wood float. Being flat and rigid this tool cuts down the bumps and pushes the excess material into the hollows. In performing this operation the arm is swept in a circle. The blade of the float is kept in contact with the surface at all times.

Floating cannot begin until the surface has stiffened enough to prevent the float from sticking. The proper time can be judged by noting the condition of the freshly applied material. When the gloss (a watery sheen) disappears the surface is ready for its rough floating.

Final floating, with a rubber faced float is the next operation. See Fig. 7-32. The rubber faced float enables the plasterer to bring up a uniform surface. Never use the rubber float alone to do both the rough and final floating. The rubber float does not straighten the work because it conforms to the humps and hollows present. For this reason, the work is first floated with a wooden float. Then the work is floated with the rubber one for final smoothness.

Until a few years ago plasterers used strips of carpet tacked onto a wood float to create a fine grained surface. This did the job, but the carpet had to be renewed frequently, which was a bother. All of the materials that had been tried before were discarded with the introduction of sponge rubber, because it makes the perfect finish float. The first sponge rubber used was round in form, later the flat sheet sponge rubber became available, and now the plastic foam.

All sand finish should be applied in a smooth, unbroken plane. Although joinings can be made a vague line remains to reveal the break.

Machine Applied Sand Finish (Spray Finish)

This uses the same mix as for float finish and is applied in the same manner as putty finish except for the drawing up and troweling. After the first and second coats have been applied and all angles straightened, the plaster is ready for the spray finish. As with any material being sprayed on, care must be taken to protect all adjacent areas of the surface to be sprayed. If spraying a ceiling, it is very wise to wear eye protecters because some of the sand and lime being sprayed will bounce back off the surface. The size of the sand used will set the texture of the finished product. This is true for the hand-float finish but when it is sprayed, by

either the gravity feed hopper gun or the large rotor type gun, three things will affect the texture of the finish: liquid state of material, size of orifice or nozzle tip, and air pressure.

Before spraying material, always spray a little clear water. This will lubricate all parts that the material will come in contact with and make the operation much smoother.

Liquid State of Material. For the ordinary sand finish texture the liquid state of the material should be like medium-thick cream. This condition is obtained by taking the regular mix and adding more water until it comes to the desired consistency. The point to remember here is the thinner the mix, the finer the

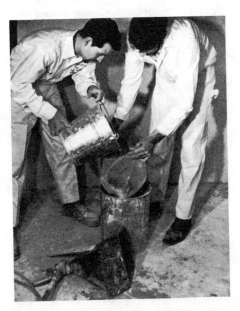

Fig. 7-33. Screening mix for texture spraying.

pattern. Never thin the mix below the consistency of heavy cream. It is also a good practice to screen this mix through a screen that will pass no particles larger than the aggregate you are spraying. See Fig. 7-33. This will eliminate lumps or small stones that will cause imperfections or clog the sprayer.

Orifices or Nozzle Tip. The orifice or nozzle tip is the opening through which the material is forced. These are usually in sizes $\frac{7}{32}''$, $\frac{1}{4}''$, $\frac{5}{16}''$, $\frac{3}{8}''$ and $\frac{7}{16}''$. To a certain extent the size of the orifice used depends on the size of the aggregate in your mix. For instance, if your material contains $\frac{3}{8}''$ aggregate, the three smallest sizes can't be used. If the air pressure and mix consistency of liquid are not changed, the smaller the orifice or nozzle tip, the finer the spray.

Air Pressure. The air pressure breaks up the material as it flows out of the nozzle. If no other changes are made, the higher the air pressure, the finer the spray. This is because the higher air pressures break up the material into smaller particles. Too little air pressure will fail to break up the material and you will get globs. Always adjust air pressure starting from low and increasing gradually until desired pattern and texture are reached.

Because there are so many things that can affect the pattern of spray, there can be no definite outline ex-

Fig. 7-34. Testing spray gun for spray pattern before applying material to wall.

plaining the pattern. Therefore, before beginning to spray, test the pattern and make the necessary adjustments to give the desired texture. It is best to mix the material stiffer than needed. Spray it on a test panel changing orifice and air pressure until the desired pattern is obtained. See Fig. 7-34. If no setting of air pressure or orifice gives the desired pattern, then try changing the consistency of the mix by adding water. When spraying, it is best to spray first in one direction and then in another direction crossing the first direction at right angles. Fig. 7-35 shows how the pattern is made.

Keenes Cement Finish

Keenes cement finish is used when a surface of special hardness is required. It can be brought to a high polish by troweling. Because of its

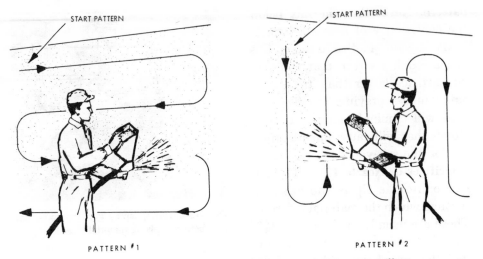

START PATTERN

START PATTERN

PATTERN #1

PATTERN #2

Fig. 7-35. Texturing by machine, moving spray gun in a cross pattern.

slow rate of set, Keenes cement finish must be worked somewhat differently than putty coat finish. The difference between the two finishes, however, is only in the troweling period.

In putty coat finish the work is troweled as soon as it is applied and it is troweled until it is set. One, two, or three trowelings may be required. These trowelings are accomplished within a short space of time because the plaster has been retarded for a certain given time. In Keenes cement work, however, the trowelings must be spaced apart so that the slow-setting material can harden. Repeated troweling of Keenes cement, while it is in the process of setting, only causes the surface to become blis-

tered and and tends to retard the setting action further.

Keenes cement has a natural ability to be retempered. Troweling, especially with water before the material is well set, will start the retempering process and thus cause the material to return to a state of plasticity. In all other respects, other than that of the troweling time, the procedure used for putty coat and Keenes cement finish is the same.

Keenes cement finish is mixed as outlined in Chapter 4. When it is thoroughly mixed, apply the material in the same manner as putty coat finish. After the work has been laid down, give it a light troweling using very little water.

On completion of the first troweling, let the work stand undisturbed for about an hour. When it has set hard, it can be troweled again to a dense, smooth surface. This troweling will require some effort, but the results obtained will be rewarding.

Finishing with Keenes cement is a special job. To hurry the work by troweling too soon is harmful. "Spiking" (adding an accelerator) with white plaster mortar is a practice condemned by the trade. A finish so adulterated will be no better than ordinary putty work.

The skilled plasterer lays Keenes cement finish over as great an area as he will later be able to trowel. For example, he may apply the plaster to the top of a room on all sides, then do another room. When this is finished, he may return to the first room, trowel the coat to a good finish, then trowel the second room. This procedure may be repeated for the lower parts of both rooms when the scaffolds are removed. Thus, the tops of two rooms can be done in the forenoon and the bottoms in the afternoon.

Until a few years ago, it was customary to apply the Keenes cement finish continuously throughout the morning and trowel it in the afternoon. This was necessary because the period of hardening of this material extended over four hours. Today, the plasterer can buy a Keenes cement that enables him to return in two hours or less and trowel up the work.

Never apply Keenes cement finish in greater thickness than the normal ⅛" finish coat. If the browning is bad, the slack places must first be straightened out by filling in the hollows with additional brown mortar. If the hollows are slight, finely screened sand may be mixed with some of the Keenes cement finish and the resulting sand finish used to fill in the hollows.

If Keenes cement finish is applied too thickly, it will shrink and check crack (another name for chip cracking). Chip or check cracking is caused by shrinkage of the material due to extreme suction in the previous coats of plaster. Once checked, it is almost impossible to bring the material back to a good finish. The surface will look good, but the cracks will show up when paint is applied later. A few scoopfuls of white silica sand added to the gauging will help to avoid the shrinkage cracks produced by suction.

Portland Cement Work

Portland cement plaster is one of the best materials to use wherever surfaces will be subjected to hard use, or where a condition of dampness is normal to the area. No other material applied with hawk and trowel is as strong or as resistant to moisture as this plaster mortar is.

A ratio of 3 to 1 (3 parts sand to 1 of cement) is used for all coats of Portland cement plaster, whether for interior or exterior application. Mix the materials as detailed in Chapter 4. Note that in cement work, suction in the base is controlled by dampening the base instead of changing the mix, as when lime or gypsum is used. The base is dampened because Portland cement plaster must be as poor a mix as possible to prevent the formation of shrinkage cracks. Moreover, any mix made in proportions less than 3 to 1 would be too hard to spread.

With Portland cement more time is required to bring the finish coat to the required texture than with most other plastering materials. More time is required because, first, the base over which the mortar is applied must be damp to insure a good bond. Thus, the floating is delayed. Secondly, Portland cement sets slowly; therefore, the setting action is of little use in bringing the surface to the proper finishing stage. The experienced plasterer can, however, cover about as much ground with this material as with any other material. See Fig. 7-36 which illustrates the application of the finish coat of Portland cement mortar.

Various types of finish textures may be used. Float finish is the most popular. Two types of floats are commonly used. The wood float is used to compact and to straighten the work. The rubber-faced float, Fig. 7-37, is used to produce a fine-grained and uniform texture. Trowel finish, Fig.

Fig. 7-36. Applying finish of Portland Cement.

Fig. 7-37. Fine floating of cement, using rubber-faced float.

7-38, is obtained by first floating the surface and then troweling it to a smooth, dense body. Blisters will

Fig. 7-38. Troweling cement to smooth finish.

form unless the surface is troweled up gradually.

Portland cement finish can also be spray textured; this is done in the same manner as explained for sand finish.

Colored Finishes

Colored plaster finishes have been in use for many years. Many beautiful effects can be obtained with the use of color in plaster and cement. It should always be remembered that only a high quality mineral color should be used in coloring plaster and cement. These colors should be guaranteed by the manufacturer to be resistant to lime and to be weatherproof.

When coloring a quantity of material make sure there is sufficient material to cover the room or area it is intended for. This eliminates the possibility of having to mix more material whose color might be slightly different from the first batch. Many of the plaster manufacturers make prepared colored finishes which are available in many pleasing colors. The advantage of these prepared finishes is that the color is always uniform and stable. The prepared colored finishes will probably cost more than the on-the-job added color finish, but the hazard of mixing too much or too little is eliminated.

Problems with Colored Finishes. Much thought and effort have been put to use in order to make colored plasters more practical and easy to use. Colored plasters have some weaknesses which should be understood and realized by the plasterer so he might know of their limitations.

1. Water used in troweling colored finishes has a tendency to wash out some of the color from the finish. This opens the possibility of leaving a slightly streaked finished face on the completed job. In order to produce a good smooth job, however, troweling with the use of a wet brush is necessary. Remember though that when troweling a colored finish, the minimum amount of water should be used in this operation.

297

2. Joinings also are a problem with colored finishes. When joinings are left on the ceiling or walls they will invariably show up when the job has dried. As the first gauging which has started to dry out is joined with the next gauging, the troweling or floating at the joining line will produce a different shade of color which is readily distinguishable. The answer to this is the entire application of the whole ceiling or the entire wall leaving no joinings. Extra help usually is needed to do this successfully. Scaffolds should be placed so that the walls may be done from top to bottom.

3. Mixing the color on the job many times leads to producing various shades of color due to inaccuracy in the measuring or the proportioning of the materials in the mixture. For maintaining the proper shade of color even the exact amount of water for each batch of mortar is important.

4. Remember that fingerprints and other marks made by the other tradesmen on the job, such as carpenters and electricians, will not wash off the finished walls.

Advantages of Colored Finishes. There are, however, many advantages in the use of colored finishes, some of which are:

1. The cost of the colored finish is very small when compared to the cost of painting the room or the job. It is two operations in one. Time can be saved in the occupancy of a house or job by using color plasters, thus doing the plastering and the decorating in one operation.

2. The application of plaster with color adds to its ease of application as the color adds plasticity to the material. If desired, various colors may be combined at one time to produce a blended effect. Mix batches of different colors on separate boards, then put a trowelful of each gauged color on the hawk and proceed to apply as though it were one color. The trowel will blend the colors producing a marbleized effect. Not more than three colors having a harmonious blend should be blended together at one time.

3. Colored sand finish is a much better working medium than the smooth finishes. Because of the floating done to create the texture, several colors can be floated in together. However, practice is needed to achieve a pleasing effect with mixed colors. Ordinarily only a single color is used in both sand and putty color finish. By careful control, mixed colors can be applied with putty coat. Experience, however, is necessary to do this effectively.

4. The use of color in the finish eliminates the necessity of painting until the walls and ceiling become dirty and in need of redecoration. If funds are limited at the time of construction, the use of color plaster will produce a decorated, attractive

home. By the time the color plaster is ready for redecorating, funds for painting the interior may be available.

5. During the period before paint is applied to the plaster surface, the lime is aging properly. This results in a stronger, harder finish. Lime becomes stronger when exposed directly to the air (the process of recarbination). When the finish coat on a job is painted shortly after the job is completed the aging and the hardening of the lime is retarded. Thus a period of aging for the lime before any paint is applied is beneficial to the plastering.

6. During this period before the painting of the plaster, there is the possibility of building settling. Settlement cracks in the plaster usually appear in the first year or two. Thus, any cracks that appear can be repaired before the interior is decorated. With the aging of the lime and the repairing of any settlement cracks which have appeared, the first job of paint decoration will last much longer than if it is decorated shortly after the plastering completion.

Suggestions for Application. There are many ways of using color on walls. Sometimes the color is mixed with the mortar immediately and a solid color is produced for the entire surface. In other cases, colors are blended on the wall surface in such a way that several shades can be seen and can form a pattern. The plasterer's skill and ingenuity will dictate the effectiveness of the colored plaster finish. Closely related to the colored plaster finish subject is the use of textured surfaces. Color is used particularly in plaster work that imitates brick, stone and wood textures. This texture is taken up in Chapter 8.

The technique of applying colored finishing mortar is the same as that described for putty coat. Only in the final finishing operations is there any difference. As mentioned before, an entire wall must be done at one time. Water should not be used in troweling the work as it tends to wash out or to bleach the color. The result will be a spotty surface of unpleasing appearance.

Suction also affects the color. It should be as uniform as possible. The best way to insure uniformity of suction is to work over slightly seasoned brown mortar.

Various colors may be combined at one time to produce a blended effect. Mix the batches of colored materials on separate boards. Then put a trowelful of each gauged color on the hawk and proceed to apply it as though it were one color. The trowel will blend the colors into a variegated or marbelized effect. Proper choice of colors is important to insure a harmonious blend. Not more than three colors should be combined at one time.

A spattered effect may be obtained

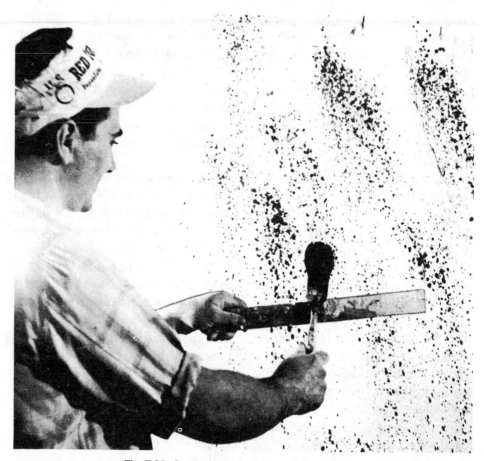

Fig. 7-39. Spattering color onto soft finish coat.

by splashing colors over freshly applied putty coat, as shown in Fig. 7-39. Colors are mixed with water to form a paint that is spattered on right after the work is laid down. Troweling then works the color into the surface, creating various effects. Experimentation will develop many blends and patterns. Work of this character may also be done with Keenes cement. This material insures a harder surface and better control because of the slow and uniform set achieved.

Texture Work

Decorative trends appear and disappear. The popularity of textured plaster work has, at times, been great and then, at other times, has declined. But textured plaster work, when suitably employed and skillfully executed, adds beauty and charm to a room. Treatment of the plastered wall is limitless and each texture requires certain materials, specific tools, and specific techniques of application.

Four basic tools are needed to create most of the textures commonly applied: the brush, sponge, float and trowel. Interior work is usually done in putty coat. Sand finish is used to obtain some effects. Only practice and natural creative ability will enable the plasterer to achieve good results in texture work.

Various Tools and Techniques. One plasterer may use a sponge to create a certain texture. Another will obtain the same or a similar effect using a brush. It is difficult, therefore, to establish rules or methods for doing work of this kind. The best rule to follow is to make samples of the different textures using various tools and devices to create each.

Then use the method that seems most suitable to you for the particular application. You must remember, however, that a sample will appear very different from the same texture seen completed on full scale work. Always tone down the work or be more conservative when applying finish over a large area. This is done by lightly brushing or sponging over the texture when set, but still wet. This will remove some of the obviously ex-

treme texture that appears on the surface. If the plasterer uses a lighter touch when texturing, this will also help.

Stippling. In Fig. 7-40 the plasterer is using a stipple brush. To use the brush stipple most effectively, it is necessary to gauge the putty and plaster into a soft mixture, using plenty of retarder and plaster. Scratch the area with a trowel. Then, using either a hawk or a small pail to hold the soft plaster, dip the stippling brush into the mixture and daub it on the wall or ceiling.

The type of brush used, the consistency of the material, and the technique of the plasterer all unite to create the texture. The stipple finish should be uniformly applied. Irregular textures or designs are hard to maintain.

A variation of any of the regular

Fig. 7-40. Texture formed with stippling brush.

Fig. 7-41. Another effect is formed by moving brush in circular motion.

Fig. 7-42. Sponge texture being brushed out.

textures can be achieved by brushing them out as the material begins to stiffen. A soft brush is used so that the stippling is merely modified and made a softer texture.

The stippling brush may also be used to create another design. By moving the brush in a circular motion, you can create a swirl design. See Fig. 7-41. Dip the brush into the soft plaster and apply to the scratcher surface. As the brush touches the surface, give it a sharp twist. The swirls may be placed in one direction, or you can vary the effect by twisting the brush first in one direction and then in another. Brushing out will soften the effect.

Rubber Sponge. The use of a rubber sponge creates the effect

shown in Fig. 7-42. The sponge is dabbed against the surface and is then lifted from it. The type and texture of the sponge will determine the finished effect. Wool or natural sponges may also be used. The sponge must be washed now and then to maintain a uniformity of the finish. Fig. 7-42 shows the use of a soft brush to tone down the finish.

Travertine Finish. In Fig. 7-43 the plasterer is shown creating a travertine effect. This plaster texture is called travertine because it imitates travertine stone used for building. To create the effect of travertine, the plaster mortar is laid on thickly. Then the surface is jabbed with a whisk broom or wire brush to form a series of ridges and valleys. When

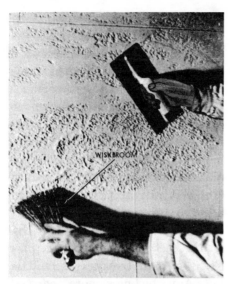

Fig. 7-43. Travertine finish formed with whisk broom and trowel.

Fig. 7-44. Trowel finish formed by arcing strokes.

the material begins to stiffen somewhat, it is troweled down. This action forms a smooth surface broken up by a number of fissures or splits occurring in parallel veins.

The travertine texture is usually produced in a light tan or beige tone to simulate the natural travertine stone. A somewhat similar and equally pleasing texture can be produced by stippling the surface with a sponge instead of the whisk broom. Then the surface is flattened with the trowel. The trowel must be kept clean and held very flat. Do not start troweling until the material begins to stiffen.

Trowel Finish. One of the simplest texture types to create is the one developed with the use of the trowel. The ordinary putty coat is scratched in and then doubled up. In the second coat, the material is laid on in short, arcing strokes, each stroke taking a different angle. See Fig. 7-44. The texture is created by the toe of the trowel as it is used to spread the finish coat. Ridges are formed as the trowel is moved across the surface. The size of the ridges can be controlled by the amount of pressure applied to the toe of the trowel.

Hand Finish. The crude method of applying plaster by smearing it on with the hands, as it was once done, is now practiced in imitation of that old practice. The attempt is to make the modern work look like the work produced by old-timers, who did their work with few tools but with a keen eye and with common sense.

To obtain the desired effect,

303

scratch in the area. Then put on rubber gloves. Pick up a handful of plaster from the hawk and rub it on the wall or ceiling. Spread it around with the fingers. When the material begins to stiffen, rub the surface with the palm of the hand. Later use a soft, clean brush to bring the plaster to a smooth surface.

Acoustic Plaster

There are several brands of acoustic plaster on the market. Generally, these plasters obtain their acoustic properties either because of pores in the set plaster or because of minute spaces between the binding material, the fibers, and the aggregates used. Sometimes the pores of the aggregates used help to produce acoustic efficiency.

Most of the brands of acoustic plaster manufactured today make use of a foaming agent. The foaming agent used is similar to baking soda because it causes tiny air bubbles to form in the wet plaster. When the plaster dries, these bubbles burst, forming many pores that soak up sound.

These materials can either be finished with a hand texture or spray finish. Be sure to follow the manufacturer's directions, since some differences do occur in the recommendations for application. Watch the water content and mixing period as these factors are important.

Acoustic plaster should not be ap-plied so that it is too thick in one place and too thin in others. A uniform thickness is desired. Because of the granular nature of acoustic materials it works quite like sand finish.

Two coats of acoustic plaster are required to insure correct thickness. Each of the coats of acoustic plaster should be $\frac{1}{4}''$ in thickness. The two coats of acoustic plaster are put on over previously applied scratch and brown coats of gypsum plaster.

Manufacturers have proven that for the best results in acoustic plasters the total thickness should be $\frac{1}{2}''$. Any thickness more than $\frac{1}{2}''$ will give very little additional sound absorbing quality in comparison to its additional cost and effort. On the other hand, any thickness under $\frac{1}{2}''$ will fall short in the ability to absorb sound adequately.

Applying Acoustic Plaster. Acoustic plaster must be applied to a clean, firm base. The best base substance is a gypsum mortar that has been properly applied and straightened. Next, the surface of the mortar is cross-scratched to construct a good key for securing the first of the two coats of acoustic plaster. Suction is important. The base must be dry and of sufficient thickness to provide continuing suction, or else dropping or separation will occur.

Low suction bases, such as concrete with a bonding agent, generally will require a three coat application. The first coat, $\frac{1}{8}''$ thick after drying, provides suction for successive coats. Wire ties and metal should be well covered with the base coat to minimize rusting that may bleed through the finish with some types of acoustic plaster. Cover visibly exposed, untreated steel (in the concrete), including nails and rods, with a water insoluble rust inhibitor, such as red lead or resin type paint. Paints containing fish oils are not recommended by the manufacturers.

Most acoustic plasters are applied over a fairly dry brown coat. Some manufacturers, however, produce acoustic plasters that are recommended to be placed over a set, cross-scratched brown coat that is thoroughly wet. Follow the manufacturer's directions as well as your own knowledge based upon experience. Generally, however, plasterers have found that acoustic plaster will fall when applied over a very wet brown coat.

Acoustic plaster is applied in two coats, generally of ¼″ thickness each. Applying two coats in this manner permits the material to dry between applications. One coat of ½″ thickness would be so wet that it would drop or separate from the base. Since only a small amount of adhesive is contained in acoustic plaster, it will not bind or adhere if too much is applied at one time.

Apply the first or scratch coat firmly. Lay on a thickness of about ¼″. Flatten the work with a trowel.

Fig. 7-45. Applying scratch coat of acoustic plaster.

See Fig. 7-45 where the plasterer is shown applying the first coat of acoustic plaster. Then straighten out the scratch coat with a plastic, aluminum, or magnesium featheredge or darby. A new wooden slicker may also be used.

Use new or clean tools. Old tools are likely to cut or drag the coat. Cutting, dragging or floating motions tend to open up the material and pull it loose. Such actions cause the acoustic plaster to pull loose because of its fibrous nature. In Fig. 7-46 the plasterer is shown straightening the first coat of acoustic plaster by using a darby.

After the work has been darbied, roughen the surface of the first coat of acoustic plaster in order to build

Fig. 7-46. Darbying scratch coat of acoustic plaster.

Fig. 7-47. Applying second coat of acoustic plaster.

a good key for the coat to be applied next. A wire scratcher or coarse-fiber broom are means by which the surface may be roughened.

Apply the second or finish coat when the first coat is dry. In Fig. 7-47 the plasterer is applying the finish or second coat of acoustic plaster, laying it on ¼″ thick. A dry base is needed for most acoustic materials to insure a good bond. See the manufacturer's directions for details.

Lay the second coat as evenly as possible. If ridges are not smoothed out immediately they will show. The trick, then, is to lay the material on and to flatten each trowelful as it is applied.

Use as little troweling as possible to lay the material down. Very little good is done by troweling if it is attempted after a period of time has elapsed since the initial application. This is because of the spongy nature of the plaster. If troweled, the mortar will return more or less to its original state once suction begins to make the applied coat firm. Troweling, therefore, must be accomplished during application or as soon after application as possible. Fig. 7-48 shows a plasterer troweling the second coat of acoustic plaster.

If it is planned to use an acoustic plaster that requires a float finish, use a cork or rubber faced float. It will be found that the floating opera-

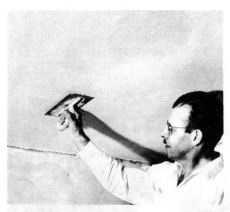

Fig. 7-48. Troweling acoustic plaster to a smooth finish.

tion must follow closely the application of the acoustic plaster if a good texture is to be achieved.

By common procedure, one plasterer applies the material and a second man follows closely with the floating. On large areas enough plasterers should be used to make it possible to complete the section without interruption.

Certain types of acoustic plasters containing chemicals that act as foaming agents are produced by their manufacturers. The foaming agents cause air bubbles to form within the plaster. These air bubbles eventually burst and leave pores within the plaster. In this way such products are able to soak up sound. Other acoustic plasters require the use of stipple or texture brushes to form a roughened surface texture that will also absorb sound readily.

When acoustic materials are used

that require a stipple finish, a technique similar to that explained for the float finish is followed. In place of the float, however, use a large fiber brush, as shown in Fig. 7-49. To achieve a fine, dense, stipple finish, punch the entire surface in a uniform manner.

The next day, trowel the work to flatten the ridges slightly. Then use a nail stippler, Fig. 7-50, to punch indentations into the plaster surface. The nail stippler contains sharp nails set into a wooden grip very much the way bristles are set in a handbrush. The indentations made by the nails provide thousands of pockets for absorption of noise.

One of the major difficulties the plasterer has to overcome in using acoustic plaster is the inability to join the material when continous application is interrupted. Each ceiling, section, or panel must be done in a continuous operation. However, when machine texture is used, this problem is eliminated. To do this many men may be required and scaffolding must be amply provided.

The use of the stippled texture already described tends to overcome this difficulty. This type of finish assists in the blending of a joining, provided the work is done by an experienced plasterer. Other finishes may be joined if the surface is to be painted.

The main difficulty with finishing acoustic plaster is not that presented

Fig. 7-49. Applying second coat and stippling with fiber brush.

by the joinings but that of keeping the color texture uniform when fresh material is joined to the old. Anticipating developments may eliminate this trouble.

The machine applied texture is usually done after the required thickness has been applied and after this material has dried sufficiently. It is troweled smooth to remove any blemishes that could show through the spray finish, such as ridges, waves, and rough areas. The procedure of spraying the texture is the same as that described for spray sand finish.

Advantages of Acoustic Plaster. Acoustic plasters are fireproof. All

NAIL
STIPPLER

Fig. 7-50. Indenting surface using nail stippler.

of the materials that go into acoustic plaster are noncombustible. Most of these plasters are highly resistant to moisture. The durability and monolithic construction make acoustic plaster dustproof and serve to prevent "breathing" of wall and ceiling surfaces. Acoustic plasters are light in weight. Consequently, excessive weight is not added to the wall structure when they are used. These attributes, plus their good insulation qualities, make acoustic plasters an excellent sound absorbing material.

Some manufacturers produce an acoustic plaster that can be sprayed directly to any sound surface providing it is reasonably clean and free of dirt, grease, oil, rust, scale, or other matter. The condition of the base should be checked for any ridges or offsets which should be leveled off with a trowel coat of the special plaster material because they may show through later. The spray should be applied to the specific thickness in accordance with the manufacturer's recommendation.

Drywall Taping

Gypsum wallboard is usually erected in sheets four feet wide and eight feet long. The joints of these sheets are reinforced by a paper tape pressed into and coated with a special plaster manufactured for this purpose. The erection of the board is usually the work of another trade. However, the taping of the joints and the covering of nailheads with this special plaster material is done by the plasterer in certain areas. The wallboards are available in ¼", ⅜", ½" and ⅝" thickness, and in lengths of 6 ft. to 10 ft. with longer lengths available on special order. The widths are usually never over 4 ft. The long edges of the wallboard are made with various shapes: tapered, square, beveled, or rounded. The tapered edge is the only edge requiring taping. The other types produce decoratives effects.

The tapered edge is recessed back two inches from each edge allowing for the thickness of the tape and the joint cement to conceal the edges. The application of two or three thin coats of the joint cement brings the tapered edge flush with the face of the wallboard, thus producing a smooth unbroken wall or ceiling surface. Single layer construction of ½" wallboard with tapered edges and concealed joints is the most commonly used "drywall" construction. The use of "laminated drywall," which means two or more sheets of wallboard nailed and glued together, is used in some special cases.

Laminated drywall also requires taping of the joints. Corner reinforcement (similar to a plaster corner bead), metal trims for edge protection, and trim around doors and windows are used. The plasterer applies the joint cement to these metal trims and finishes them flush to the beads. This is similar to plastering to a corner or casing bead in conventional plastering.

A plasterer can do a very satisfactory job of taping with the use of his regular plastering tools. However, there are tools recommended which are classified and used primarily for taping drywall.

Before beginning to tape make sure there will be good ventilation and adequate heat. Inadequate heat causes uneven drying, leading to bond failure, delayed shrinkage and joint discoloration. For best results temperatures should be at least 55°F.

First, check all nails. Protruding nails should be driven home leaving a dimple in the paper to receive the cement. Then butter the joint cement into the channel formed by the tapered edges of the wallboard; fill the channel or hollow fully and evenly; avoid heavy fills. Now take the tape and center it over the joints of two boards and press it down into the fresh applied joint cement. Then, using the drywall trowel, trowel along the tape with enough pressure to remove excess joint cement and leave sufficient joint cement under the tape for proper bond. As soon as the tape has been applied and troweled, cover the tape with a thin coat of joint cement to fill any recess between the tape and the board surface.

Now apply the first covering coat of joint over all nailheads.

Butter both sides of the angle on inside corners with joint cement, using the angle trowel. Fold the tape along the center and crease and feed the tape into the angle. Then, with the flexible angle tool, trowel the tape to remove excess cement and be sure to leave sufficient joint cement under the tape for proper bond. With the same tool, cover with a thin layer of joint cement.

After the inbedding and covering coat is completely dry or hard (at least 24 hours) apply a second coat, feathering the material a little beyond the edges of the first coat, usually about 1½".

Spot nailheads again. If the manufacturer specifies 3 coats, sandpaper the second coat lightly after it dries and apply a thin finishing coat. Feather the material an inch or two beyond the second coat edges. When everything is completely dry, sandpaper all the areas where the joint cement was applied.

On external corners, the same procedure is followed as for flat joints except tape will not usually be necessary and the material will finish flush with the edges of the beads.

Machine Application for Multi-Story Building

When using the plaster machine for applying gypsum or Portland cement mortars careful planning is required. Further planning is necessary when we use the machine for multi-story buildings. We should first select an area large enough to accommodate mixing and pumping units, materials, bagged goods (enough for one day's operation plus), and a stockpile of sand (aggregate). Select a permanent location for the entire job. This area must be readily available for further delivery of materials. Store bagged material on pallets or planks to prevent water damage. Plan to enclose your area with a temporary shelter consisting of a wooden frame covered with plastic or water proof tarps. If it is going to be used in cold weather, some type of heat will have to be used, such as the kind using propane. This will keep the water, sand, and equipment from freezing. Water will be in continuous demand for mixing and cleaning, so check the supply. A place to dispose of dirty water must be found.

Two 2½" diameter pipes, should

be installed vertically from the ground floor to the top floor of the building. One pipe serves to deliver material to each floor as it is sprayed. This riser pipe should be located at the most advantageous point from which the hose can be run to reach and spray the entire working area. Most often these pipes are placed on the outside of the building but can sometimes work as well if erected in elevator shafts. The other pipe is installed for cleaning out the hoses and pipe by returning dirty water.

In joining lengths of pipe and hose, all connections should be covered with a non-setting pipe joint lubricant. Pipe and loose fittings should be sufficiently tightened so as to avoid leaks. The pipes should be fastened to the building very securely to eliminate possible vibration. Pipe bends should be made with long radius ells, because right angle fittings restrict flow of material. This is illustrated in Fig. 7-51. The hose connected to the delivery pipe on the floor of spraying operation should be 2″ in diameter, regular high pressure rated. Connected to this should be 1½″ regular high pressure rated hose to which the spraying nozzle is attached.

Clean the pipes and hoses by running a sponge through the lines at least once a day, and twice a day if the machine is cleaned and shut down for lunch. To prevent separation of sand from gypsum or cement,

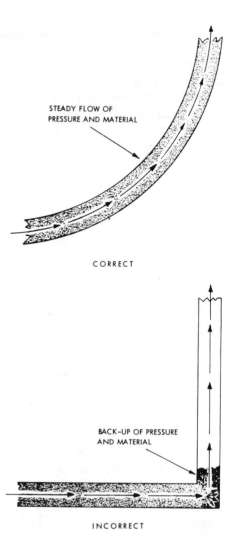

STEADY FLOW OF PRESSURE AND MATERIAL

CORRECT

BACK-UP OF PRESSURE AND MATERIAL

INCORRECT

Fig. 7-51. Pipe bends should have long radius ells in order to convey plastering material with minimum restriction of flow and back pressure.

you should first disconnect hose from the machine, insert the sponge, reconnect, and then flush with water behind the sponge.

After cleaning equipment each day, inspect ball valves, valve seats, O-rings, hose fittings and couplings. Ball valves, valve seats, and O-rings should be replaced as needed from a stock of spare parts which should be kept on the job.

Inaccurate proportions of aggregates and other ingredients, and under-measured water, can produce extreme pressure when pumping materials. This pressure can break hoses. Improperly mixed fireproofing materials also cause high pumping pressure, so it is especially important to follow the manufacturer's recommendations precisely.

When using the machine, it is a good practice to wet the interior of the hose by pumping some water through it before beginning to spray with the plaster. When applying mortar with the gun the nozzle should be moved with a steady, even stroke, laying on the proper thickness with one pass and overlapping successive strokes. The angles of the nozzle to the wall should remain uniform; if the nozzle turns out toward the end of the strokes, the material will build up much more heavily directly in front of the operator. Around door bucks or window frames (which should be protected by covering) the nozzle should be moved up to within a few inches of the surface and pointed away from the area where overspray will occur. This will reduce overspraying.

Checking on Your Knowledge

The following questions give you the opportunity to check up on yourself. If you have read the chapter carefully, you should be able to answer the questions. If you have any difficulty, read the chapter over once more so that you have the information well in mind before you go on with your reading.

DO YOU KNOW

1. Why does a plasterer use a scratch coat when plastering over metal lath?
2. Should plaster be applied over tar or bituminous material?
3. Should plaster be applied over smooth concrete surfaces?
4. What should a plasterer do if the plaster is to be applied over a high suction base?
5. How can a plasterer increase the fire retardation for columns?
6. How can a plasterer safeguard his work after he has finished plastering over hot water radiant heating panels?
7. What safeguard should a plasterer use when using Portland cement?
8. How should the surface be scored when applying a cement scratch coat?

9. How can a plasterer form an arris with Portland cement?

10. When using the plaster machine for a multi-story building, what planning should be used?

11. How does a plasterer straighten an angle when applying a finish with putty coat?

12. Why does a plasterer finish putty coat by painting water over the surface?

13. How does a plasterer make a "sand finish"?

14. How does a plasterer spray finish?

15. What are some of the weaknesses of colored plasters?

16. How does a plasterer apply tape for drywall?

Special Finishes

The special finishes that will be covered in this chapter are those that will aid the plasterer to imitate various materials such as brick, stone, etc. The growing number of new materials that are being added to the building industry for decorative purposes or fireproofing will be discussed. Many of these new materials require completely different application techniques than those discussed in Chapter 7.

Common to all exterior finishes is the need to provide the proper protection against the entrance of water behind the finish materials and from the effects of freezing and thawing cycles. For a detailed explanation of flashings, control joints and termination members, refer to Chapter 5.

The end of this chapter will cover acoustic tile and grid work. In most parts of the country this work is being done by the carpenter, but it is wise for the plasterer to have some knowledge of this work. Many plastering contractors have broadened their bidding to cover acoustic tile and grid work. As a result, one international organization has changed its name to reflect the fact that installing tile and grid systems is now a regular part of the acoustic work being done by its members.

Imitation Brick, Stone and Wood

Imitation brick, stone and wood may be used either for interior or exterior plastering. In the last few years, plasterwork that imitates masonry has had a marked revival. Imitation plasterwork was developed from Italian sgraffito work. *Sgraffito* is an Italian word meaning "scratched". Although the word sgraffito had become a forgotten one in the trade, recently both the word and the work is represents have been brought back into circulation.

Sgraffito work is accomplished by laying various coats of colored mortar one over the other. While the coats are still soft, portions of the top coat may be removed, exposing the color beneath and allowing some of the surface color to remain untouched. Work of this nature is usually done with a stencil in order to produce a pattern. Proper base coats and curing, combined with good colors, will make a strong, lasting job.

The mixing and application of the materials are regular plastering operations. Interior work may be done using putty coat, sand finish or Portland cement. Portland cement is used for interior plastering when dampness is present. For exterior work, Portland cement is the proper material to use.

Imitation Brick

To produce an imitation brick effect, mix two batches of material. Mix one batch of regular or white mortar. Then mix a batch of mortar in the brick shade desired. Apply the white mortar over a properly browned base. Usually ¼" thickness is about right. Next, apply the brick-colored

mortar to a thickness of ¼". See Fig. 8-1 showing this application.

The two coats bond into one, producing a strong job. The plasterer then floats the surface to straighten and compact the mortar.

Fig. 8-1. Applying brick-colored mortar over undercoat of white mortar.

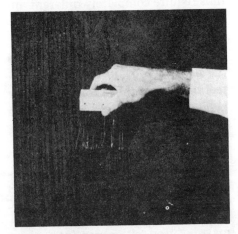

Fig. 8-2. Scoring the surface using a wire brush.

The next step is to comb or score the surface to reproduce the texture of the brick to be imitated. This operation is shown in Fig. 8-2. A single row of wires set in a wooden handle is used in this case. Other textures can be used, but this is the texture most commonly applied and the one easiest to create.

To lay the brick joints, various methods are employed. One of the devices used consists of a large sheet of aluminum in which slots are cut to represent the horizontal joints of the brick. This guide sheet or template (a pattern) is suspended from a strip of wood nailed at the top of the wall. The joint cutter is then drawn through these slots to form the joint.

When the section is finished, the template is moved over and the cutting is repeated. Working from left to right, move the form along until the width of the job has been cut. Now lower the form until the top slot lines up with the bottom joint just cut; then repeat the operation until all horizontal joints have been cut in the same manner.

Fig. 8-3. Cutting horizontal joints, exposing undercoat color.

A similar template is used to cut the vertical joints. The slots run the length of the joints and are staggered in imitation of actual brick joints. When you use both templates, the work progresses at a fast rate. It is almost impossible to make a mistake. Templates are expensive to construct, and only firms that do a considerable amount of imitation brick use them.

A simple and inexpensive guide can be made as shown in Fig. 8-3. Light wood strips are nailed to form a square frame and a row of nails is spaced out evenly on the two upright strips of the frame. Each space is equal to the thickness of a brick plus the width of one mortar joint. A wood strip with a piece of sheet metal attached at each end is used as a guide strip to give a straight line when cutting the joint. The guide strip rests on parallel nails protruding from the upright strips of the frame as shown in Fig. 8-3. The cutter is now run over the guide strip as shown, and the joint is cut, exposing the white mortar.

The cutter consists of a short length of wood to which a round loop of flat sheet metal is nailed. The width of the joint is set by the size of the metal loop. The depth at which the cutter will score the plaster is controlled by the notched part of the handle.

In the top and bottom strips, as in the uprights, nails are arranged in a

Fig. 8-4. Cutting vertical joints to resemble brick.

Fig. 8-5. Shading various bricks to show contrast.

row. The nails are spaced a distance equal to one half of a brick plus a mortar joint. Thus, as the straight-edge is set against the nails, alternate joints are cut. This phase of the process is shown in Fig. 8-4.

Exercise care in using the cutter on the vertical joints. It is very easy to cut too far and cut into the horizontal joints. The procedure to follow is to cut downward almost to the point where the horizontal joint will meet the vertical joint that is being cut. Then start the cutter at the bottom and work up. This will prevent the chipping of corners or cutting beyond the vertical line.

After the area has been cut, the surface is brushed lightly with a soft brush to remove any loose mortar left clinging to the joints and to soften their hard, straight edges. Color is applied to a few bricks to provide the variation in color found in real brickwork. Fig. 8-5 shows the plasterer applying color.

To achieve a perfect imitation of real brick in color and permanence many firms use crushed brick as the aggregate in mixtures of the brick-colored mortar. Only enough color is then added to tint the cementing material added to the aggregate.

Imitation Stone

A discussion of the two common ways of producing imitation stonework is necessary to any full treatment of the subject of textured plasterwork. Both methods of producing imitation stone give a certain degree of satisfaction to the beholder as a decorative feature.

The first method to be discussed below is similar to the one used to produce artificial brickwork (that is, the application of various colored mortars in layers). The second method, called Perma-Stone, is produced by using a series of molds into which the mortar is placed. The molds are then transferred to the wall surface. Each of these two methods is excellent for the production of surfaces resembling roughly cut stonework.

The roughly cut imitation stonework produced by the two methods introduced above can be used for interior work. However, imitation stonework is gaining in popularity for exterior use. Finishes used on exterior surfaces require the use of Portland cement as the cementing material.

Two-Coat Stonework. In the first of the methods used to produce imitation stone, the plasterer mixes a batch of mortar and colors it to match the base color of the stone being copied. A natural mortar color batch is mixed at the same time. As in brickwork, the undercoat is applied first. Then the colored mortar is spread over it. If a smooth-faced stone is to be reproduced, a thin layer is applied and floated. Fig. 8-6 shows the result.

Fig. 8-6. Imitation smooth-faced stonework.

Fig. 8-7. Applying mortar to form imitation rough stonework.

If rough-faced stone is wanted, the mortar is laid on irregularly in crude gobs. See Fig. 8-7. When the mortar begins to stiffen, the rough gobs of mortar are brushed out. A fairly coarse brush is used for this purpose. This operation harmonizes the humps. Do not brush out all of the rough spots, as some must remain to make the finished job look natural.

While the mortar is still soft, for either smooth or rough face brick, spots of color are placed here and there. Dark shades are put in the hollows and some lighter shades are applied at one side or the other of the projecting portions. Using a small, soft brush, work these colors into the mortar. The knack of blending the colors properly can be acquired with practice. Study natural stones to see how the colors merge and harmonize.

The joints are now cut through the colored coat into the coat of mortar beneath. Lay out the joint to correspond to the type of stonework being done. In Fig. 8-8 the plasterer is seen cutting some joints. The joint cutter is the same as that used for brickwork. Pay little attention to the humps and cut right through them. Note that a level is used here (Fig. 8-8) as a guide for the lines.

When all the joints have been cut, the stones are shaped so that the humps are contained within the individual stone and do not seem to break through the joint and carry into the adjoining stone. Use a pointing trowel to trim down the humps as the joint is approached, then brush over the area to soften the break.

To develop grain or veins in the

Fig. 8-8. Cutting joints of imitation rough stonework.

Fig. 8-9. Using color to shade stonework.

stones, use a wire brush. Run the grain vertically for the square stones and run it horizontally for long, narrow ones. To develop more life and a natural effect, dent the mortar here and there, using a small, wire brush for the purpose. This denting gives a weathered appearance to the stone. To highlight veins and weathered spots, color may be stippled in, as shown in Fig. 8-9.

On stones having a light tan or yellowish color, the shading is done with deeper tones, such as those of the lighter shades of brown. Then, to give the stones a natural look, specks of red and green can be worked into the hollows. Only a few spots of color should be used. A tiny spot worked in every third or fourth stone will do the trick. Nature is deceptive. A stone that appears dull at

first will, upon examination, usually reveal a variety of colors blended into a harmonious whole.

Perma-Stone. Perma-Stone is a type of stone-work that is formed by use of metal molds. This system is patented and the work is done by licensed contractors. The metal molds are shaped to imitate the natural stone-cut surface. The mold is filled and then it is applied to the wall. This is usually applied to a set Portland cement scratch coat which will give the proper suction and bonding qualities. When the contents of the mold are released, the result is an effective representation of a block of stone.

To prevent mortar from sticking

SHAKERS FILLED WITH
VARIOUS COLORS

SHAKERS FILLED
WITH VARIOUS
COLORS

Fig. 8-10. Sprinkling colors over the wax paper in metal mold. (Perma-Stone Co.)

to the mold the plasterer first places a sheet of waxed paper in this mold. Colored material is then sprinkled over the wax paper in the mold. The molds are of different sizes, each producing the effect of one stone block.

The coloring material is mixed with cement and silica sand or with marble dust; it is then put into the shaker cans, which resemble oversized salt shakers. To insure a variety of hues, usually two colors are sprinkled into the mold simultaneously. Colors are alternated in order not to bring stones of the same color close together. Fig. 8-10 shows a plasterer (on the left) placing waxed paper in the mold. The plasterer to the right is sprinkling color over waxed paper already in the mold. After color is sprinkled on the waxed

Fig. 8-11. Placing filled mold against wall.

paper, the mold is filled with cement mortar. The plasterer then strikes the mortar off even with the edges of the mold.

In Fig. 8-11 a plasterer is shown placing a filled mold against the wall,

Fig. 8-12. Home covered with Perma-Stone. (Perma-Stone Co.)

ready to slap it firmly in place. A quick motion is necessary; otherwise, the mortar will slip out of the mold.

Suction in the base coat holds the soft mortar in place and the waxed paper lining allows the mold to be removed almost immediately. When the mold is emptied the waxed paper remains in place. It is left on for a short time to allow the cement to harden.

As each stone is formed, it is trimmed up and the edges are tooled into the stone previously set in place in order to form a smooth, natural-looking mortar joint. A home covered with Perma-Stone is shown in Fig. 8-12. Note the variations in shading and shape of the stones.

Proper arrangement of the stones is important. Large and small stones must be arranged to look right and to appear as though structurally possible. Joints must never run one above the other. A grouping of small stones should be separated by a large stone.

Imitation Wood

Imitation wood has many uses and possibilities. Wood members of nearly every kind may be reproduced. The half timber construction of early English and French architectural styles is artfully imitated. To construct structural members of this kind, stucco and brick are used. Tree trunks and logs may be copied so realistically that only a knife cut

will convince the onlooker that they are not "the real thing."

Simulated wood, being fireproof, rotproof, and termite proof, is ideal wherever timbers are required as units of architectural design. For interior use, putty and plaster are used, as these materials are better adapted to fine detail. For exterior uses, portland cement may rightly be considered the only safe material. In exterior work, the base color—usually a raw umber—is added to the mortar. The tones are blended in later. In interior work, the coloring is done after the work is shaped and finished. Imitation wood is treated just as though it were real wood.

The first step is to build up a mass of material resembling the general shape of timber. This is done, as shown in Fig. 8-13 with high gauged putty and plaster. As the plaster be-

Fig. 8-13. First step in building up imitation wood.

gins to set, it is trimmed and shaped to the required size. Knots are outlined, as shown in Fig. 8-13, and are placed at random. The knots should vary in size and in shape.

The next step is to form the grain of the wood. To do this, a flat wire brush is drawn across the surface of the imitation wood. The grain should take a natural curve around knots and knurls. In Fig. 8-14 this operation is illustrated. Notice that the knots are cracked and that they have been depressed. This was done to make them look very much like the real thing.

To age the timbers and to give them a weathered look, the surface is pricked at various points with a stiff wire brush. Notches and ax marks can be cut in the corners at this time.

The next step is to build up the pegs that hold the timbers together. Study the timbers in an old house or barn to learn how the pegs should be placed. Remember that the peg holds the mortise and tenon joint together. Keeping this fact in mind will help you in placing the pegs.

The last step is the coloring of the timber. A stain made from linseed oil and turpentine mixed with some raw umber is first brushed over the timber, after which the knots are rubbed off with a rag to lighten them. Next, the grain is highlighted by brushing in some stain prepared as just directed, but this time using burnt umber instead of raw umber. Rub the work down as soon as it is applied. Thus the dark stain will penetrate the deep grain. The stain can be wiped from the outer surface with a rag. As shown in Fig. 8-15, the next operation is to darken the deep-etched places and timber joining, using an even darker color than was used before.

Fig. 8-14. Developing grain in wood, using wire brush.

Fig. 8-15. Dark stain applied with knots shaded lighter in color.

325

After the whole timber has been stained, it is rubbed down. This is done to make the smooth places, such as knots and high spots, shine. Some practice will be necessary before good results will be obtained. Remember that timbers are massive, and, to look true to life, units representing them must be of substantial size. Also, because of their large size, timbers are usually cracked and windshaken. Large cracks, therefore, must be put in close to the ends of the simulated timbers. Proper color blending will do wonders for the job.

The creation of imitation wood is not limited to simulated wood timbering. Very nice imitation wood clapboarding can be produced also. Combinations of imitation wood and stone can also make up a fine appearance.

Textures for Portland Cement Finishes or Stucco Work

Stuccowork can be finished in accordance with individual tastes and in ways appropriate to the particular purpose. Certain textures are uniquely suited to the climate and prevailing style of architecture of given regions. In the South and West, float finishes are popular; California and Florida, however, favor the swirled and irregular textures done in color. The colors used are sometimes of vivid shades. More often, however, they are subdued shades like some of the pastels. (The addition of color to stucco mixtures was discussed in Chapter 4.)

Midwestern, northern states and Canada seem to prefer the rugged pebble-dash finishes. All of the various finishes applied are, of course, found in all sections of the country regardless of their apparent traditional association with a particular region.

There are many stucco textures. Space does not permit discussion of all of the popular texture styles. Some of these textures, however, are shown in Fig. 8-16.

Whatever the style of architecture used, choose a stucco texture that will harmonize with it. Your choice of the kind of texture will determine the harmony of the overall architectural design. Each of the textures has been given a name. Usually these names have been derived from the kind of architecture with which they are identified. For example, the style called "Modern American" is well fitted to colonial buildings. See Fig. 8-16, center left. Irregular, feathered finishes, such as "English cottage," Fig. 8-16, bottom left, are in harmony with houses of English design. Other textures fit the French (Fig. 8-16, top left), Spanish and Italian designs. (Fig. 8-16, top right).

The method for applying one of these textures, called *travertine*, (Fig. 8-16, bottom right) is shown in the series of pictures. See Fig. 8-17. The plasterer is beginning to apply

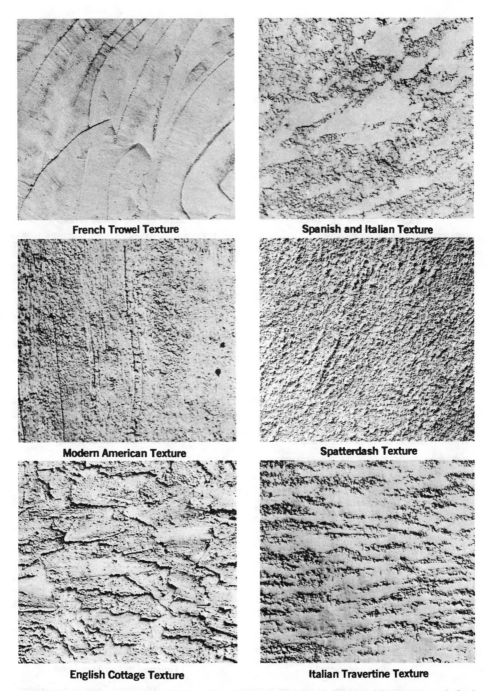

French Trowel Texture

Spanish and Italian Texture

Modern American Texture

Spatterdash Texture

English Cottage Texture

Italian Travertine Texture

Fig. 8-16. Traditional texture finishes for plastered surfaces. (Portland Cement Association)

Applying mortar ⅜" thick

Stippling with whisk broom

Holding whisk broom at angle

Appearance after stippling

Troweling horizontally

Finished Italian Travertine

Fig. 8-17. Steps for applying travertine texture.

(Portland Cement Association)

Italian travertine stucco texture in Fig. 8-17, top left. Apply a very thick coat of mortar (about ⅜″) to a well-dampened brown coat; the thickness will retard hardening and permit texturing operations. Trowel the finish coat fairly smooth. Then stipple deeply, using a whisk broom or wire brush to pull up an irregular texture. See Fig. 8-17, top right.

Hold the brush at an angle as shown in Fig. 8-17, center left. Stippling should be irregular in appearance. Finally, trowel horizontally to smooth out the higher portions of the texture. The depressions made by stippling remain rough and furnish the veined effect.

The finished texture is seen as a stippled, troweled finish. Using greater pressure on the trowel will give a finer veining. The finish may then be cut with a jointing tool to create joints. The result will duplicate travertine stone blocks.

Fig. 8-18 shows a plasterer about to apply *spatterdash* to a wall surface. For this texture, regular sand and cement mortar is made up and kept soft. Use a stiff fiber brush to apply the material. The stucco mortar is thrown at the wall surface. To prevent lumping and the possibility of an uneven texture, it might be well to snap the brush against a stick held in the other hand. This action spreads the mortar material into a thin, even spray. The finished result is shown in Fig. 8-16, center right.

Fig. 8-18. Applying spatterdash to wall surface.

Fig. 8-19. Applying pebble dash to wall surface.

Fig. 8-19 shows the application of *pebble dash*. This texture requires a mixture of graded pebbles (usually about ⅛" to ¼" in size) and Portland cement. No sand is used. Make the mix very soft and stir it at frequent intervals to prevent separation of the stones and the cement. Proportions for this special finish are 1 to 1.

The texture is applied with a small, flat, square-nosed scoop. Place the material either in a pail or a flat trough. Pick up a scoopful of material and, with a sweeping motion, throw the mortar against the surface of the wall. The motion is about the same as that made in throwing a baseball by the sidearm delivery. A certain amount of the stones flung at the wall will bounce back and drop. This cannot be avoided altogether. However, practice will enable the plasterer to hold this wasted effort and material to a minimum.

The knack to be acquired in doing this work is that of flinging the stones in such a way as to accomplish uniform coverage of the base and at the same time create a texture. It is best to start at the rear of the building and work around to the front. Such a procedure gives the plasterer time to work into the "swing" of the texture.

Dry dash texture is similar to pebble dash. The main difference is that dry stones are used. Two men are needed to accomplish this job. One man trowels on a mortar coat of sufficient thickness to hold the pebbles, and the second man dashes stones onto it. The second man uses a scoop similar to the one seen in Fig. 8-19. The action of the arm is also the same as that used for pebble dash. Considerable force must be used, however, to make the stones adhere to the base. To bring the surface to a more uniform, flat plane, the stones may be patted down. For this purpose, a clean wooden float is used. Doors and windows must be protected and covered during the dashing operation.

Only practice and experience will enable the plasterer to produce a good job in the various types of textures. Personal likes and abilities permit one plasterer to do a good job with one process. Another plasterer will achieve equally good results with another and perhaps totally different procedure.

Exposed Aggregates

Generally, exposed aggregates fall into two groups: one group consists of a mix or emulsion of binder and aggregates that are simultaneously troweled on to a surface; the other group consists of a matrix which is applied to a surface and then seeded (either by gun or by hand, depending on the matrix) with aggregates. The manufacturer's specifications as to how these materials are to be applied should always be followed.

Normally, exposed aggregate will be applied to newly constructed backing made of block or brick, concrete (pre-cast or poured), cement plaster and cement asbestos board, or existing painted surfaces.

Surface Preparation

Block or brick should be free of oil, dirt, etc. The walls should be plumb and true, with no projecting blocks. All excess mortar should be rubbed off, and it is most advantageous to have the joints struck flush. In new work these conditions can be had by informing the mason contractor before he begins his work. If he neglects this, it will mean that the plasterer will have to prepare his surface by chiseling any protruding blocks or bricks that may project beyond a point of a flat surface. Cement plaster in any uneven surfaces, especially at corners where the wall was not erected plumb, will have to be corrected. The wall will have to be scraped of excess mortar that may have been splashed onto the surface during the erection of the wall. If the joints are not flush, it may in many cases require a coat of cement plaster over the entire wall. This would be more evident in the emulsion type exposed aggregates, which are troweled on. With the matrix type material, the use of the matrix troweled over the joints will make the surface flat prior to application of the bed coat.

Concrete (pre-cast or poured) should be free of oil, separating agents, dirt, etc. If oil is present a solution of tri-sodium phosphate should be used for cleaning, followed by thorough rinsing. Slight imperfections of concrete lines will be hidden by the materials applied, but any concrete that is noticeably out of line should be brought to the attention of the architect or owner, to arrange for the surface to be put into shape.

Cement plaster should be rodded straight and plumb and float finished; it should be held back from the finish line to allow for the thickness of the exposed aggregate to be used. If the matrix for seeded type aggregate is to be used, the plaster should be held back $\frac{1}{4}''$ to $\frac{3}{8}''$.

Cement asbestos boards should be clean and free of dust remaining on the surface after cuts have been made in the boards. They should be securely fastened by a method that does not allow popping out of the fasteners, such as common nails. The asbestos board should be at least $\frac{1}{4}''$ thick with rust-protected screws countersunk into the backing on 12'' centers. Asbestos boards are sold in sizes similar to plywood. It is always necessary to make sure that the architect and builder understand that the joints formed by butting of two or more asbestos boards should never be covered over with material, because in no case can the covering up of these joints be expected not to crack. In all asbestos board construction each board should be laid out as an individual panel, leaving at least ¼" control joint between separate panels. With this information the builder and architect can predetermine what the board sizes should be cut to in order to make the panels look symmetrical.

On Existing Surfaces.   If the existing surfaces have been coated with one or more coats of paint, they should be sand blasted to the bare original surface. Then, depending on the type of back, the previously described procedures are followed. When the backing is prepared, then proceed with the application of the exposed aggregates.

Emulsion Type

Emulsion type exposed aggregate is a trowel-applied material consisting of aggregates combined with specially formulated binders. (The binder is a thick, milky material which on drying exposes the aggregate.) The manufacturer's recommendations should be followed when applying this type of material. Gen-erally, the work should not be applied when temperatures are likely to drop below 40°F before material dries. Check the surfaces to make sure they are in an acceptable condition. Remember that regardless of how strong the applied material is it will only be as sound as its backing.

Most manufacturers recommend that the alignment of the backing should not require the application of their material in excess of ⅛". In such cases, a leveling coat of Portland cement plaster should be applied. Use plastic containers instead of the customary mortar boards to keep the material in as you use it. Mortar boards will subject the material to the air, thus starting its hardening action. (This material is like a glue, it air dries to a hard state. As much of the material as possible should be kept unexposed to the air before using it.) Apply the undercoat or binder and sealer coat (if so specified by the manufacturer) no more than 1/16" thick even when filling voids.

After the undercoat has dried to a sufficient hardness, the marblecoat may be applied. The required drying time for the undercoat varies with weather conditions. Usually 1 to 2 hours will be enough.

Before applying the marblecoat, however, check the possible weather conditions. This material should not be subjected to rain until it has cured. Rain on a newly completed application can wash out enough binder to prevent proper long term performance, even though the surface will appear to be undamaged. In hot weather, when temperatures are 85°F and over, the job should be planned so as to avoid working in direct sunlight wherever possible. The increase in temperatures on the surface of the material will hasten the hardening of the material thus making the finish troweling more difficult.

Next, check the size of the area to be covered. With experience the plasterer will be able to judge the number of plasterers necessary to complete the area in one continuous operation. With this type of material joints must not be left for any length of time. The edge of applied material should be in a wet state when joining a new surfacing because lap marks will result if the material cannot be worked into the joinings. The number of plasterers necessary will increase as the area covered increases and the temperature increases (causing faster drying). Like glue, this material will harden faster when exposed to high temperatures.

The right number of men will insure a wet edge at all times. Also, sometimes under certain conditions the finish troweling will have to be done by another plasterer directly behind the plasterer or plasterers applying the material. Such cases would be when an area is in direct sunlight or high-velocity wind.

Be sure enough material is mixed (or on hand) to cover the entire area. The best trowel to use is one made of stainless steel. This type of trowel avoids discoloration of the material when final troweling is done and it also is easier to keep clean. Allow no interruptions until the area is completely covered. Begin applying the material at the top of the panel and work down.

Keep the material in plastic containers and scoop out a hawkful at a time as needed. Apply it to the wall, using several short strokes compressing the material with the trowel as it is applied. See Fig. 8-20. After the hawkful has been applied, turn the trowel on edge and draw it across the applied material in both directions. It is necessary to cut off excess materials to keep the finish straight. Now flat trowel this material in both directions with a whipping motion,

Fig. 8-20. Plasterer applying troweled-on exposed aggregate. (Cement Enamel Development, Inc.)

making sure the trowel blade is always flat to the surface. (With experience, this will become very easy. The reason it will seem difficult at first is because the trowel is used at the opposite angle to which the plasterer is accustomed.) This whipping motion will bring some of the emulsion to the surface. Then flat trowel the surface in one direction. All voids and trowel marks must be worked out immediately because once the drying process has started these imperfections cannot easily be repaired.

The binders of this material tend to create a film on the trowel which must be removed frequently in order to avoid drag. A second trowel may be used to scrape off film. Trowels must be kept clean at all times, otherwise a rough, uneven finish will result.

Proceed with another hawkful and continue until the surface is completely covered. When the material begins to show its finish color, the final troweling should begin. Flat trowel the surface in one direction using a perfectly clean trowel. This troweling will leave a flat, smooth surface. It is absolutely imperative that the trowel for final tightening be kept completely clean by regularly passing a wet brush over the face of the trowel. Excessive troweling should be avoided, as it tends to create a greasy appearance. Water is not used on the surface when troweling as is used in troweling other materials such as the putty coat.

In overhead work, or ceilings, gravity has a tendency to pull the material off before it becomes hard, so special precautions should be used.

1. Backing must be thoroughly dry before undercoat (if it is called for) is applied.

2. Undercoat must dry thoroughly before finish work begins. Allow twice the ordinary drying time whenever possible.

3. Finish coat must be applied in a thin, well-compacted coating. The finish will sag and drop off if it is applied too thick. Do not try to trowel it back into place if this occurs. Remove it, dry the area with a clean cloth and reapply a thinner coat.

4. Consistency of the material should be a little stiffer than what is normally used for wall application.

Large Unbroken Areas. Occasionally an architect or owner will insist on large unbroken areas, but most often they will understand that this requirement will be extremely difficult to meet. It is just not feasible to have enough equipment and plasterers on one job just to make sure the edges or joinings are made satisfactorily. For example, assume the job is a wall 15 ft. high and 100 ft. long. It will be necessary to determine how wide an area can be applied carrying a wet edge. If the plasterers can complete a section 20 ft. wide and 15 ft. high without interruption, then this would be a good place to divide the wall into five equal parts separated by using special beads recommended by the manufacturer.

The panels can also be separated with taped joints. Over dried undercoat, if called for, or if color contrast of similar finish material is wanted, the contrasting color is first applied at the point of division and allowed to harden and dry. See Fig. 8-21 top left. A plumb line is chalked over either one to mark the joints, using white chalk which can be easily brushed off. After the lines have all been chalked take tape cut to the width of desired joint and place it along the chalk line. See Fig. 8-21,

Fig. 8-21. Four step procedure for separating panels with taped joints. (Cement Enamel Development, Inc.)

top right. The best tape for this is ⅛" thick rubber tape, although masking tape can also be used. When the rubber tape is used, do not stretch it because it will contract later, causing the tape to fall off. Now trowel on the finish using tape as a ground. If rubber tape or regular masking tape is used (for liquid binder), apply as little material as possible over it. See Fig. 8-21, bottom left.

After the material has been applied, and just before final troweling, in the case of regular masking tape, place a clean trowel flat on the finish material at the edge of the tape. Now, with a free hand, pull the tape down and toward the edge of the trowel as if you were cutting it on the trowel. Continue pulling this all the way to the end of the joint. Align any irregularities by pushing the material to a straight line. Finish trowel the entire panel.

If rubber tape is used strip off the tape as soon as the finish has dried hard. See Fig. 8-21, bottom right. Rubber tape must be stripped before it adheres or vulcanizes, so to speak, and discolors the joints. Where work is being done in very high temperatures and direct sunlight, it may be necessary to apply masking tape under the rubber tape to prevent it from sticking and discoloring the joints.

Exterior Corners. These can readily be made to a sharp true line.

The simplest procedure, of course, would be to use a special bead designed for this purpose. Another procedure is to use wood strips that are tacked along one side of the corner extending ⅛" beyond the corner. The troweled aggregate is then applied on the side opposite the wood, using the wood strip as a ground. The wood strip will be found to work best if it has been beveled to a ¼" edge on the side which is used as the finished edge. See Fig. 8-22. The reason for bevel edge is to make the troweling on of the material easier. There is less surface for the material to build up on. In other words, the smaller or sharper the surface edge, the cleaner the work will finish. The reason it is turned away is for a flatter nailing surface. The reason we don't make it razor edge is that in han-

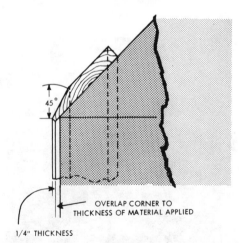

1/4" THICKNESS

OVERLAP CORNER TO
THICKNESS OF MATERIAL APPLIED

45°

Fig. 8-22. Use of wood strip to establish mortar thickness on exterior corner.

dling it would be too easily nicked. The strip should be covered with masking tape first, then with petroleum jelly over the tape so it will be more easily removed. The strip should not be removed until the material has dried and hardened. This will usually be the next day. After removing the strip, scrape the side that was next to the strip so as to remove any of the petroleum jelly that may have remained. Now apply the remaining side of the corner, working the material to a sharp arris.

Where cement plaster backing is used, regular exterior-type plaster beads may be used effectively to obtain a sharp true corner. See Fig. 8-23. The cement plaster coat is held

Fig. 8-23. Troweled-on aggregate applied at a corner using a regular exterior type plaster bead. (Cement Enamel Development, Inc.)

back from the finish line so that the surfacing can be installed to meet the edge of the bead.

Inside Corners. No bead is required at the inside corner returns, but a trowel cut is often made in the surfacing at the corner to assure a neat, sharp break if any differential movement occurs.

Work Suggestions. If the following steps are followed most problems will be avoided.

1. Make sure there are enough plasterers and equipment to do the job.

2. Protect the surface from rain at least 24 hours.

3. Avoid application on irregular backing.

4. Protect adjoining surfaces before work begins.

5. Use designed joints to break up large areas; try to keep unbroken, unjointed areas to 50 sq. ft. whenever possible.

6. Mix material thoroughly.

7. Always work with a "wet" joint. Do not wet with water but with soft material. It is no longer wet if the surface has started to dry.

8. Never apply when the temperature is below 40°F.

9. Make final trowelling to tighten up surfaces in one direction only.

10. Keep trowel clean at all times.

Matrix Aggregate Seed Type

There are three kinds of this type of exposed aggregate. They are classified by the composition of the matrix: (1) liquid binder or special cement, (2) marblecrete, (3) epoxy.

Liquid Binder or Special Cement

This kind of exposed aggregate uses a special liquid binder. It should be used as directed by the manufacturer. The condition of the backing must be checked to see that it is in an acceptable condition to receive the exposed aggregate. Most manufacturers recommend that the temperature be 40°F and rising before any material is applied.

A presealer is applied to all the surfaces to be covered; a brush, roller or sprayer is used. Apply a sufficient amount of pre-seal so as to completely coat without any misses, including any voids. A missed spot the size of a half dollar may show when trying to seed the aggregate into the applied matrix. This is because the suction underneath has sapped all the *open time* out of the matrix at this spot and the aggregate will not adhere well. (Open time is the time after application that the matrix will easily receive aggregates.)

This pre-seal should be allowed to dry sufficiently before any further application of the exposed aggregate is started. Some manufacturers recommend the pre-seal be allowed to dry overnight, others say two hours.

After the pre-seal is dry, or while it is drying, certain preparations should be made. Check temperature, size of panels and number of men available. This should be done so as to know how much area can be covered before stopping, because once a panel has been started it should be carried out to completion with no interruptions or delays. An estimate of the area that can be covered will also give an idea of how much matrix and aggregates will be needed at one time.

A rule of thumb for rough estimating is 1 bag of Portland cement with all the ingredients to make the prescribed matrix will be needed to cover 100 sq. ft. Aggregates, on the other hand, will vary as to size, but a guide is 2, 100-lb. bags of aggregate or 6, 12-quart pails full will be required to do 100 sq. ft. (Note: This quantity includes the amount of aggregates that will fall and not remain on the wall.)

The aggregates are kept in buckets, which should be placed so as to be available throughout the entire panel. For instance, if the panel is of any height there should be enough aggregates at each level of scaffolding to properly seed that area before the application of the bed coat begins. (The "bed coat" is not the same as "pre-seal." Bed coat is the applied matrix that will receive the stones or seeded aggregates.) This precaution is necessary because there will only be a limited length of time during which the stones can properly be seeded into the matrix bed coat after it has been applied. This time will be shortened with a rise in temperature and increased velocity of wind.

Preparations should be made to catch the aggregates that fall. To reduce waste, these aggregates should be re-screened before using again. Usually a clean tarp (called a "catcher") is used for this purpose. If an aggregate gun is to be used, make sure there is enough extension cord to allow the gun or guns to move freely over the entire panel. It is also wise to check the operation of the gun before applying the bed coat.

Scratch Coat. First scratch a *tight coat* (very thin coat) over the entire panel. See Fig. 8-24. Do not try to straighten or fill any voids, keep the coating tight. This coat is applied to give a little controlled suc-

Fig. 8-24. Applying a tight scratch coat over the entire panel.

tion for the bed coat. In the case of weather-struck mortar joints in masonry backing, it will prevent joint marks from showing throughout the panel when the job is finished. This can happen because there will be slight sags at the horizontal joints if this tight coat is not applied. The coat can be tested for dryness by placing a finger onto the surface and pulling it away. If the material tends to cling to your finger, it is not ready for application of the bed coat; if it does not cling, then you can start to apply the bed coat.

Bed Coat. There are two methods that can be used successfully in doing the next two steps. In one method the bed coat is applied to the entire panel before seeding the aggregate. This method would, in many cases, require many men and the use of several aggregate guns.

The other method requires fewer men. The bed coat is first applied at the bottom of the panel and then proceeds up the wall. Application of this bed coat is followed directly with the seeding of the aggregate. In seeding, the bottom of the panel is started first to prevent the aggregates from falling into the matrix that is being applied below. This method would make the aggregates caught in the catcher dirtier due to the dropping of matrix by the plasterers applying the bed coat.

Using either method, it is well to test the surface for suction and dry-

ang conditions. Use a small area where the aggregate can be readily scraped off with a trowel. This will indicate the conditions under which the work will be done, such as the consistency of the material and how far ahead the bed coat can be applied before seeding of the aggregate becomes difficult.

In the latter method (working bed coat from bottom up, followed by aggregate), one or two plasterers will apply a sufficient amount of matrix to hold the aggregate to be seeded. Normally the average size aggregates are ¼″. The bed coat should be cut back slightly at the bottom of any panel having a horizontal bead or stop. This will allow for the slight slump that will occur when the aggregates are applied. Due to the weight of stones being added the material will sag over the intended stop bead. The bed coat should be slicked or smoothed out with the trowel before proceeding further up the panel. This area is then seeded with aggregates.

Seeding. The aggregates can be broadcast onto the surface by hand or with the aggregate gun. The aggregate gun, in most cases, is the most practical for this type of material. Fig. 8-25 shows the use of an aggregate gun. The proper air volume setting on the gun will vary with different size aggregates and the wetness of the matrix to which it is going to be blown on. The air setting should be just enough to allow the

Fig. 8-25. Plasterer applying aggregate with an aggregate gun. (Portland Cement Association)

aggregate to be imbedded about one third of its thickness without too much loss of aggregate. Excess pressure will drive aggregates too deep or possibly cause them to bounce off the backing. Not enough pressure will result in stones rolling off without adhering to the surface. The best procedure when starting to blow aggregates onto the bed coat is turn the air volume low and then increase it until it reaches the best setting.

The aggregate gun should be held square with the wall at all times. See Fig. 8-26. Aggregates striking at an angle will not adhere as well. Even worse, they will leave a slight tear as the stones enter the bed coat. Even though this defect may seem very

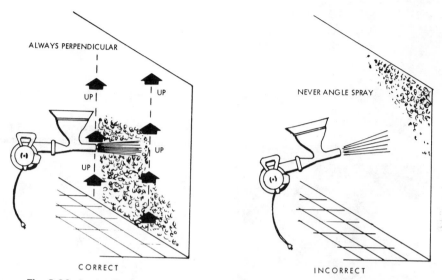

Fig. 8-26. Angle of spray should be perpendicular to area covered at all times.

slight at the time the stones are applied, it will become more noticeable when the whole panel is completed. These slight irregular pattern marks will usually show more noticeably at the scaffolding heights where the applicator has a tendency to point the gun up when he nears the limits of his reach or he points it down when beginning each scaffold height.

The aggregate gun should not be directed at the wall when empty, as the air pressure will further dry the surface of the matrix, making it difficult for the aggregates to adhere. When applying the aggregate, the plasterer should move the aggregate gun slowly. This is done so as to cover the surface with as much ag-

gregate as possible the first time the gun is moved over an area. Aggregates will not adhere as well the second time they are blown at an area, or they may not adhere at all.

If the corner arrises have been formed by hand, the aggregate gun should not be used to apply the aggregates closer than 2" to 4" from the arris. The aggregates should be *carefully* applied by hand at this point so as not to disfigure the formed arris. See Fig. 8-27. If a small area (a square inch, for example) does not accept aggregates while the area is being seeded by the gun, the plasterer applying the aggregate should push a few stones into the bed with his fingers without disturbing the

Fig. 8-27. Lightly hand seeding the arris previously formed by the outside corner tool.

surrounding area. After the aggregates have been applied, they should be tamped lightly with a trowel or, preferably, a soft rubber float. See Fig. 8-28. The rubber float will tamp the assorted size aggregates that may be used, whereas the trowel would only tamp those aggregates that protrude the furthest from the surface. The purpose of tamping is to imbed the aggregate more firmly and bring the surface to a more even plane. Some plasterers use a paint roller with a heavy nap, such as lambs wool. The roller is *lightly* rolled over the aggregates in one direction, starting from the bottom up, making sure not to use too much pressure. After the

aggregates have been applied, they should remain protected from rain for several hours. Once the matrix has set, water will not harm it.

Large Unbroken Areas. In working with large unbroken areas, refer to the section covering this subject earlier in this chapter under "Emulsion Type." The difference will be in the thickness of the beads or tape used to form the joints, which must correspond to the thickness of the matrix applied. Masking tape cannot be used.

Corners are usually handled in the same manner as for troweled-on aggregate, allowing for greater thickness. Another method can also be

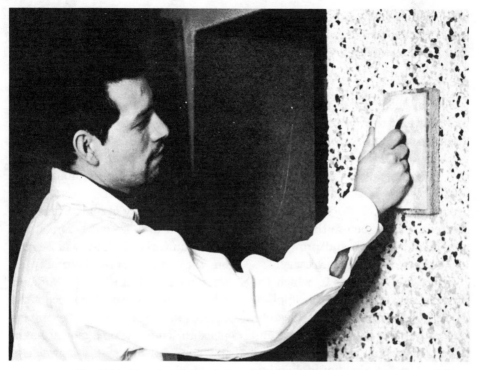

Fig. 8-28. Tamping assorted size aggregate with a rubber float.

used. This is the freehand forming of the arris using the outside corner trowel. See Fig. 8-29. Some plasterers use a rod as was explained in Chapter 7 for forming the arris when browning Portland cement.

Sealer. Many manufacturers recommend using a sealer after the exposed aggregate has dried sufficiently. A sealer is sometimes used with the liquid binder. It is applied with a brush or heavy nap paint roller. The plasterer must make sure to cover all areas uniformly, especially where a colored matrix has

Fig. 8-29. Forming an arris with an outside corner tool.

343

been used, because any uneveness of sealer on the matrix will appear as slightly different shades of the color.

Marblecrete

The application of marblecrete's bed coat and the seeding of the aggregates is almost the same as for the liquid binder exposed aggregates. One difference is that marblecrete is applicable only to those surfaces that are acceptable for Portland cement plastering. Also, before the bed coat is applied a scratch and brown coat of Portland cement is necessary.

There are two ways in which the brown coat is readied for the application of the bed coat of marblecrete. One is to let the brown coat cure and dry; then, using a brush roller or sprayer, one of the liquid bonding agents for Portland cement is painted over the brown coat. The second method is to brown a section and then apply the bed coat for marblecrete as soon as it becomes stiff enough to carry the weight of the bed coat and aggregates. This normally takes about 1 hour.

Epoxy Type

Epoxy exposed aggregates are very different from the other exposed aggregates in application and composition. The manufacturer's instructions for application should be followed, of course. First check to see that the aggregates are available, that a catcher for the aggregates has

been provided, and that there is a source of electricity if the aggregate seeder gun is used, etc. The surface to which the bed coat is to be applied must be perfectly dry. The aggregates should never be dampened. It is not necessary to complete the entire section before stopping as is the case with the other exposed aggregates. With a little care unnoticeable joinings can be made.

Mix a quantity of matrix that can be applied and seeded before it becomes unworkable. This will depend on the temperature. Normally at temperatures from 50° to 70°F, it will remain soft enough to accept the aggregate. Immediately after mixing place the contents on a mortar board and spread it out with a trowel. See Fig. 8-30.

Fig. 8-30. Mixed epoxy matrix spread on mortar board will be used immediately. (Ceram-Traz Corp.)

Fig. 8-31. Plasterer applying epoxy matrix. (Ceram-Traz Corp.)

In most cases aggregates from sizes #2 to #5, or a mixture of these, will require a bed coat thickness of ⅛″ to support their weight. Larger aggregates will require a thickness of ³⁄₁₆″ or more.

This type of material cannot be rodded. To insure the proper thickness use a stainless steel trowel notched as follows: for ⅛″ thickness use a ¼″ notched trowel, for ³⁄₁₆″ thickness use a ⅜″ notched trowel, and for a ¼″ thickness use a ½″ notched trowel. Now apply the matrix with the properly notched trowel

and then level it to the proper thickness with a smooth stainless steel trowel. See Fig. 8-31. Finally, apply the aggregate by hand or with the aggregate seeder gun. If done by hand use cotton gloves and pat down; do not roll the aggregate. This method has been found to be very successful. See Fig. 8-32. Fig. 8-33 shows the finished work.

Remember to clean your hawk and trowels before the epoxy becomes hard. Use hot, soapy water or lacquer thinner, because once the matrix has set it will have to be chipped

Fig. 8-32. Plasterer hand seeding aggregates. (Ceram-Traz Corp.)

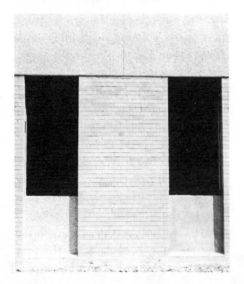

Fig. 8-33. Epoxy exposed aggregate. (Ceram-Traz Corp.)

off, and in most cases this is not feasible. Some manufacturers require a sealer. It is applied after the epoxy

aggregates have cured at least 24 hours and have permanently set.

Murals or Designs

Murals or designs can be made using any of the exposed aggregates or Portland cement finishes. The work consists of using the architect's or artist's sketches or drawings and projecting it onto the wall in a predetermined size and then applying the specified materials in the color or colors chosen by the architect or artist. The only restrictions are limited colors available.

Simple designs can be made very easily by using the rubber tape or beads in the manner explained earlier in the discussion of large unbroken areas in the section "Emulsion Type." Fig. 8-34 shows an example.

Mural Design

Troweled-On Aggregates. Mural designs for the emulsion type troweled-on aggregates, can be applied in two ways. One is the *superimposed mural*, the other is *flush-jointed mural*.

To construct the *superimposed mural* first apply a finished background of emulsion type troweled-on aggregates. After the background has dried, draw the mural outline (with pencil or chalk) as shown in Fig. 8-35, top left. Then apply any type masking tape around mural areas. Contact (adhesive) paper is sometimes used. See Fig. 8-35, top right. Next, apply troweled-on aggregate in contrasting color or shade as shown in Fig. 8-35, bottom left. As soon as the material has dried strip the tape as shown in Fig. 8-35, bottom right.

To construct the *flush-joint mural*, experience in using emulsion-type troweled-on aggregate will be necessary, for the troweling in of the final jointing is a difficult skill to master. First apply the undercoat and allow it to dry completely. Then, using chalk or pencil, draw the complete outlines of mural on the wall. Mark

Fig. 8-34. Simple design made by using rubber tape. (Cement Enamel Development, Inc.)

Fig. 8-35. Four step procedure for constructing murals. (Cement Enamel Development, Inc.)

areas of the mural to show what color will be added in various places. Example: red, yellow, green, etc. Numbers could be used, such as : red, No. 1; yellow, No. 2; green, No. 3, etc. (This is similar to the popular paint-by-number kits.) Apply tape to areas throughout the whole mural that will use the same color because only one color is applied at a time and allowed to dry completely before an adjoining color is applied. After this area has dried thoroughly, strip the tape that was used to outline the areas we just applied and then take masking tape and cover the edges of the already finished color area so as to protect them from the next color of emulsion-type troweled-on aggregate to be applied. Now apply the troweled aggregate up to the line of tape. Avoid building up too heavily a coat over the tape's edges. Just before the final troweling, remove the tape and work in the jointing as smoothly as possible.

Seeded Aggregates. For the seeded aggregates the superimposed method of constructing a mural is not recommended because of its irregular surface. A method similar to the flush joint is used. The difference, of course, is that the *joints* are not troweled in and the tape used will have to have the same width as the thickness required for the bed coat. (The juncture of two adjacent colors is called a *joining line* or *joints*

because this is a bedcoat; the aggregates are seeded, thus the material is not trowel finished. We cannot trowel in the joints as was just shown for the emulsion-type troweled-on aggregate flush joints.) It also will not be necessary to wait until an area dries completely before an area adjacent to it is applied. Usually only an hour is necessary, for this will allow the surface to tighten up enough so that aggregates being applied at the jointing will not adhere to the previously applied area.

There is also a means of dividing off different areas using a permanent

Fig. 8-36. Divider strips are used to separate colors in murals made up of seeded exposed aggregates. Top view shows a sketch on a wall to which divider strips were nailed. Bottom view shows the completed mural.

divider that is similar to metal dividing strips used in terrazzo work. Fig. 8-36, top, shows a sketch on a wall to which divider strips were nailed.

Fig. 8-36, bottom, shows completed mural. This divider strip is designed specifically for murals using seeded aggregates. It is made of 29 gauge

Fig. 8-37. Applying a divider strip to a line that will outline a mural.

Fig. 8-38. Example of a mural design developed by using a combination of troweled and seeded aggregates.

galvanized steel or aluminum. It conforms simply and easily to any shape. It comes in ¼″, ⅜″ and ½″ widths. It has nailing brads every 6″. Fig. 8-37 shows a strip being applied to a line that will outline a mural.

The imagination and ingenuity of the plasterer can often help him in devising methods to develop the architect's or artist's murals or designs. Fig. 8-38 shows such a design developed by using a combination of troweled and seeded aggregates.

Another unusual design is shown in Fig. 8-39. The basic design was constructed by the lather with the plasterer applying Portland cement to the lines formed by the lather. Fig. 8-39, top left, shows the prepared surface. After the basic design was finished with the proper three coats, the lettering was cut out of styrofoam as shown in Fig. 8-39, top right. The cutout styrofoam was then applied to the wall. In this case it took more than one thickness of the cut styrofoam to bring the letters out to the desired thickness as shown in Fig. 8-39, bottom left. The jointings were troweled flush with finish material

Fig. 8-39. Unusual design basically constructed by the lather with the plasterer applying Portland cement over the lines. View at top left shows the prepared surface. Lettering was then cut out of styrofoam, as seen at top right. Letters of several thicknesses of styrofoam were applied over the basic design, as shown at bottom left. After plastering and texture spraying, the final result appears at bottom right. (Lathing & Plastering Institute)

and then all the surfaces were texture-sprayed with the finish. The finished work is shown in Fig. 8-39, bottom right.

Swimming Pools

Many formulations and methods are used in plastering swimming pools. Most pool companies specify the formula and method of application. One successfully used formula and method will be discussed here.

The base material that forms the shape of the walls to which the plasterer applies the formulated cement is "gunite". (This is a material composed of cement, sand or crushed slag, and water. It is forced through a cement gun by pneumatic pressure.) The base material that forms the floor is poured concrete, to which the plasterer applies the finish. The two base materials are installed by another trade.

Fig. 8-40. Mixing setup for swimming pool. (Huron Pools Co., Inc.)

When the "gunite" and concrete have cured or hardened sufficiently, usually a couple of days between the application of these bases, the plastering application can begin.

Application

Have all the necessary materials and equipment available. Fig. 8-40 shows a swimming pool site. Check the temperature so as to know how much calcium chloride should be used. Calcium chloride is used as an accelerator for the setting of the cement. High temperatures also hasten the setting action of the cement. Thus, as temperatures increase, the need for calcium chloride decreases. Also, when the cement hardens quickly more men are needed on the job because the cement must be in a trowelable state when finishing in order to obtain the smoothest surface possible.

Walls. First, wet down the walls using a water hose. Wet the walls from the top, letting the water run down the sides. When the walls have been thoroughly dampened mix the brown coat using a mix of 1½ bags of fine white marble dust, 1½ bags of coarse white marble dust, 1 bag of white cement and a 1 lb. coffee can full of calcium chloride solution (when temperatures are normal). The calcium chloride solution is made as follows: Using a 5 gallon container, fill it ⅔ full with calcium chloride and add water until full.

Fig. 8-41. Applying swimming pool brown coat mix. (Huron Pools Co., Inc.)

Apply the mix with trowels as quickly as possible to all the walls, including the cove at the bottom of the walls, to a thickness of about ⅜". See Fig. 8-41. Use a swimming pool trowel to shape the material as you proceed around the pool. The brown coat should be left fairly smooth, so as to avoid trouble in applying the finish coat. Wait for the brown coat to stiffen sufficiently to receive the tight finish coat.

Mix the finish coat using a mix of 1½ bags of coarse white marble dust, 1½ bags of fine white marble dust, 1½ bags of white cement and a

Fig. 8-42. Applying the finish coat. (Huron Pools Co., Inc.)

Fig. 8-44. Wet troweling the finish coat. (Huron Pools Co., Inc.)

Fig. 8-43. Drawing up the finish coat. (Huron Pools Co., Inc.)

Fig. 8-45. Working into the coves at the base of the walls. (Huron Pools Co., Inc.)

1 lb. coffee can of calcium chloride solution (for normal temperatures).

Under normal weather conditions the application of the finish to the walls can be done with two men in the following sequence: One man applies the finish to the desired finish thickness, usually about ⅛″ if the brown coat was applied properly. See Fig. 8-42. The second man follows behind him and draws the cement up (tightening it in and further eliminating any trowel marks or "cat faces"). See Fig. 8-43. Just before the cement of the first applied finish sets up he drops further back and with a brush he wet trowels the finish, leaving the first man to draw up the finish himself as he continues

around the pool. See Fig. 8-44. The first man, after he has reached the point at which he started, begins to wet trowel the walls for the second wet troweling.

Floor. Apply the brown coat to the floor of the pool to a thickness of about ½″. See Fig. 8-45. Work the material into the edge of the cove at the base of the walls, being careful not to over-trowel. Over-troweling at this intersection can cause a black mark.

As soon as the floor has taken up enough to walk on without marring it too much, wet trowel the brown coat twice.

Then apply the brown coat to the steps of the pool in the same manner

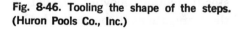

Fig. 8-46. Tooling the shape of the steps. (Huron Pools Co., Inc.)

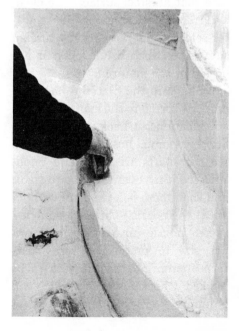

as the floor. See Fig. 8-46. Usually templates cut from plexiglas are used to aid in the shaping of the steps.

Note: Only one-coat application is used for the floor and the steps.

Direct Applied Fire Insulation

Manufacturer's instructions should always be followed in applying cementitious, mill-mixed material using the plastering gun. The materials that use the regular plaster gun have either a gypsum or a cement and are not to be confused with those that use a type of glue to hold the material together.

Make sure that the temperatures will be warm enough so the material will not freeze before it is set and dry. Also see that there is proper ventilation to promote drying. Make sure that all steel surfaces are clean and free of dust, soot, oil, grease, water-soluble materials, or any foreign substance which would prevent adhesion of the material.

Make sure the material has been properly mixed to the manufacturer's specifications. Check the machine hopper, pump, hose and nozzle to see that parts are clean and unobstructed, that connections are tight, and that there is a proper supply of air at the nozzle. Then set pumping speed, nozzle air volume, and orifice size to produce a spray pattern appropriate for the area to be worked. Choose a pattern which will minimize over-spray and waste.

Avoid unduly high pumping speed (volume output) and high nozzle air pressure. Too much speed and pressure make application of uniform thickness difficult, reduce the plasterer's efficiency, and generally produce excessive waste.

Application to a cellular and corrugated steel surface. Set spray machine and nozzle to produce a spray pattern slightly finer and narrower than normally used for browning plaster work. Hold nozzle about one inch from the edge of the corrugations or flutes at such an angle that the spray does not overlap the adjacent cell bottom. Direct the spray parallel to one side of the rib and spray a length that can be comfortably done by swinging the nozzle to the left and then right, applying a good even thickness. Spray with a steady motion, moving parallel to the corrugations so the mortar builds up in the flute to a level flush with the cell bottom. Move in a straight line backward and do the same to the next rib. Continue doing this for several feet and then reverse and do the same to the other side of the rib. All of the surface should be evenly sprayed to make sure all the metal

Fig. 8-47. Fireproofing of structural steel building. (Check local codes on the use of fireproofing material.)

Complete one side first, then the other in the same manner. Finish off flange bottom and then edges in one operation. Avoid unnecessarily high nozzle air pressure. Straighten with hand tools, if required, using very light pressure with the trowel.

Mineral Fibers Fire Proofing and Sound Insulation

Special machines, such as the one shown in Chapter 2, Fig. 2-71, are used for spraying on fire proofing material.

The application and equipment should be in strict accordance with the directions and specifications of the manufacturer.

Continued breathing of the dust that accompanies working with this material is very hazardous to the health of the applicator. He should always use some type of respiratory mask while working with this type of material. It is also advisable to have some type of protection for his eyes. See Fig. 8-48.

is covered to the proper thickness. Fig. 8-47 shows fire proofing being sprayed on.

When flutes have been filled, use the machine as for normal "browning" plaster work. If a smooth surface is required, allow mortar to "take-up" or stiffen and then level with darby or straightedge.

For *uncaged beams* use a fine pattern, spray ¾ of the specified thickness on bottom of flange and then over the web and inner flange surfaces until it is at specified thickness. Sweep nozzle from side to side to achieve a gradual, uniform build up.

Surfaces to receive the sprayed-on material must be thoroughly cleaned of dirt, dust, grease, oil, loose paint and other material which would prevent good adhesion of the spray-on fibers. When it is to be applied to metal lath, high-rib lath should be used. The lath must be drawn tight and be firmly tied. For application to gypsum boards, all joints should be taped or sealed. For all exposed finish applications to lath,

357

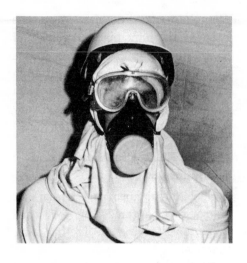

Fig. 8-48. Plasterer properly protected for fireproofing application.

a scratch coat of sand and gypsum is usually preferred.

Before the fiber is applied, check equipment to make sure there is no foreign material in the hopper that may cause an obstruction to the movement of the stator. Make sure all hose connections, both water hose and material hose, are secured tightly. Turn on the water that supplies the pump and fill the drum that will hold the water. Now turn on the machine and then the water pump. Set the water pressure gage to the required pressure for the type of nozzle to be used. Set the volume control on machine for the desired output of fiber. Now turn off the pump and then the machine. The plasterer tender should be ready to turn on the equipment and add material to the hopper at the plasterer's signal. Before signaling, make sure the water is turned off at the nozzle.

After the machines (water pump and blower) have been turned on, turn on the water. The atomized water spray should form a complete circular pattern. This will demonstrate that the nozzle is in working order. With some types of machines the applicator will have complete control of the machine at the nozzle end. (This will include the output as well as the on/off switch of the machine.) Then turn on the switch to feed the material. Take a sample in your hand and squeeze it firmly. If a small amount of water exudes between your fingers, the material is being wetted enough to apply.

If an adhesive is required it should be sprayed on the metal surfaces just before applying the fiber. The fiber must be sprayed while the adhesive coat is wet and tacky. If no adhesive is required, the surfaces are wetted by shutting off the output and spraying with the atomized water. Only wet far enough ahead so that it will not dry before the fibers are applied. This will insure a sound bond of the material to the metal surfaces.

To apply to cellular or corrugated surfaces direct the spray parallel to one side of the rib, spraying a length that can be comfortably done by swinging the nozzle to the left and then right applying a good even thickness. Move in a straight line backward and do the same to the next rib. Continue doing this for several feet and then reverse and do the same to the other side of the rib. All of the surface should be evenly sprayed to make sure all the metal is covered to the proper thickness.

Application to Beams. Direct the spray parallel to the beam from left to right starting at the top of the beam and working down to the bottom flange. Direct the spray down along the top of the bottom flange, then move sideways and do the same. Continue this operation for a short distance and then walk along and cover the edge of the bottom flange. Do the same to the other side.

After both sides have been coated, direct the spray toward the bottom of the beam. Continue this until the entire beam is covered. (If the finish calls for the material to be tamped, use a wooden float about the size of a trowel and tamp the material to the required thickness). Apply a light spray of water over the applied material. (If a sealer is required, this is sprayed over the applied material in place of water. This is usually applied where flaking or dusting must be prevented.)

Machine Care

The machines used for these materials are usually equipped with electrical fuses. It is advisable to have spare ones available on the job. It is also a good idea to have a roll of insulator's tape on hand to make minor repairs to material hose, and to tape the connections. If the nozzle is used that causes the wetting action to take place inside the nozzle, the operator should occasionally clean out the inside of this nozzle to remove any built-up material. If the blower is turned off, the operator should make sure the water is also turned off so it will not slip back down into the hose and wet the material. If material is allowed to become wet while in the hose, and the machine is turned off for any length of time, it could compact and cause an obstruction. It is also advisable to caution the plasterer tender not to feed any excessively damp material or to allow any foreign matter to fall into the hopper.

If the holes in the nozzle should become clogged, the manufacturer's recommendations for clearing the obstruction should be followed.

Acoustic Tile and Grid Ceilings

Noise reduction through the use of sound-absorbing, soft or porous tile is a development of the last thirty years. Today, a great variety of tiles of this kind are available. Manufacturers offer many types and sizes of acoustic tile of rated noise-reduction ability.

Some mention should be made of insulating tile, which is very similar to acoustic tile. Insulating tile, of course, is produced to prevent the passage of heat or cold. Acoustic tile is produced to present a high degree of sound-absorbing capacity. Consequently, insulating tile and acoustic tile may often be similar in appearance. Insulating tile is, however, less expensive than acoustic tile.

Both acoustic and insulating tile can be nailed or clipped on, or they can be stuck to the surface to which they are applied with mastic. Each of these methods of application is suited to given conditions. It often happens that noise reduction is achieved as much by the use of the right suspension method as by the tile itself.

Tile Sizes. Certain types of tile, such as fire-resistant tile and grease-resistant tile, are made to suit specific applications. Sizes conform to architectural needs. Standard sizes are 6″ x 6″, 6″ x 12″, 12″ x 12″, 12″ x 24″, 24″ x 24″, 16″ x 16″, 16″ x 32″, 16″ x 48″, 24″ x 48″, and 48″ x 48″. Thicknesses vary according to the amount of noise reduction the job calls for. Some of the thicknesses are ½″, ⅝″, ¾″, ⅞″, 1″ and 1⅛″. Each thickness has a rated sound absorbing ability.

The smaller sizes of tile are usually acoustic tile. Acoustic tile seldom exceeds 12″ x 24″ in size. The number of companies making acoustic tile for general use are few, but these manufacturers offer a total of more than twenty different types.

Methods of Attaching Tile. Most of the acoustic tile erected is stuck to the base with mastic. See Fig. 8-49. This method of erection is suited to bases such as concrete slab, gypsum board, and plastered metal or gypsum laths. Plaster and gypsum lath provide the best all-round base materials because they are straight, strong, fireproof and dustproof.

It is extremely important that conditions permit a good bond between the tile and the adhesive and the surface to which the tile is cemented. To insure a satisfactory job check the conditions of the backing. No oils, residue from concrete forms or other foreign matter should be on

Fig. 8-49. Applying acoustic tile with mastic. (Insulation Board Institute).

the surface. Surfaces should be level to ¼″ inch in 12 feet, with minimum irregularities. All new concrete surfaces should be primed with a primer recommended by the adhesive manufacturer. Surfaces must be dry. A moisture test should be used if in doubt.

To test for moisture place about ¼ teaspoon of calcium chloride crystals in a small transparent container and tape the open end of the container to the surface to receive the tile. See Fig. 8-50. Be sure to keep it airtight to the surface throughout the test period to prevent room humidity from affecting the crystals. Watch for signs of moisture. If present, the crystals will become damp, stick together and eventually dissolve, or drops of water will form on the side of the container. Tests require from one to three days.

Tile may also be nailed or stapled

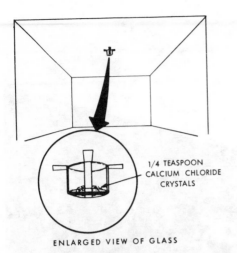

1/4 TEASPOON
CALCIUM CHLORIDE
CRYSTALS

ENLARGED VIEW OF GLASS

Fig. 8-50. Testing for moisture.

Fig. 8-51. Applying acoustic tile by nailing.

to wood furring strips. See Fig. 8-51. This is done on some remodeling jobs and on new work when the ceiling is attached to the joists. This system is subject to seepage of dust through the joints. Moreover, it is not rated as fireproof, since the strips are of wood. Furring strips may also be nailed or tied to suspended metal channels.

Metal clips and special channels also are used for suspension of acoustic tile. These systems provide fire-safe, positive-grip methods of erection, but they are not proof against filtration of dust through the joints.

Tile Layout and Patterning. In applying acoustic tile first lay out the centerlines of the ceiling. All tile work is started from the center of the room in order that a balanced border may be obtained. Centerlines are best established by the bisecting method for squaring a room described in Chapter 6. (See Fig. 6-40 of Chapter 6.) The bisecting method is the most accurate way of establishing straight centerlines.

After the centerlines have been established, the next operation is to determine if the tiles should be started with one edge of the tile on the centerline or whether the tile should be offset with the centerline running directly through the centers of the tiles.

Fig. 8-52 shows a row of tiles (running from left to right) that has been set so the center of each tile is on the center line; the tiles running up and down the figure are set to the edge. In Fig. 8-53 the tiles have been set with each of the edges of the top row of tiles on the center line. In this case you can see that the layout as shown in Fig. 8-53 is less satisfactory than the one shown in Fig. 8-52 because one of the tiles, near the border, is just 7½″ wide at the wall line. Consequently it does not have as good an appearance as the layout in Fig. 8-52 presents. By offsetting the tiles as was done in Fig. 8-52 a large section of tile is left at the border.

Tile placement depends upon the

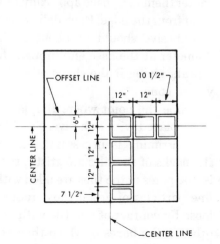

Fig. 8-52. Tile plan with tiles offset.

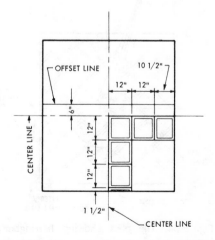

Fig. 8-53. Tiles set with edges on center line.

size of the ceiling you have to work with. It is best, therefore, to do some measuring after the centerlines are established to see how many full-sized tiles will fit. Then decide whether it is best to set the first row of tiles over the centerlines or whether to set the tiles' edges next to the centerline.

Many variations of tile patterns may be used. One interesting tile layout might be called the "herringbone" because it resembles the cloth weave also called by that name. See Fig. 8-54. The herringbone design is one of the hardest to lay out, but it is a very interesting design, especially when using tiles that are twice as long as they are wide.

Five possible starting points are shown in Fig. 8-54. After the center-

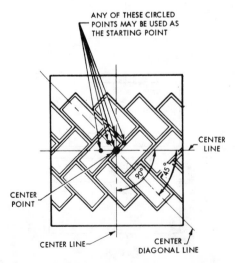

Fig. 8-54. Tile plan showing herringbone layout.

lines are placed, a line is established at a 45° angle cutting through the center point. This 45° line is a guide for the long dimension of the tiles to be attached.

Only by laying out a scale drawing of the ceiling can the proper starting point be found. Remember that in this pattern if the design is shifted backward or forward, it affects the size and the sides will work out with unbalanced pieces.

When applying the tiles in a diagonal design, care must be used to prevent locking them in a position that will make it hard to place the next row of tiles. Only by working back and forth along one side can this difficulty be avoided.

Applying Tile with Acoustic Tile Adhesive. When applying acoustic tile, use four dabs of adhesive per 12″ tile as shown in Fig. 8-55. Place the dabs of adhesive near the corners and center them in a space approximately 2″ in from the edge. Each dab or spot of adhesive should be about 2″ in diameter at the base and brought to a peak, giving it the shape of an inverted cone.

Prime the point where the dab is to be placed with adhesive. See Fig. 8-55. Priming is necessary because the backs of the tiles are gritty; that is, the pores of the tiles are filled with fine particles of grit or dust rubbed loose by contact of one tile with another in the boxes in which they are packed. (Adhesive priming is the rub-

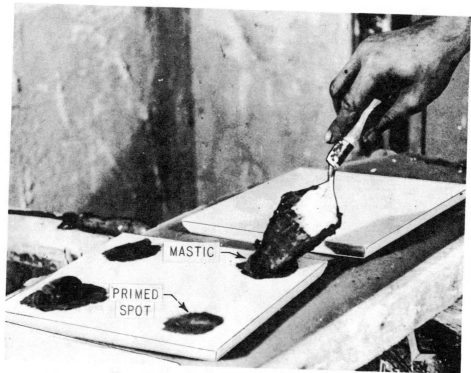

Fig. 8-55. Applying mastic to reverse side of tile.

bing of a very small amount of adhesive into the tile where the dabs of adhesive will be added. Fig. 8-55 shows the primed spot.)

A pointing trowel or a putty knife is used to rub a small amount of the adhesive into the tile. Over this primed area place the dab of adhesive. Use a twisting motion while applying the adhesive to prepare a well-formed dab. Winding the pile of adhesive into a pigtail also will do the trick.

Some plasterers make an applicator of a section of 1½″ pipe 16″ long.

They cut away about one-half the circumference of the pipe for a distance of about 10″ of the length of the pipe. They then round the end of their trowel to fit into this hollow portion of the pipe. They use it to scoop up the adhesive, then wipe off the excess amount with the trowel. Finally, taking their trowel and setting it in the groove, they push off the adhesive onto the tile. They find that by doing this all their adhesive dabs are the same size. Fig. 8-56 illustrates the manufacture and use of this holder.

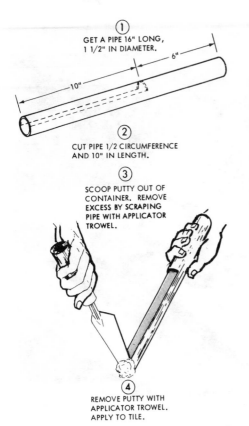

① GET A PIPE 16" LONG,
1 1/2" IN DIAMETER.

10"   6"

② CUT PIPE 1/2 CIRCUMFERENCE
AND 10" IN LENGTH.

③ SCOOP PUTTY OUT OF
CONTAINER. REMOVE
EXCESS BY SCRAPING
PIPE WITH APPLICATOR
TROWEL.

④ REMOVE PUTTY WITH
APPLICATOR TROWEL.
APPLY TO TILE.

Fig. 8-56. Handy putty applicator for setting acoustic tiles.

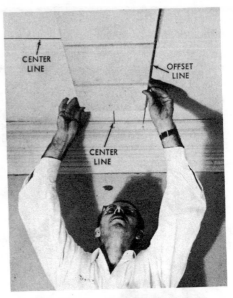

CENTER LINE

OFFSET LINE

CENTER LINE

Fig. 8-57. Pressing tile into place.

Do not butter many tiles in advance of installing them; a skin will form on the adhesive dabs and this will weaken the bond. Clean or chalk your hands or wear white gloves before handling the tiles.

As the tile is placed in position, it should be moved back and forth gently so as to insure a good bond. The tile should be in position and level after working it into place. Fig. 8-57 shows a tile being pressed into

place. Note that the mechanic is placing the tile with its edge on the offset line. Press the tiles in place using the edges. Never press the tiles in at the center when putting them in place. Temporary distortion may be produced in this manner. Then later, as the tile moves slowly back into a flat plane, the adhesive loses its hold and the piece of tile may fall.

If tiles are used that call for splines, drop back after installing the first row of eight to ten tiles and insert these splines, setting them in the kerfs of two adjacent tiles. See Fig. 8-58 showing installation.

Some types of tile are subject to swelling or shrinking because of moisture conditions in the room. The

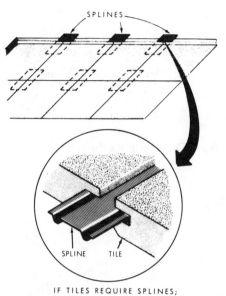

SPLINES

IF TILES REQUIRE SPLINES;

① AFTER INSTALLING A ROW OF TILES, RETURN TO STARTING POINT.

② INSERT SPLINES INTO KERF OF TWO ADJACENT TILES.

③ REPEAT PROCEDURE UNTIL CEILING IS COMPLETED.

Fig. 8-58. Installing tiles with splines in edges.

Fig. 8-59. Nailing acoustic tile to furring strips.

presence of excessive moisture will cause the tile to swell. Unusual dryness will cause shrinking. To avoid either of these two conditions, do not bring tiles to the job and do not erect them until the humidity in the building is about normal.

Place the tiles in the room in which they are to be used a few days before they are applied. This permits them to adjust to the normal atmospheric conditions of the building.

Applying Tile by Nailing. Acoustic tile may also be nailed to furring strips that are spaced far enough apart to meet the width of the size of tile used. In perforated tile, the nails can be placed in holes located near the corners. This hides the nailheads, which are driven into position with a nail set.

Plain tile must be nailed at the edges. Drive the nails in at an angle to prevent each tile from working loose. Use four nails per tile, placing each near one of the corners. (Note that a hand stapler is also sometimes used for this.) In Fig. 8-59 a mechanic is nailing plain acoustic tile to furring strips. Note that there are three strips supporting the tile. In Fig. 8-51 of this chapter only two furring strips per row of tile are used. Three furring strips, of course, would

367

Fig. 8-60. Marking tile for cutting.

give a more solid base for the tile; three are, however, more expensive.

Cutting Tile. To fit tiles to border lines around the ceiling and near the walls, often each tile must be cut to size. To achieve a good tight fit requires careful marking and cutting. In Fig. 8-60 the mechanic is holding a tile in place to mark it. Note that the tile is bottom side up. The tile is held bottom side up in order that it may be more accurately measured and marked.

When the tile has been marked at both edges and cut, it is placed in position top side up. If care was used in marking and cutting the tile, it will fit perfectly even though the wall surface is not straight or parallel. Bevels may be formed by sanding. Use sandpaper wrapped around a block of wood for this procedure.

If molding is to cover the border joints, cut the border tile ¼″ shy of the wall joint.

Suspended Grid System—Acoustical Ceilings

There are many installations incorporating the suspended grid system, which may fall under exposed grid (direct-hung suspension system) or concealed grid. These are for acoustical ceilings only. (The ex-posed grid system will show metal. Concealed zeebar is a system which shows no metal; only tile is visible from the floor.) In all cases the recommended procedure for installation should be followed.

Study the blueprint to determine the correct height for the ceiling. Check electrical prints for light fixture layout. Using a water level, place bench marks on each of the walls at the corners. These marks will provide an accurate starting point from which to measure to the desired ceiling height. Now measure up from the benchmarks to the specified ceiling height and then add the height of the wall molding to be used and mark at both ends of each wall. See Fig. 8-61, top. Snap a chalk line using these two marks as points to hold the end of the line. Note: It is important to set the chalkline level with the intended height for the top edge of the molding, so that none of the chalk line will be visible after installation is complete. Fasten angle moldings to walls, making sure the top edge of the molding touches the chalk line at all points on each wall. See Fig. 8-61, bottom.

Exposed Grid

After squaring the room, determine border panel sizes; hang lines as guides (nylon is the best to use) in positioning the first main runner and cross runner. See Fig. 8-62. Install the main runner line approximately parallel to the centerline and

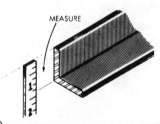

MEASURE

① ADD MEASUREMENT OF MOULDING TO REQUIRED CEILING HEIGHT. (USUALLY 3/4" TO 1 1/2")

② PLACE MARK AT BOTH ENDS OF WALL AND SNAP A CHALK LINE FOF INSTALLATION GUIDE.

① FASTEN MOULDING TO WALL WITH SCREWS OR NAILS.

② MAKE SURE MOULDING FITS SNUGLY TO WALL FOR ELIMINATION OF SOUND LEAKS.

③ MITER CORNERS FOR A TIGHT FIT OR USE CORNER CAPS.

Fig. 8-61. Suspended grid system for installing acoustical tiles.

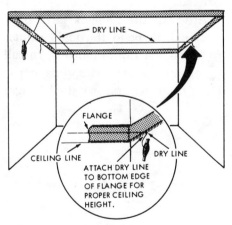

DRY LINE

FLANGE

CEILING LINE

DRY LINE

ATTACH DRY LINE TO BOTTOM EDGE OF FLANGE FOR PROPER CEILING HEIGHT.

① PLAN TO INSTALL MAINRUNNERS PARALLEL TO WALLS AND AT A DISTANCE EQUAL TO DESIRED BORDER PANEL WIDTHS.

② STRETCH DRY LINE TO INDICATE WHERE TO HANG MAINRUNNERS.

③ DRY LINES SHOULD INTERSECT AT RIGHT ANGLES TO OTHER LINES.

NOTE: NYLON IS USUALLY USED FOR THE DRY LINE

Fig. 8-62. Procedure for constructing exposed grid.

at a distance from the wall equal to the desired width of border panel. (These *panels* are also called pads, lay-in panels or ceiling boards.) The line should be level and set 1½″ above the flange of the wall molding, for it will also be used to measure the lengths of hanger wires. Then install the second guide line (dry line) at a right angle to the other line where the first cross runner will be installed. Install the first hanger at the point where the two lines cross and install all others four feet *on center* (center to center). Using the long line as a guide, mark each hanger where the line touches before moving the long line over four feet to install more hangers.

After all the hangers have been installed, suspend the main runner from the first row of hangers. Install it so that the nearest end runner is hung from the first installed hanger

wire. In order to place slots for cross runners in proper location for the layout, establish the length of the runner in each row. Cut the runner, if necessary, so that after lining up the cross runner slot in correct position for the cross runner, the main runner will rest on the wall molding. If length of room in the direction the main runners are installed is greater than the length of these runners, splice two or more runners together. Various types of splices are supplied by the individual manufacturer. After the ends of main runner rest on the wall molding of two facing walls, then loop all hangers in the rough insert holes and over the top of the runner. Twist wires and secure. See Fig. 8-63.

Alignment and leveling of main runners is all important in installing grid ceilings. Install the first cross runner directly above the dry line,

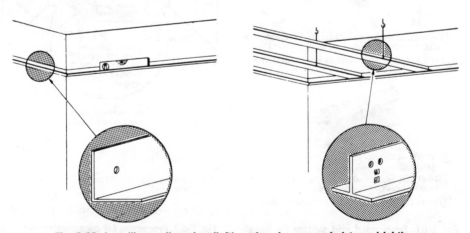

Fig. 8-63. Installing wall angles (left) and main suspended tees (right).

Fig. 8-64. Steps in installing suspended ceilings.

measure the distance between the main runner and the wall molding and cut the cross runner to this measurement. Lay the end that was just cut onto the wall molding so the main runner and cross runner are perpendicular to each other, then lock them together with a pin. Fig. 8-64, top, shows a cross runner being installed. It is recommended that every other cross runner be pop riveted to the wall molding. This helps to stabilize the ceiling. Continue in this manner until all the cross runners and main runners have been installed.

Now install ceiling panels as the job progresses. See Fig. 8-64, bottom. It is good practice to wear gloves or

rub the hands with white chalk before handling the pads. (Most tiles are white and white chalk will not mar surfaces.) Cut border tiles with a razor knife to fit as required. Make certain all pads lie flat and snug in the grid. Sometimes hold-down clips are used to secure the pads tightly.

Concealed Zeebar

The concealed zeebars with splines provide an unbroken, monolithic appearance which is extremely popular today. This system uses $1\frac{1}{2}''$ channel which is suspended by the lather.

The zeebars are attached with clips at right angles to the $1\frac{1}{2}''$ carrying channels (see Fig. 8-65) at intervals specified, usually 12" on center, wherever the zeebars come in contact with the channel, generally

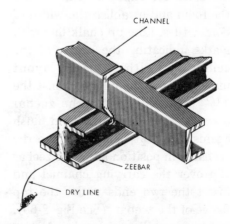

Fig. 8-65. Concealed zeebar. (1) Center zeebar over dry line; hold flush against channel. (2) Attach zeebar clip over channel; insert both ends under top flange of zeebar. (3) Fasten additional clips to zeebar if it intersects with channel.

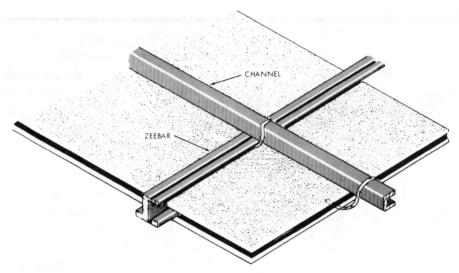

Fig. 8-66. Supporting tiles during installation. Use temporary clips when possible to provide support for tiles until next zeebar is inserted. This prevents bending of zeebar flange.

four feet on center. Wall moldings are installed in the same manner as described for exposed grid. Square the room and establish the width of border tiles. Line up chalk line with marks indicating the center of the ceiling; or, with adjusted layout marks, snap a chalk line against the 1½″ channel. Center the zeebar over the dry line and hold flush against the 1½″ channel. When the zeebar is in position hang the zeebar clip over the carrying channel and insert the two ends under the top flange of the zeebar. (See Fig. 8-65).

After installing about ten feet of the first zeebar, begin placing tile into position. Line up the first tile with the centerline, running at right angles to the zeebar. Ease the kerf

of each tile gently over the zeebar flange, without forcing. Insert splines into all tile kerfs that are at right angles to the zeebar. Temporary clips should be hung over carrying channels to provide support for tiles until the next zeebar is inserted into the fourth kerf (fourth side of tile) of these tiles. This prevents the weight of unsupported portions of tiles from bending the zeebar flange. See Fig. 8-66.

For installing border tiles between zeebars running at right angles to the wall molding, cut tile to the desired border width. Insert the tile into the flange of the previously installed zeebar. Insert the next zeebar into the kerf of this tile on the opposite edge. Place a spring steel spacer between the molding and the border tile. Let the border tile rest on the molding. The spring steel spacer will keep the tile from drifting.

For installing border tiles between the last zeebar and the molding running parallel to it, cut tile to the desired width. Lay the tile on the wall molding and compress it against the spring steel spacer. Let the tile spring back out and ease the kerfed edge into the zeebar flange.

Checking on Your Knowledge

The following questions give you the opportunity to check up on yourself. If you have read the chapter carefully, you should be able to answer the questions. If you have any difficulty, read the chapter over once more so that you have the information well in mind before you go on with your reading.

DO YOU KNOW

What are the steps used by a plasterer to produce an imitation brick surface?

2. Describe three ways of producing imitation stonework.

3. Name and describe the means used to produce four stucco textures.

4. How does a plasterer prepare the backing for exposed aggregates?

5. How does a plasterer plan for the application of exposed aggregates?

6. How does a plasterer apply the emulsion-type exposed aggregates?

7. Name three types of exposed agggre-gates using a matrix to which aggregates are seeded, and how are they each applied.

8. How does a plasterer develop murals or designs?

9. How does a plasterer apply the plaster finish to a swimming pool?

10. Describe how a plasterer applies one type of direct-applied fire insulation.

11. How does a plasterer prepare a tile layout and patterning?

12. Describe the method of installing a direct-hung suspension system for acoustic grid.

Ornamental Plaster

This chapter will cover the basic areas of ornamental plaster work and will endeavor to give the reader an understanding of the various processes involved in developing each area.

There are four subdivisions within the general heading of ornamental plastering. (1) Plain cornice work, either run or stuck in place. (2) Enriched cornice work: combining a plain cornice with cast ornament embedded in it. (3) Cast ornamental work: running ornament, coffers, strap work, columns, caps, etc. (4) Modeling and casting: making original models and reproducing them in glue, wax, rubber and plaster piece molds. Each of these four subdivisions has many specialized areas. This chapter will cover plain and enriched cornice work. The more advanced work of casting and modeling may be studied by reference to spe-

cialized texts in the field. (Appendix B lists reference texts for basic and more advanced ornamental plastering.)

The terminology used in discussing ornamental plastering will be explained as the items are discussed. Keep in mind that various areas of the country use different terms or words to describe a particular step, operation or method. Most of the general terms used will be covered.

Basic to all work of this kind is the architect's drawing of what is required. In some cases the modeler and plasterer must make shop or working drawings from these architectural drawings to do their particular work. For example, when a cornice is shown on the architect's drawing, he details only the actual members. If it has ornamentation he will show that in profile (cross section) and elevation (face view).

The modeler then must develop a cross section detailing the pockets (recesses) to be formed in the cornice to receive this ornament and the additional recess required to provide room for the plaster used to stick (glue) the ornament in the cornice.

The plasterer working from the architect's drawings produces mold drawings to suit the particular conditions found on the job site. He may have to adapt the same mold to a number of situations changing the method of *running* the mold to suit the conditions. (*Running* means to form the cornice with plaster, using the mold in repeated runs or passes to develop the members of the cornice to shape.)

The caster is the craftsman who works from the original models made by the modeler and makes glue (gelatin), rubber or plaster piece molds to reproduce as many pieces of ornament as will be required for the job. The plasterer then sticks or places these ornaments in the cornices or other areas on the job.

Some classes of work may require the services of the architectural sculptor (an artist who produces original art, usually statuary or similar art forms, in clay or other media. The difference between the sculptor and the modeler is that the modeler usually works in developing the basic ornamental forms such as running ornament (ornament that repeats its motif at regular intervals), geometrical forms and natural elements of nature such as leaves, flowers, etc. One is classified as an artist; the other a craftsman.

In many areas of the country cast ornamental work is called *staff work* to separate it from plain cornice work run in place. It was called staff work because, in many cases, the pieces were supported in place with sticks or staffs until the plaster used to stick them in place had hardened or set.

Plain Cornice Work

Making a Cornice Mold

Step number one in any cornice work is to make the mold or metal form which is the exact reverse of the cornice to be run. On the blueprint or drawing showing the full size profile of the cornice, line out a distance of about 2″ from the last member of the cornice on the ceiling and about 1″ from the bottom member on the wall.

These marks are the top and bottom ends of the mold. (See Fig. 9-1, top left for the "ceiling line" and bottom right for the "wall line.")

Place the drawing over a piece of 26 gage galvanized sheet metal and fasten both down on a smooth work

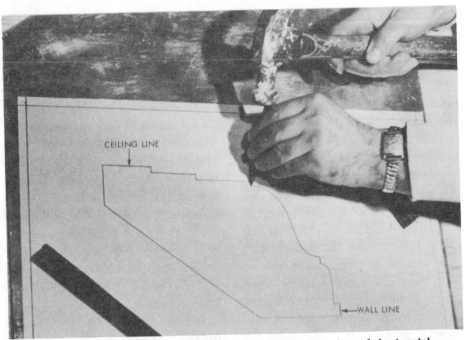

CEILING LINE

WALL LINE

Fig. 9-1. Transferring cornice profile from the drawing to a piece of sheet metal.

bench by nailing a piece of cornice strip on each side.

Using a sharp prick punch, as shown in Fig. 9-1, pin point light punches through the drawing and into the sheet metal. Do not make holes, just visible pin points; do this at regular intervals about ¼" apart on the curved lines and on the straight lines only at the corners. In Fig. 9-2, the drawing has been removed and the profile (outline of the cornice) is shown punched in the metal. With a soft lead pencil carefully trace the profile from point to point so that it is clearly indicated on the metal. Some plasterers use

carbon paper to trace the profile directly from the drawing. Another method used is to transfer the profile from the blueprint to a heavy paper and paste this on the metal.

Using a pair of sheet metal snips, cut out the profile following the line accurately; do not cut beyond the lines or too far into the corners. Some plasterers prefer to use aviation snips which can be obtained in both right and left hand cutting types. These snips make it easier to cut around some of the intricate shapes. Fig. 9-3 shows the snips cutting the metal on a curved line. Notice how the metal is held in one hand and, as the

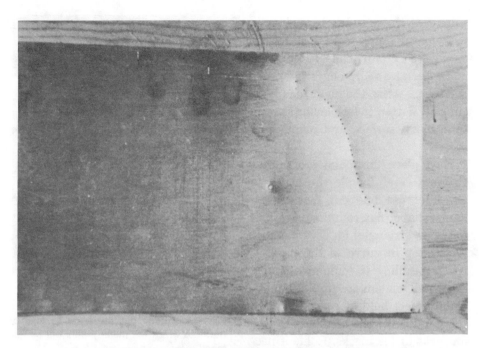

Fig. 9-2. Sheet metal profile outlined and ready to be cut.

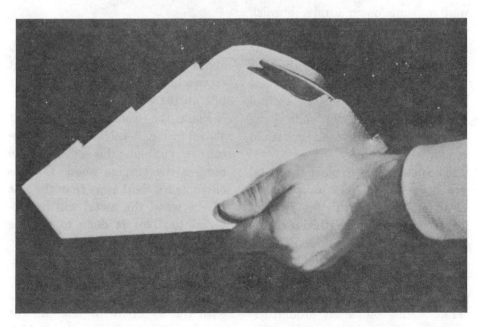

Fig. 9-3. Cutting out the metal profile using sheet metal snips.

snips cut, the metal is turned so as to help the snips work around the curve. Be careful while cutting the metal because sharp edges can cause deep cuts. After the metal has been cut out, flatten it by placing it on a solid metal surface and hammering it flat.

Next place the metal profile in a vise and carefully file the metal to the line; use fine cut files of the proper shape (flat, round, square, half round and triangle) for each member (each step or curve in the profile). After the filing is completed, run a large nail or similar round smooth metal object held at right angles to the profile over the members to test for nicks or file marks. Refile if necessary. Then, using the same tool and pressing down firmly, burnish the metal edges to a final smooth surface. Tipping the nail from side to side while drawing it over the edge will help to remove any burrs left on the sides by the file. Be careful not to let the nail slip while doing this operation, as a serious cut could result.

The next step is to place the metal profile over a ¾″ thick piece of good clear pine lumber with the grain of the wood running lengthwise to the longest part of the profile. This is done to provide the greatest strength to the mold in the line of the greatest stress when running the cornice. Trace a line completely around the metal on the wood. At the ceiling

Fig. 9-4. Cutting out the wood backing or horse to which the metal profile will be fastened.

line or top of the mold, mark off about 1″ and then from this point trace a line ¼″ back from the profile line on the wood.

Place the wood into a vise as shown in Fig. 9-4. Now, using a coping saw, cut along this second line to remove the surplus wood. Do this cutting on a slant away from the side of the wood the metal will be fastened to. This is done to provide more clearance behind the metal so plaster will not build up there when running the cornice. The metal is normally applied to the wood backing with the metal to the left and the wood to the right when the mold is

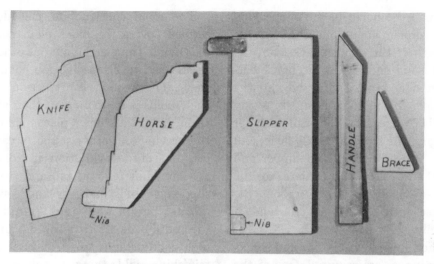

Fig. 9-5. Individual parts of the mold ready to be assembled.

held vertically with the top part up. When the mold is completed it will move from right to left and is, therefore, a right-handed mold, normal for most plasterers.

Fig. 9-5 shows the various parts of the mold ready to be assembled. The knife or metal profile is now ready to be nailed to the wooden horse. Use the prickpunch again to punch a series of holes for the ½″ or ⅝″ brads that will be used to nail the metal to the wood. These holes should be made about ½″ back from the edge of the metal; make the holes about ½″ apart along the member edge and approximately 2″ apart for the rest of the mold. At the ceiling line of the horse nail a 1″ wide x 1½″ long strip of sheet metal across the top of the wood. This strip will act as

a glide permitting the mold to slide smoothly over the ceiling screed without cutting into it. This metal strip is called a *nib* or *shoe* by the trade.

Now nail the knife to the horse, making sure the ceiling line is flush with the ceiling nib and the bottom line is flush with the bottom of the horse. The member edge or forming edge of the knife will then project ¼″ beyond the wood everywhere except at the ceiling nib.

The *slipper* or bottom section of the mold is a piece of ¾″ thick pine about one third longer than the longest part of the horse and wide enough to support the horse securely. Nail a nib or shoe at each end about ¼″ back from the ends. (See Fig. 9-5.) Mark a square line (to use as a guide

379

to nail the horse to the slipper) on the slipper about one third of the length of the slipper back from the left hand end. Place the horse bottom side up in a vise and nail the slipper to the horse using six penny nails. Make sure the wall member edge of the knife is exactly flush with the face or edge of the slipper and that the slipper is square to the horse. Some plasterers rabbet the horse into the slipper to hold it firmly.

For a mold of this size and shape, two braces are required. One of the braces serves as a handle. See Figs. 9-5 and 9-6 for size, shape, and how these braces are attached to the slipper and horse. Make sure the horse is set square to the slipper when nailing the braces in place; use a steel square to check for squareness. Also make sure the braces do not cross beyond any of the knife's members and check to see that there is ample room for the hand that will be placed around the brace when pushing the mold. Larger molds and various other designs will require more braces plus different overall construction.

Fig. 9-6 shows all the parts of the mold and the completed mold as seen from the back. The slipper nibs, where they cross the slipper, should be hammered into the wood so their sharp metal edges will not cut into the wall screed as the mold is pushed over it.

The mold construction just described covers the basic plain cornice mold. There are numerous other types used to fit specific job conditions. The general construction of such molds is the same, but their attachment of the slipper, placement of the bracing and other supports varies. Some of these variations in molds are named as follows: tail and twin slippered molds, hanging and soffit molds, bench and panel molds, hinged molds, raking molds, pin molds, adjustable splay molds, diminishing molds (hinged or sliding), miter molds. Some of these molds will be shown in actual use forming the work they were specially designed to do. With certain modifications some of these molds are used in numerous situations, each suited to a particular job or condition. The plasterer must be alert to these variations and be able to pick out the type best suited to each job condition.

For all cornice molds the proper bracing and stiffening of the mold is most important. A poorly braced mold will vibrate and *chatter* (violent vibration creating ripples in the cornice face) due to the swelling of the setting plaster when the cornice is in its last stages of completion. A strong mold will resist this action. A mold must be studied carefully during construction to spot any weak points. All weak points should be properly braced.

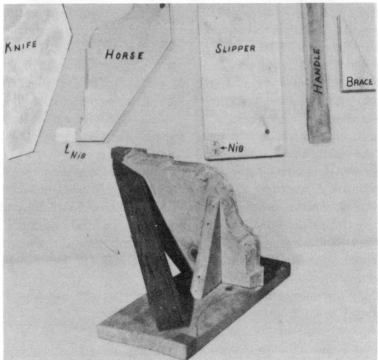

Fig. 9-6. Completed cornice mold and all the individual parts.

Dotting and Screeding

Dotting and screeding for cornice work use the same basic techniques covered in Chapter 6 on dotting and screeding a room for the brown coat. Instead of using brown mortar for the dots and screeds, the plasterer uses putty and gauging plaster mixed approximately fifty-fifty or half putty and half plaster. This mix, called *high gauge*, provides the hardness required to permit a cornice mold to run over the screed without cutting into its surface.

When a cornice is to be run in a room or area, the browning is not done, in most cases, until the cornice is completed; then the screeds used for the cornice are used as the guides for the browning operation. There is an exception to this rule and that is in residential and light commercial fields. In these fields the area is browned out first and the cornice is run afterwards, because the degree of accuracy is generally not as important as it is in larger installations.

Most cornice work is done in areas where the plaster base is metal lath. A scratch coat is applied and then the dotting and screeding can proceed. Common practice is to plumb dots around the walls, screed to these dots, then establish the *slipper line* (underside of the slipper, bottom of the mold or the cornice strip line) by placing the cornice mold on the wall screed down from the ceiling about ½″, then marking the slipper line directly under the horse section of the mold. This point is used to prevent errors which could occur if the slipper is not set perfectly level on the screed.

If the work calls for the cornice or ceiling to be a specific height then the mold must be set on the screed at this point. Water level this point around the room, check for accuracy

by cross leveling and returning exactly to the starting point. Lines at the corners must intersect perfectly or there will be serious trouble when the *mitering* (completing the cornice into the corners where the mold could not run because the slipper's projection) is being done. Use a hand level to carry the water level lines into the corners.

There are many occasions when the plasterer must produce screeds on extremely long walls where there are no grounds to plumb up from or the grounds may not be lined straight enough to use for in-line dots. For these conditions the method of setting small or pin dots to a line is used.

When setting dots to a line, the wall line must be established at each end of the wall either by plumbing up from an established point or set from centerlines if the room is to be perfectly square. Fig. 9-7 shows how

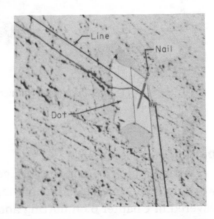

Fig. 9-7. End dot with nail placed under line.

a small dot or pin dot has been established at one end of the wall and a chalk line fastened behind it and stretched over it to the other end of the room. Notice that a nail has been placed between the line and the dot. This is done at each end, permitting the line to clear all the intermediate dots. Each intermediate dot is then checked for being in-line by testing it with a similar nail as shown in Fig. 9-8.

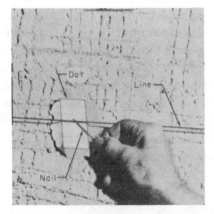

Fig. 9-8. Line should clear dot the thickness of a nail.

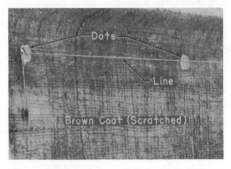

Fig. 9-9. Dots placed in line to establish a straight screed.

Fig. 9-9 shows how the intermediate dots are set to the line. These dots are spaced so that the long straightedge will cover from one dot to the next one. When all the pin dots are in place the dots are enlarged by building up additional material below or above the pin dots and pressing a *plumb dot* (a dot approximately 6″ long pressed vertically to the pin dot) using a hand level and a piece of paper over the dot to prevent the level from sticking to the plaster. Form screeds around the room to these dots and finish them to a true smooth surface.

When the water level lines have been established and leveled into the corners, snap chalk lines from corner to corner or to intermediate points on the screeds. Now nail cornice strips ($\frac{5}{16}$″ x $1\frac{5}{8}$″ x 16′ clear pine strips) to this line keeping the top of the strip exactly even with the line. Space nails about 18″ apart. Butt the strips tight into the corners and at joinings. Make up enough high-gauged plaster and daub spots of plaster over the lower part of the strip and on to the screed. Space these daubs about 12″ apart to support the strip firmly and to keep it to the line while running the cornice.

The next step is to press ceiling dots using either the mold itself or a gage made specifically for this purpose. For a single room the mold can be used to save time. If identical cornices are to be run in various areas, it is better to make a gage. Use of the gage will make the work of pressing the dots easier and more accurate.

When using the mold for this purpose, nail a short piece of ¾" x 1" wood about 6" long across the top face of the mold and set even with the ceiling nib and centered on it. Now tack a piece of cornice strip about 12" long vertically on the face of the mold so that it is exactly parallel with the vertical members of the mold. This will be the guide against which a short level will be held while pressing the ceiling dots. Fig. 9-10 shows how these strips are fastened to the mold.

With the mold prepared, place it on the wall strip and check around the room using the hand level held against the vertical strip to plumb the mold to see that there is clear-ance all around the room for the screeds. Cut strips of newspaper about 2" wide and 8" long to place over the soft plaster of the dot so when the dot is pressed into position the mold or gage will not stick to the plaster.

Now prepare enough high-gauged plaster to press the required dots on the ceiling. With the trowel, form wedge-shaped dots at the screed line spaced a straightedge length apart. Place a strip of paper over the dot and, holding the mold on the cornice strip and the short level against the vertical plumb strip, press the mold's top strip into the soft plaster dot until the level's plumb vial shows the mold to be perfectly plumb. Repeat the operation at each dot, then pull the paper off the dots and trim off the surplus plaster that has squeezed out from the sides.

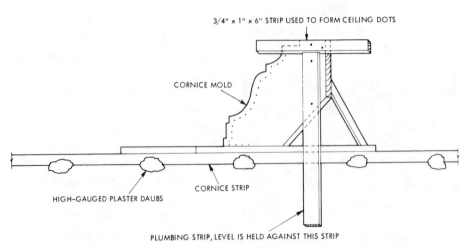

3/4" x 1" x 6" STRIP USED TO FORM CEILING DOTS

CORNICE MOLD

HIGH-GAUGED PLASTER DAUBS

CORNICE STRIP

PLUMBING STRIP, LEVEL IS HELD AGAINST THIS STRIP

Fig. 9-10. Strips are fastened to cornice mold to form ceiling dots and hold level for plumbing.

Check each dot again with the mold and level to make sure the dots did not sag when the pressure of the strip was removed. A slight over-pressing to compensate for this action can be learned with practice.

The same operation can be performed using a gage made to the same projection and height as the cornice mold. With the gage it is more accurate to fasten the level on the horizontal strip, as this usually provides a longer distance to level and is easier to watch as the gage is pressed upward into the dot.

With the dots pressed, trimmed, and checked, the next step is to form the ceiling screeds. Use the same procedure used to form the wall screeds. Because the cornice mold has only one nib which runs on the ceiling screed, this screed must be very carefully finished so there are no holes, scratches or other imperfections. The ceiling nib of the mold will dig into any screed imperfections, gradually enlarging them and creating a kink in the cornice. The need to produce a perfect ceiling screed is extremely critical to the final outcome of the finished cornice.

There is another method used to screed for cornice work, as mentioned previously. This is used primarily in the residential and light commercial field. For this work, it is common to brown the walls and ceilings. When this coat is dry, screed a thin coat of putty and plaster over the brown coat, straightening out the screeds with a featheredge. Darby the screeds with a second coat of putty and plaster to bring them to a smooth finish.

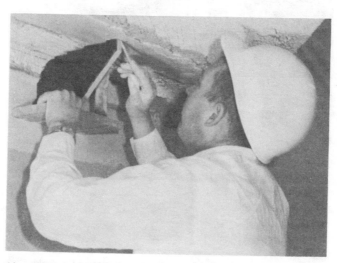

Fig. 9-11. Marking the top and bottom lines of the mold to establish the cornice strip lines.

Fig. 9-11 shows the plasterer setting the cornice mold on the screeds. The mold is placed as squarely as possible to both screeds. Fig. 9-12 shows the plasterer using a level to check the squareness of the mold as set to the screeds. Mark the top back edge of the mold on the ceiling screeds as shown. Also mark the bottom of the slipper on the wall screed. Do this near each corner of the room on the two opposite screeds only.

Now use a hand level to carry the wall marks into the corners and level around the corner on to the other walls. Set the mold on these marks drawn on the walls opposite to those previously marked, and mark the ceiling lines. Snap chalk lines to all these marks and nail the cornice strips in place on the wall line, using only enough nails at this point to hold the strips in place temporarily.

Nail the strips securely to the wall line in the corners now and make sure they are in perfect alignment, then place the mold on the wall strip near the corner and start to slide it along the screed, watching the ceiling line (used to produce a straight cornice) to make sure the mold is on the line. One man pushes the mold and watches the ceiling line while the other nails the strip at about 12″ intervals as shown in Fig. 9-13. If the mold's top edge moves *inward* over the line, then the man nailing pushes the strip upward, forcing the top of the mold outward until it meets the line again. If the mold's top edge moves *outward* over the line, the wall strip is pushed down until the mold

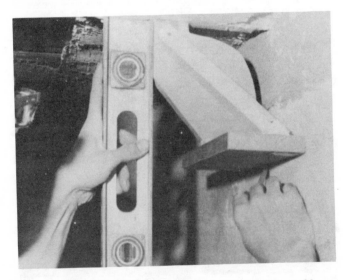

Fig. 9-12. Plasterer checking the squareness of the mold.

Fig. 9-13. Nailing the cornice strip to the position set by the mold as it follows the ceiling line.

Fig. 9-14. Strips nailed in place and the mold set on the strips ready to run the cornice.

moves back to the line. Continue this operation around the room, nailing the cornice strip securely as the mold is moved along. Fig. 9-14 shows the strips nailed in place and the mold set on the strips ready to run the cornice.

The technique involved here will

develop a straight ceiling line while the variations in the two screeds is taken up in the wall line. The ceiling line of any cornice will show screed defects to the eye quicker than the wall line will show them. If there should be too much movement needed in the cornice strip to overcome screed hollows or bumps, the screeds must be corrected in that area before the cornice can be run.

This method will produce an acceptable cornice, appearing straight to the eye. For most classes of work, if the screeding has been done with reasonable care, this will be considered a good job.

Running a Cornice

Running a cornice means producing a finished product by filling in and repeatedly passing the mold over the area until the members of the cornice have been formed completely and the cornice is polished to a smooth surface. The cornice shaped by the mold is developed by blocking out the run; that is, filling in the cornice area with high-gauged plaster until the members of the cornice begin to develop. For blocking out, the gauging is made quite stiff so the material will hang in place readily.

Blocking Out. Two methods are used by the plasterer to block out a

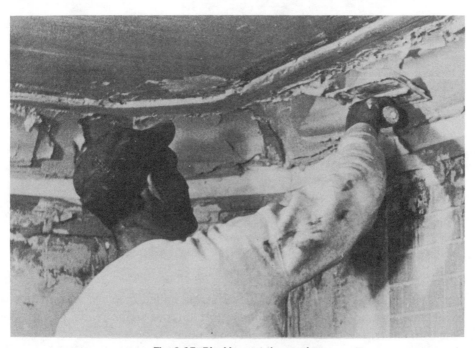

Fig. 9-15. Blocking out the cornice.

cornice. In one method material is applied with the trowel in small strokes across the scratch-coated cornice outline, as shown in Fig. 9-15, until the mold begins to form some of the members as it is run over the screeds. Fig. 9-16 shows the mold forming the cornice members. One man usually pushes the mold while the other catches the surplus material with his hawk held at the front of the mold as it is moved along.

Fig. 9-17 shows the plasterer spotting hollow places with some of the plaster the mold has just cut off. This operation is repeated a number of times until the material becomes too stiff to use. By this time the cornice should be roughly shaped, or as it is called, "blocked out."

The other method used to block out a cornice requires the plasterer to use a full trowel of material. This is thrown onto the scratch coat with a snapping action of the wrist, gradually covering the area with thick daubs of plaster. The area is built up this way until the mold, when run over the screeds, begins to show member formation. This method is particularly effective where the cornice is large or there are many members and large projections.

This second technique requires much practice and is dangerous to the eyes if glasses are not worn. However, it is fast and will cover hard-to-get-to places with ease.

Finishing. When the blocking out is completed, a job that may re-

Fig. 9-16. Mold forming the members of the cornice.

Fig. 9-17. Spotting hollow places in the cornice.

quire more than one mixing of material, the next step is to prepare a new batch or mix of putty and plaster. This should be so *retarded* (set of the plaster slowed) that the material can be applied a number of times to the cornice (to bring out the members to their true shape) before the material is too stiff to use.

Two methods are used to apply this finishing material. Both methods are called *stuffing the mold*, and consist of applying the soft plaster against the face of the mold and the rough members of the cornice while pushing the mold over the screeds until the cornice is completely formed and the surfaces polished.

The most common method is to hold the soft material in place with the trowel while the mold is pushed against it. The pressure applied to

the trowel forces the soft plaster to fill the holes and slack places in the roughly formed cornice. This operation is usually done by one plasterer, who replenishes his trowel with more material as needed from his hawk, while his partner pushes the mold as shown in Fig. 9-18.

The other method is to wear a large gauntlet-type rubber glove and to stuff the mold while holding a handful of plaster against the cornice, as seen in Fig. 9-19. One man can stuff a small cornice this way quite readily by pushing the mold with his right hand and stuffing with his left. Do not attempt to reuse the material too often, because it will soon stiffen enough to start dragging off the members rather than forming them.

Make up small batches so fresh material will be available to stuff the mold. One technique is to soak enough plaster in a ring of putty to complete the stuffing operation, but to mix only a part of the batch at a time. Placing a board or darby across the ring and cutting off what is to be mixed, will prevent the plaster from running out of the opened ring.

Stuffing and polishing is continued until the cornice is finished. The last run of the mold can be made to give a final polish to the cornice. To do this splash water ahead of the mold with the plasterer's large brush. This run removes all the slobbers and drippings that may have stayed on

Fig. 9-18. Stuffing the mold, using the trowel.

Fig. 9-19. Stuffing the mold, using a rubber glove.

Fig. 9-20. Final run of the mold to finish the cornice.

the members during the stuffing operation. Fig. 9-20 shows the clean mold being run over the finished cornice to give it that last polishing operation.

Precautions. A number of precautions must be taken in running a cornice. First, the blocking-out operation must be completed and the stuffing procedure well under way before the first batches of plaster begin to swell due to the setting action of the gauging plaster. The first batches must be retarded just enough to permit the blocking-out operation to be completed, with the material in a setting stage but not to the point of swelling. If the blocking material starts to swell it will cause the mold to chatter and may spoil the cornice.

If this should happen, the condition can be overcome by letting the total swelling action take place after running the mold over it as often as possible to cut off all the swelling material. Then loosen the three nibs on the mold and slip *shims* (very thin wood or metal pieces $\frac{1}{16}''$ thick or less) under them to create a slight clearance between the mold and the hard material. The finishing operations are then completed as quickly as possible to avoid further trouble.

At all times the back of the mold (the $\frac{1}{4}''$ space between the metal profile member edges and the wood backing) must be kept clean. Material gathers here which, if left in place after a run, will drag on the members being formed, preventing them from being brought out to their true shape. Cleaning all parts of the mold at regular intervals will help to produce a perfect cornice in the shortest time.

Another rule that must be followed at all times is to hold the mold firmly against both screeds with a uniform pressure. The mold must be held down onto the cornice strips and pressed inward at the same time to hold it solidly against the ceiling screeds. Check both screeds constantly for any material gathering where the nibs run over the screeds. In applying the plaster to form the cornice, some of this material will at times cover parts of the screeds. This material must be scraped off before the mold is pushed over it. Otherwise, it will gradually gather in thin layers and cause the screeds to become crooked or bumpy.

Watch particularly near the corners on the wall screeds, as here the front of the mold will be full of excess material and it will tend to build up on the wall screed at these points. Check the screeds after each run, looking also along the ceiling line of the cornice. Any growth on the wall screeds will be indicated by a rapid curving outward at the corners.

As the cornice is being blocked out, fill in surplus material into the miter areas (corners or intersections) keeping it cut back enough so it will not interfere with the mitering later on. This procedure will save a lot

of time and trouble in the mitering process, particularly when the cornice is large and has a number of projections.

Many other techniques are used in running special types of cornices. These specialized items will be covered when these types of cornices are explained later in this chapter. However, all cornice running uses the same basic operations outlined above. Each plasterer has his own variations on how and when to do certain steps, but the final overall procedures will be the same.

Mitering

Mitering, or developing by hand the members of the cornice into the corner or intersection, is an art that requires much skill and practice. No other plastering operation taxes the plasterer's skill as thoroughly as mitering.

After the cornice is run (completed) and set, the intersections at the corners are checked for straightness and alignment, using a miter rod or joint rod as it is usually called. The length of the rod to use depends upon the size of the miter or the distance the joint rod must span, plus sufficient bearing on the cornice to provide guidance in projecting the members of the cornice into the corner. Usually 6″ of bearing on the cornice is needed to insure a true projection. Thus a 12″ joint rod would be used for a miter whose

Fig. 9-21. **Blocking out the miter, using the trowel to roughly shape the plaster into the miter.**

greatest distance from the run cornice to the corner is 6".

If any of the cornice members do not line up with each other across the corner, or there is a curve outward at the end of the cornice, carve off the members using the joint rod as a knife. Shave thin slices until the members are straight and true to intersection. This checking will also show up any surplus material that might be in the corners; remove this excess by chopping it out with a sharp hatchet.

Mix enough material to block out the miter. Use the same proportions as used for the cornice. As shown in Fig. 9-21, fill in the miter area, using

the trowel to roughly shape the soft plaster into the rough outline of the cornice. Using the joint rod with the long thin edge against the cornice, and working up from the bottom of the cornice, start to roughly shape the plaster to the contours of the cornice on each side of the miter. Use a combination pressing and sliding motion with the joint rod to squeeze the soft plaster into shape. Do not at this stage try to cut or drag downward with the joint rod, or you will pull the soft plaster out of the miter. Fig. 9-22 shows how to hold the joint rod. With the hand closest to the miter, the thumb goes on top and the finger underneath; the other hand is

Fig. 9-22. Shaping the members of the miter, using the joint rod.

held on top or over the joint rod. These positions of the hands permit full control of the various motions the joint rod must make in developing the miter.

One technique many plasterers develop is to let the ends of the fingers of the hand holding the rod (where it is bearing on the cornice) project over the rod and lightly touch the cornice members. This action permits the rod to float over the members of the cornice without exerting too much pressure on them and prevents the rod end from cutting into the freshly run cornice. Even though

the cornice material is set and relatively hard, it is easy to damage the members with the rod if care is not used.

With the miter roughed out, start again at the bottom and, with the point of the rod projecting only to the edge of the run cornice, trim off any material that is built up over the members during the first operation. Use a cutting and backward sliding action with the rod to do this trimming. This operation cleans the cornice of material built up in the filling-in process and permits the joint rod to glide freely.

Now proceed to develop the miter an inch or two at a time across all the members, working tight to the run cornice with the back end of the joint rod. Work all sharp member edges by pushing inward and upward with the rod from each edge and at the same time sliding it back and forth on the cornice.

Fig. 9-23 shows the miter blocked out, with each member developed to its true shape and in line with the run cornice. Notice that the miter is not smooth and polished at this point, and that the intersection line has some gathering of material left there by the joint rod. The technique is to develop this intersection

Fig. 9-23. Miter blocked out, ready to be finished.

as closely as possible without digging into the opposite side and leaving a void.

The next step is to gauge a small amount of fresh material, keeping it quite soft. With a small tool brush, brush on a thin coat over the members on one side at a time. For a large miter, such as Fig. 9-24, only a few members at a time may be brushed.

Again, work the joint rod over the members using a troweling action to smooth each surface; work away from each member edge. Slant the face of the rod downward and upward as it trowels the members to work the material into the holes and voids. Clean out the brush in a water bucket as soon as the material has been brushed on. This prevents the plaster from hardening in the hair of the brush.

As the operation proceeds, a little water is thrown over the area being

Fig. 9-24. Brushing on finishing material to complete the miter.

worked on to lubricate the action of the joint rod and to keep the material soft enough to work with. Repeat this operation until the members are all true in shape and have a smooth surface.

The miter is now ready for the *pointing-up* (filling and forming the final point of the intersection of the two sides of the miter). Use an ornamental tool (see Chapter 2 for ornamental tools) to finish this intersection, filling in any holes and carving off any surplus. This work requires a steady hand and careful manipulation of the ornamental tool to prevent damage to the members already formed. Fig. 9-25 shows the plasterer using an ornamental tool to form the cove member of the miter intersection. All members are carefully joined at their intersections. Finish this operation by brushing clean water over the entire miter, using a soft clean tool brush; this develops a smooth clean miter. All the members

Fig. 9-25. Pointing up the intersection of the members, using a small tool.

should be sharp and true to shape, with all surfaces polished smooth.

Fig. 9-26 shows the plasterer using a pointing trowel and joint rod to form a perfectly straight line down the intersection.

Similar sharpening or emphasizing the interior angles of the members can be done with the joint rod, sliding it back and forth in each angle. Make sure the rod is clean before doing this. This operation is approved by some plasterers and disapproved by others who claim it fakes the true miter intersection. In most cases, if it is done correctly and lightly, the final effect will not be noticeable, but the miter will be sharp and true.

There are other phases of mitering which may be used in certain conditions. For very large cornices, with overall width of 3 ft. or more, the mold may be cut down by removing the upper section after the cornice is run. Then the lower section, with a shortened slipper, is run into the

Fig. 9-26. Cutting a straight line through the intersection.

large miter so it will reduce the extremely long miter left by the large mold.

Most cornices of this size have some large flat projecting members roughly near the center of the cornice which can be extended into the miter and used as a ceiling screed for the shortened mold. Fig. 9-27 shows a typical large cornice with a reduced

section of the mold outlined to show how the mold can be cut down.

Miter Mold. The miter mold is a much discussed but seldom used item. Few plasterers consider the work involved in making the mold and the problems encountered in running a cornice with it worth the time and effort. It is included here only to show that it is possible to run

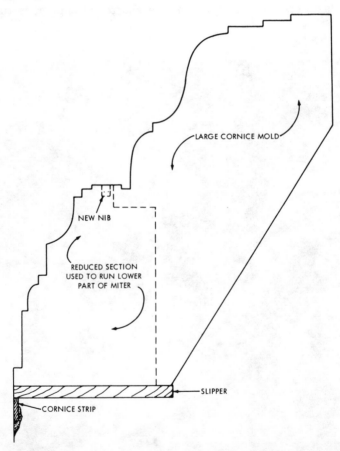

Fig. 9-27. Outline of a large mold, with dotted lines indicating how a reduced section can be cut out.

Fig. 9-28. Miter mold run into the corner.

Fig. 9-29. Miter mold turned, ready to run into the next miter.

a cornice and the miters in one operation. Fig. 9-28 illustrates the miter mold run into the right side of the miter. It must then be pulled out of this side and turned as shown in Fig. 9-29. The mold now faces into the next miter. Fig. 9-30 shows the mold pushed into the left hand side of the miter where it will again have to be pulled out slightly before it is turned and the operation repeated.

The main problems with this method is that it is almost impossible to build the mold so perfectly square that when it is turned the members or profile of the mold will fit exactly. There is usually a slight difference in alignment. Also, entering the miter and keeping the mold tight to the screed with material on the face of the mold presents problems. It can be done, but few plasterers will spend

Fig. 9-30. Miter mold pushed into left-hand side of the miter.

the time required to project the regular mold profile on to a 45° angle and cut a new profile. Also, it is slow running and there are finishing problems caused by the profile being set on a 45° angle.

Running Breaks in Place

In many cases when running a cornice, the plasterer finds he must work it around a break in the wall. This could be a pilaster, fireplace breast or any other projection interrupting the continuity of the wall line. There are a number of methods used to run *breaks*, as these interruptions in the straight line of the cornice are called.

Two basic methods are in common use. Each one is suited to a specific condition and each has some variations to suit special needs. The two basic methods are: (1) run-in-place or actually forming the break on the spot, modifying the mold and screeding to suit the condition; and (2) benching the mold and running a

section of the cornice on a bench or table, then cutting it into sections and *planting* (sticking the cut piece in position) the sections in place as needed.

The plasterer makes his choice of which method to use based upon the size of the break or projection. If the break in the wall will permit the mold to be run on it with some additional temporary screed development, then the run-in-place is the better and quicker method. If the break is narrow and would not permit any part of the mold to run on it, or the cornice has very intricate members, then running the cornice on the bench and sticking the cut-to-size pieces in place is the better method.

Fig. 9-31 shows a typical break, run in place, set up for a cornice to be run around a pilaster. Notice how the 1″ x 4″ wood strip has been fastened and supported on the face of the pilaster. The only modification

Fig. 9-31. Setup used to run a break in place.

on the mold is the addition of a rabbet nailed to the bottom of the slipper. The rabbet is any strip of wood (usually a piece of cornice strip) nailed to the bottom of the slipper back from the wall line the thickness of the strip the mold will run on, in this case ¾".

Braces, as shown, support the running strip both horizontally and back to the wall. Fasten all parts of these braces with both nails and extra-high-gauged plaster (25 per cent putty to 75 per cent plaster) to hold them securely. It is customary to run the main cornice around the room and the face of the break at the same time if conditions permit. Then the sides of the break are run last.

The break face is now run, making sure to extend the run section well beyond the corner on both sides, so that when the sides are run later the mitering will be formed automatically by the mold. Because these will be external miters the mold forms the miters by cutting through the previously run face of the break.

When the breast or face of the pilaster has been run, remove the 1" x

403

4″ strip and nail two shorter strips, one on each side of the pilaster. Each of these pieces should project out far enough to permit the mold to run beyond the ceiling line of the cornice on the pilaster face. When these strips are in place and braced, place the mold on the strip and carefully carve away any part of the previously run cornice so the mold will clear and form the miter as the run is made. If the strips have been put up accurately and the carving with the mold is done carefully a perfect external miter will result. Note: The mold must be used to carve and clear the external miter before the side can be run.

The two sides of the pilaster are now run in place. To protect the ex-isting cornice from damage, it is wise to brush some raw putty over the members at the ends of the previously run cornice. When the sides are completed, any material that may have gathered on the face of the cornice will lift off without damage. Brush the area clean with a tool brush and clean water. Never let the raw putty stay on the cornice overnight or it will dry on so hard you will damage the cornice getting it off.

Fig. 9-32 shows a similar break run in place, but with a very small side projection — too small to normally run in place. Here is one of the exceptions to the rule of running a break on the bench if the area does not permit it to be run in place. With only two small breaks to form, it

Fig. 9-32. Break face run in place.

would not be worthwhile to set up a bench run. The plasterer has two choices: (1) if it is a simple mold with few members he could form the

Fig. 9-33. Temporary wall setup used to run side of narrow break.

breaks by carving them in place by hand (a procedure to be covered later); or (2) it can be set up as shown in Fig. 9-33, providing a temporary wall to run enough cornice in place so the miter can be made.

Notice that some raw putty has been brushed on the existing cornice. The temporary wall setup is a piece of gypsum lath. The short section of cornice is now run in place on the temporary background as shown in Fig. 9-34. When the cornice is finished the miter is then completed as shown in Fig. 9-35, using the short run of cornice to guide the joint rod in developing the miter.

With the miter completed, cut away a section of the temporary cornice just beyond the face as shown in Fig. 9-36. When the cut is completely through the cornice to the backboard, remove the balance of the cornice with a hatchet and take down the running strips. Now very care-

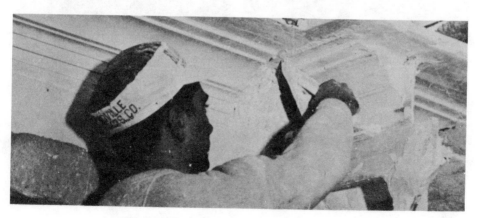

Fig. 9-34. Running the narrow break in place.

Fig. 9-35. Mitering the break, using the temporary cornice as a guide.

Fig. 9-36. Removing the temporary cornice to complete the break.

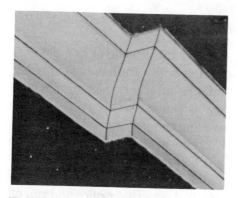

Fig. 9-37. Completed break.

fully carve off the balance of the cornice until the raw putty is reached. The remainder can then be washed off with water. Only a minor amount of pointing up of the external miter will be necessary. The completed break is shown in Fig. 9-37.

This procedure seems like a lot of work, but in actual practice it can be done quite fast. In many cases the running of the temporary cornice section is done while some of the other walls are being run.

Running Breaks on the Bench

In many cases, particularly when a number of small breaks must be made in a cornice, it is best to run a section of the required cornice on the bench. It can then be cut into both right and left hand pieces and stuck in place with high-gauged plaster.

On a work bench long enough to hold the length of cornice required, set up a backboard (usually a scaffold plank 2″ x 12″ x 10′ long) at right angles to the bench. Brace it from behind the plank with enough braces to hold it securely. Make sure both the plank and the bench are straight and smooth. Place the mold in position, with the slipper on the bench and the ceiling nib against the backboard. Square up the mold and mark the slipper line at both ends of the bench. Check by measuring from these lines to the backboard at both ends, to make sure they are parallel.

Now snap a chalk line to these two marks and nail a cornice strip to the bench at this mark. The slipper of the mold will ride on the bench and against the cornice strip while the nib of the mold bears against the backboard.

With the mold in position, check to see how wide a piece of ¾″ thick board can be fitted on an angle between the bench and the backboard while leaving about an inch of space between the projecting members of the mold and the board. When this is determined, fit the board in position at the required angle and nail it lightly at each end. This board will act to reduce the thickness of the piece of cornice to be run and will assist in removing the run section from the bench when it is finished.

Coat the whole area where the cornice is to be run with raw putty and sprinkle over this enough dry sand to cover the putty completely. Both the putty and the sand are parting agents. The putty keeps the plaster from sticking to the bench and backboard while the sand keeps the putty from sticking to the cornice. The sand will become imbedded in the back of the plaster and will help to hold the plaster used to stick the piece in place.

Make a trial run with the mold to check for clearance and smooth running of the mold. It will be helpful if the slipper and nib lines are given a coat of grease or oil to make the mold run easy and prevent the plaster from building up.

This type of run must be made using gauging or molding plaster only to provide the required strength. Soak enough plaster in a pail or tub to cover the area with a layer of plaster up to ½" thick. In some cases, when the cornice is large with flat members, it is wise to lay in this first coat of plaster some fiber or open mesh burlap to reinforce the run. This reinforcement strengthens the

pieces so they can be handled later without breaking.

Cover this first layer as quickly as possible with a fresh batch of plaster, building up the cornice members as fast as you can to keep ahead of the swelling action of the setting plaster. Keep the mold clean and tight to the bench each time it is run.

Finish up with soft fresh material splashed over the cornice and run off repeatedly until the cornice is finished. Do not reuse any of this final mix material, as it will drag off (damage) members rather than build them up. The mold must be run almost continuously over the cornice to keep the swelling plaster cut down.

Fig. 9-38 shows the completed run on the bench. Note that in this case, because the bench had a straight outside edge, the mold was *rabbetted* (guide strip nailed to slipper), eliminating the cornice strip normally used. If the work bench is rough

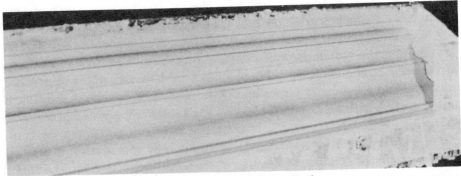

Fig. 9-38. Completed bench run cornice.

wood the rabbet can also be used to run on top of the cornice strip to provide a smooth running surface for the slipper.

The cornice being made with plaster is set hard and strong enough to be removed, turned over so that the wall line laying on the bench will face up and be the top member while the ceiling line will be on the bench. This is done to permit the marking and cutting to be done in a vertical position. Remove the nails that are holding the slanting ¾″ thick board under the cornice. Using a hammer, tap the upright back plank to loosen it from the run; also tap the bench a few hard raps to do the same thing.

Now, lifting on the slanting board at both ends and holding the cornice firmly, slide it forward on the bench away from the backboard. Remove the backboard and its braces, and clean off the bench. Pick up the board and the cornice, swing the cornice around so it is now facing the opposite way, and turn it over. The wall line will now face upward. Fig. 9-39 shows the cornice in this position. Remove the slanting board and clean off the raw putty from the back of the run. Support the cornice temporarily during the next operation with some pieces of 2″ x 4″ or similar blocking behind and under it. All cuts can now be marked and cut straight down and on a 45° angle, as shown in Fig. 9-39.

To lay out the cuts when both right and left hand pieces are needed, start from the center-mark of the run and mark off a square line down the face of the run. To do this, use two small steel squares. Hold one in the left hand, placing it on the upper member of the cornice so the tongue (short end) faces out and the corner of the square is on the center mark. The other steel square, held in the right hand, is placed upright on the bench and pushed against the first square. Now remove the upper square and mark the bottom corner of the remaining square on the bench.

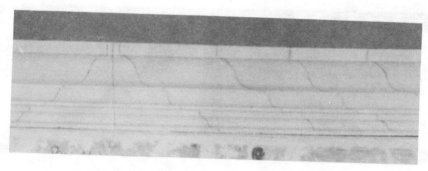

Fig. 9-39. All the required pieces marked off on the cornice.

Draw a straight pencil line from the top mark to the bottom mark using the square or a joint rod as a straightedge. From this centerline, on both sides, measure out ½″ (provides thickness for the break at the wall line) and draw lines at the top to mark the wall lines for the right and left hand pieces which will be needed to form the right and left hand breaks.

Next, using a 45° triangle held horizontally against the top wall member and a square held vertically on the members and against the triangle, mark the members downward until a true 45° line has been developed from the top to the bottom in a series of steps. Be sure the 45° triangle is held perfectly flat against the cornice, not tipping down or up, or the line drawn will be off. The square must also be held vertically to insure a true line. Mark a 45° line on each side of the centerline.

Check this 45° line, its projection to the ceiling line measured horizontally should equal the width or projection of the cornice. The projection of the cornice is the true 45° line, so these two must check out.

Measure out from each 45° line each way the length of the pieces needed and draw the required 45° lines. Fig. 9-39 shows all these lines drawn for the required right and left hand pieces for the breaks.

Use a coarse toothed saw (eight teeth or less per inch) to make the cuts. Hold a triangle in place to guide the saw. A few drops of water on the saw blade as the cutting proceeds will prevent binding. Wet plaster, when it is cut with a saw, tends to act like glue and will stop the saw in the cut.

Fig. 9-40 shows some of the cuts completed and ready to be *planted* (stuck or glued in place) in position. Wash off the cut pieces with clean

Fig. 9-40. Bench run cornice cut into various size pieces.

water to remove the slobbers left on the face of the run by the saw.

Another method is sometimes used when the cornice is small and has simple members. This is a variation of the run-in-place technique combined with part of the run-on-the-bench technique. Select a section of the screeded area where the cornice is ready to be run, and coat in a streak with raw putty which will permit the finished run to be removed when completed. Make sure the scratch coat is completely covered and the putty is troweled quite smooth, using stiff putty for this operation. Throw some sand into the putty as was done in making the bench run previously described. Run the mold over the area to make sure there is good clearance for at least ¾″ thickness of plaster to cover the area.

Gauge up enough extra-high-gauged plaster and putty to block out the required length of cornice. Fill this in over the putty and complete the blocking out. Finish the cornice with all extra-high-gauged material and when it is finished cut the run loose at the wall and ceiling line, using a pointing trowel to cut in behind the run piece to free it from the putty background. Now keep pushing the full length of the pointing trowel into the putty, gradually loosening the run cornice.

When you have cut as far in as possible at the top, bottom and both ends, carefully pry the piece loose with the pointing trowel. Be sure you hold it with one hand while you pry with the other. When the cornice has been removed place it on the bench, clean off the back, and cut the required pieces. This method saves setting up the bench for a run but it can only be used for very simple molds. Do not try this method for large complex cornices. Maximum length of a run of this type would be about three feet.

Sticking Breaks

When the required pieces of cornice, or breaks, have been cut to size, they must be *planted* (stuck or glued) in place. To support the breaks and to line them into the proper position, wall strips are nailed to the wall screeds. Set these strips level and on a line equal to the bottom part of the break. Fig. 9-41 shows the strip setup. The area behind the break to be planted should be checked for clearance and wet down so the break can be moved into its proper position before the sticking material (high-gauged plaster) has stiffened because of the suction in the background.

Check each break in position and mark a line along the top edge of the break on the ceiling screed. This line will then be used to guide the break into its proper position when it is planted. If the break is large or heavy, have some nails or a strip of

Fig. 9-41. Planting a break, showing supporting strips and ceiling line.

wood with nails started in it ready to nail on the ceiling screed to hold the break in position until the sticking material has set.

Fig. 9-42. High-gauged plaster applied to the back of a break ready to be planted.

Fig. 9-42 shows the break with the high-gauged plaster spotted on the back. It is ready to be set in place. The surplus material will squeeze out around the break and indicate that the planting area is full and will hold the break securely. Check the break using a square on the top and bottom members and held against the adjoining cornice to make sure it is in the proper square position. If needed, nail the wood strip on the ceiling screed to hold the break until the plaster has set. Wash all surplus material off the break as soon as possible.

Fig. 9-43. Long break being squeezed into position.

Fig. 9-43 shows a long section being squeezed into place with the sticking material flowing out at the ceiling screed. If the break will not move into position properly, remove it, scrape off the used plaster, and plant it again using fresh material. If the break is forced into place it may break in the effort to move it.

The small return, or dead ending, of the cornice against a plain wall is shown in Fig. 9-44. The piece is being fitted into position before planting. Each break, in position, must be checked for clearance and proper alignment before attempting to plant it. In many cases it is necessary to chop away part of the wall screed to

permit the thick wall section of the break to fit back into place. The ceiling screed may also need some cutting away to provide the required

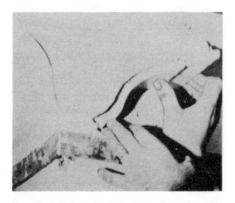

Fig. 9-44. Fitting a return used to stop a cornice.

413

clearance. Make this clearance for the screeds very carefully, using a sharp hatchet, so as not to loosen the screed where the supporting strips are nailed.

Making a Return in Place

A *return* is, as the name implies, the cornice turning in on itself and stopping at a given point. The return is usually found where a cornice must stop due to an opening such as a stairwell or similar obstacle where the cornice cannot continue. It is a neat and ornamental ending for any cornice when certain conditions prevent continuity.

The cornice is run as far as possible and then a ceiling line is selected that will provide a few inches between the end of the cornice and the opening end of the ceiling. Use a plumb bob and square as shown in Fig. 9-45 to plumb the ceiling line down to the wall line. At the same time, note on the steel square the projection of the cornice (measured from the wall to the plumb line). See Fig. 9-45. Now measure back (to the left) from this wall mark set at the ending of the cornice the projecting distance just measured and mark this distance on the wall as shown. This will establish the 45° mark to be drawn next.

Place a joint rod as shown in Fig. 9-46 on an angle from mark to mark. Pick a joint rod whose length will fit as close as possible to this distance.

Using another short joint rod and holding it plumb while standing directly in front of and in line with the mark to be made, draw the rod slowly

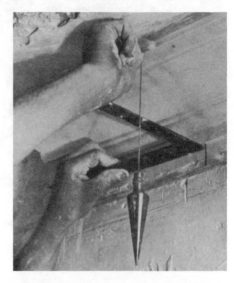

Fig. 9-45. Plumbing the ceiling line of the return down to the wall.

Fig. 9-46. Marking the miter cut for the return.

down the side of the first rod and cut a light mark into the cornice. Be sure you are directly in line with the mark to be made and that the joint rod is held plumb while it is drawn down.

Now, as shown in Fig. 9-47, carve away the excess plaster to form a square return of each member from the mark just made and back to the wall. Notice how the plasterer is using the back, square end of the joint rod to do the carving. When the carving is completed the return is finished off by brushing on some fresh gauged soft plaster, using exactly the same procedure as used to finish any blocked-out miter.

Fig. 9-48 shows another method used to make a return. The marking of the line is done in the same manner as in the first method. Then, instead of carving out the members, the entire cornice is dug out on a 45° angle so as to develop the external miter line.

The end is filled in square with some fresh-gauged plaster, again using the square end of the joint rod. See Fig. 9-49. Be careful not to dig into the miter line; this will result in an edge that is not straight.

The first method is very practical when the carving can be done as soon as the cornice is run. At that time the plaster is still soft enough to be carved quite easily and the return can be finished in a short time. The second method is used when the cornice has been run the day before,

Fig. 9-47. Carving away the excess plaster used in forming the return.

Fig. 9-48. Cutting the cornice on a 45° angle to form a return.

Fig. 9-49. Blocking out a return miter.

or earlier, and is too hard and dry to be carved in place. The return should be completed with a straight external miter line and the returned members should be level into the wall.

These two methods are useful for ordinary cornices without too many members or drips, projections and similar hard-to-carve members. For that type of return the run-on-the-bench and planted piece is best.

Enriched Cornices

In many cases a cornice is designed to receive various types of running ornament planted in the *beds* (recesses formed by the mold to hold the ornament) of the cornice. Many types and sizes of ornaments can be planted in a cornice, and in some large cornices very elaborate designs are worked out in combination with various members of the cornice.

The plasterer must fit these pieces of ornament into their proper place. In doing so he must at times stretch out or shrink the pieces to make the units fit in the space allotted. It

Fig. 9-50. Fitting and spacing ornament.

is important for each corner or miter to have the same unit or part of the ornament in it to be architecturally correct. A decision must therefore be made as to what unit should be placed in the miter as a starting point.

In Fig. 9-50 the lower row of ornament, the *dentil* (so named because the blocks are spaced like teeth), is shown with a full block or unit in each miter, both internal and external. (The upper ornament is called "acanthus" and the lower "egg and dart.") Some plasterers would prefer to start with a space rather than a block; this question has been debated for years.

Spacing. With the corners set, the ornament must now be spaced out to see how it will fit in the remaining space. Two methods are used. One method is to fit it in place against the miter section, then mark the cornice in the bed at the end of the piece and slide it along, repeating the marking until the end is reached. For small simple units this method is safe. When planting larger and more complicated sections, however, it is better to fit each piece in place and hold it temporarily with one or more nails pushed into the cornice above the ornament as shown in Fig. 9-50.

There are two good reasons for recommending this second method. One is that castings made with glue molds vary in length; as the mold dries out it shrinks, so that pieces cast on the second day will be shorter than those cast on the first day. (This condition, of course, can be used to advantage when the shorter pieces are needed.) Therefore, if the ornament is spaced out and only marked, some of the pieces used later might be longer or shorter and would thus throw off the spacing. Another problem, similar to the first one, is that there may be variations in the width of the ornament, in which case individual pieces may have to be planed off slightly (using a small block plane) to make them fit in the bed.

If the section being spaced out requires either shorter or longer pieces, a decision must be made as to which method is best under the circumstances. For most of the simple ornaments, like the dentil, it is very easy to shave off just a thin slice from one end of each piece of ornament and thus shrink an inch in a matter of a few feet of ornament. When the space to be fitted is short, the individual pieces of ornament may have to be cut into separate units and each space reduced enough to make the ornament space out correctly.

In all cases, the stretching or shrinking required can never be greater than the span of one unit of the ornament. Shrinking, whenever possible, usually requires less *pointing up* (filling in the spaces between

417

the ends of each ornament) then stretching. For some of the larger ornaments with a large amount of detail, the caster will utilize the shrinking of the glue mold to produce both short and long pieces of ornament. This saves a lot of time on the job. Also, the modeler may make up special miter pieces which have a leaf, flower or other unit molded to fit the condition. This is done when the design does not miter properly.

Sticking. When preparing the sticking plaster make a thin ring of putty on the finishing board and, after mixing in the required amount of retarder (amount varies according to how many pieces of ornament will be stuck and the size of the ornament), fill the ring with water and sift in the molding plaster, keeping it quite soft. Next, soak the ornament to be stuck in a pail or tub of water so it will absorb enough water to prevent it from drawing water out of the sticking material before it has set. The cornice should also be wetted.

Some plasterers prefer to use *glue water* (gelatin soaked in water) to retard the plaster used for sticking the ornament. There are also gums that can be mixed with the plaster, and when these are used neither the cornice nor the ornament needs to be wetted.

A look at Fig. 9-51 will show that each ornament bed consists of the overall recess for the ornament. In the back of the bed there is a second recess for the plaster used in sticking the ornament to flow, while the ornament itself slides solidly against the top and bottom ribs of the bed.

These are laid out when the cornice profile is cut and must be of the correct size and depth to hold the

Fig. 9-51. Ornament bed, showing how space is allowed for sticking material.

ornament. Some plasterers fail to provide the second or plaster recess and if the sticking plaster has a small lump or stiffens slightly the ornament cannot be seated properly, resulting in a crooked run of stuck ornament.

As each piece is ready to be put in place, butter the back of the ornament with enough plaster to fill the bed completely. Then push and slide it into place, making sure the ornament is solidly against the back of the bed.

There are, of course, many variations and techniques used in sticking certain ornaments. Large heavy ornaments are sometimes planted in run cornices and these may require sticks to be placed under them for temporary support until the sticking plaster has set. In other cases, the bed for the ornament is made to lock the pieces in place, and they must be pushed into place from one end of the run where an opening has been made to let them slide in.

Pointing. Fig. 9-52 shows the plasterer pointing up the joints in the ornament (in this case the "egg and dart" ornament). To do this he places a small amount of the sticking plaster on a pointing trowel, and using the proper shaped ornamental tool, fills in the joints plus the top and bottom edges if needed. With certain shapes of ornament it is possible to have the sticking plaster flow out at the top and bottom of the or-

Fig. 9-52. Pointing up the joints in the ornament.

nament as it is being pushed into place. This outflow is then brushed off with the tool brush and clean water as each piece is stuck. This practice eliminates the need for pointing at these places. However, if the ornament has many deep recesses brushing may create more trouble than it is worth. A judgment must be made as to which method (tooling or brushing) to use.

Carefully brush off each joint after it is filled in, using the tool brush and clean water.

When the pointing has been completed the total cornice should appear as one piece. Fig. 9-53 shows a completed section of cornice with ornament pointed up and brushed clean. The egg and dart (center ornament in the figure) is the hardest to point up of the three types of or-

Fig. 9-53. Completed section of cornice with ornament pointed up and brushed clean.

nament shown because the pointing must be done on the center of the egg. In such cases it is best to fill in the joints roughly first and then go over them a second time to achieve a smooth, true surface.

Light Troughs, Coves and Compound Curves

The modern light trough presents many problems for the plasterer. The projection of the lip of the trough creates problems in bracing the mold, and the restricted working area makes it hard to run the mold.

Fig. 9-54 shows a small light trough mold combing a cove and trough lip in a single mold. Larger light troughs would require separate molds. To make this job easier the plasterer runs the cove first as shown in Fig. 9-54. Then the cove knife is removed and the lip part of the trough completed.

In Fig. 9-55 the plasterers are running the lower part of the mold to form the light trough lip. The man pushing the mold is also stuffing it

while his partner is catching the excess material that falls off. The mold must be held very firmly and extra care must be exerted to keep the long tail end of the slipper in the strip.

There is a tendency for the mold to leave the running strip because the horse is set at the extreme left end of the slipper. For some of the larger light troughs a tail mold is sometimes used to run the lip. This type of mold needs only the flat ceiling screed and provides good bearing for the solid support of the mold.

Fig. 9-56 shows a compound-curved cove being run with the mold shown leaving the convex section and entering the concave part. Notice the way the horse is set at the end of

Fig. 9-54. The cove has been completed and the light-trough lip will be run next.

Fig. 9-55. Running the light trough lip.

Fig. 9-56. Compound-curve cove mold set on the convex curve section.

the slipper and the braces are placed to reinforce the mold. The ends of the slipper in such a mold must be rounded off so it will not dig into the screed as the mold rounds the curves.

Many of our modern buildings call for compounded curves, that is, the curves combine both concave and convex curves into one motion much like an ocean wave. To run a cove or any cornice of this type two changes must be made in the mold. First, the horse must be set at the extreme left end of the slipper so that no matter how the mold moves in or out of a curve its knife will be in contact with the screed. Also, the slipper must be hollowed out between the two nibs when the mold is set at the convex curve. Figs. 9-54 and 9-56 show two examples of this type of mold used to run on compound curves.

Developing and Hanging Coffers

A coffer is a cast unit of plain or ornamental plaster used to form a ceiling. A number of units are cast to cover the required ceiling area. The coffers are sometimes bordered with a run cornice and in other cases extend to the wall lines, forming the complete ceiling.

Coffers (recessed panels) are made up in the shop by modelers and cast by the caster. The plasterer hangs the coffers in place on the job. The units must be made to exact measurements. They are hung level (except for curved areas) and fitted to wall lines so that any wall treatment, such as marble or wood, will fit.

Coffers can be made in many sizes,

Fig. 9-57. Single coffer hung in place.

limited only to practical handling of the castings between the shop and the job. Most units are made so they can be positioned by hand but for heavy units a block and tackle or chain hoist must be used.

Fig. 9-57 shows how the coffer has been hung in position with hanger wire. Other units will now be hung adjoining it, as shown in Fig. 9-58, and they will be aligned by using strips of wood placed across the joints, both top and bottom. Tie wire is used to pull these strips tight against the coffers holding them in perfect alignment.

There are numerous methods used to hang coffers. Most common is the cradle or rack made of wood. The rack must be strong enough to support the coffer without distortion and permit the units to be moved, if necessary, to align the rows. Racks are

Fig. 9-59. Coffers supported by a rack until fastened.

either hung in place first, lined as to centers and leveled, then the coffers placed on the rack, or one or more coffers are placed on the rack and both coffers and rack are raised into position together.

Fig. 9-59 shows two coffers set on a hung rack made of 2″ x 4″ wood with the two cross bars providing a true level surface at the coffer edges. The rack was set level and in the proper location. The plasterer will now *wire* and *wad* (plaster and fiber rope shaped loops placed around the supporting channel iron and the channel iron cast on the back of the coffers) or coffer.

Fig. 9-60 shows the wires in place. These will be tightened by twisting until they support the cast without lifting it from its rack. The wires are used as double insurance that the coffers will never fall.

Fig. 9-60 also depicts how the plasterer drapes the plaster wad over the supporting channels and then bring

Fig. 9-58. Lining up the second unit with the first.

423

Fig. 9-60. Wadding the coffer in place.

the ends under and over the channels on the back of the cast. Place the wads over the supporting wires and squeeze the wad into a solid mass, using the hands to do this. When dipping the fiber into the plaster make sure that the plaster com- pletely covers all the fibers. To insure this, pull the fibers apart so that the wad is loose and roughly flat in shape. Then, after it is completely coated with plaster, form it into a rope shape as it is lifted out of the pail of plaster.

Fig. 9-61. Highly ornamented coffers, showing how the joints are covered with ornament stuck over them.

Never use a wad when the plaster is starting to set, because it will lose its strength during the placing process. Wet the back of the cast where the wads are to be placed and along the joints, which are also wadded. Enough wads must be used to insure absolute permanent support of the coffer.

There are many methods used to support racks and coffers. Space does not permit showing them all. Some of these are: strap iron hooks; turn-buckles used on either the rack or directly on the coffer, eliminating the need for a rack; and struts, usually made of wood 2″ x 4″s. Any support for the rack must be placed upon the floor or other solid base.

Fig. 9-61 shows an example of highly ornamental coffers: The joints of the coffers have been covered by ornament stuck in place after the coffers were hung. This is a common method of hiding the joining of coffers.

Geometrical Panel Construction

The geometrical panel or strap work is also called *repeating* or *diaper* ornamentation. It is found in many old English ceilings and was derived from very ancient pre-Christian decorations. It is also a favorite Arabian decoration.

This type of ornament is usually cast in individual units then glued to a smooth, previously finished ceiling. Sometimes a series of units are combined into large cast sections which are then hung to supporting channels to form complete ceilings.

The single unit ornament is placed to guide lines struck on the smooth ceiling. The layout of these lines must be very exact, as any mistakes will be compounded as the sticking of the units progresses. Start center-lines that are accurate and square to

Fig. 9-62. Geometrical panel layout and panel unit application.

each other. Lay out the pattern to find out if the unit should straddle the centerline or start on it. Remember the units must end at the cornice or wall with a perfect fit. The units usually intersect with the cornice's

425

outside member to complete the panel.

In Fig. 9-62 the unit centerlines are shown as required for the pattern used. As illustrated in this figure, when these centerlines are not accurately drawn the units will not fit. Fig. 9-62 also shows the units being stuck in place using soft plaster buttered on the back of the unit. The ceiling underneath the unit is scored lightly to provide a key for the sticking material. Soak the ornament and wet the ceiling area just before applying each unit.

For light units, as shown, quick setting plaster mixed in small batches will hold the units without supports. For heavier units, support-ing props should be placed under strips which cover the face of the unit so as to support all of it equally. One prop will usually support most of the medium size units and two props will support almost any size unit.

Many variations in the designs and method of construction are found in the geometrical panels or strap work. Fig. 9-63 shows the plasterer sticking strips of molding that were previously run on the bench. Here he does the complete job. The step-by-step sequence is: first, make a mold to run the required strips or straps; lay out the lines for the pattern wanted; then cut pieces to fit the layout, using a miter box. These

Fig. 9-63. Sticking moldings previously run on the bench to form panel layout.

pieces are then stuck in place, using plaster as previously noted.

This method would be used only for small ceilings or in case a cast shop is not available to produce complete units. Because the method is slow it would be too expensive for large ceilings and has the further dis-advantage of requiring a lot of pointing up on many joints.

However, the method shown would be useful to the modeler for making up the original unit over which he would make a mold to cast the required pieces.

Special Molds

Job conditions create problems for the plasterer in the type of mold he should use to fit the situation. There are many types, each one usually designed for a specific condition. There are variations of basic types, with almost unlimited modifications possible. These can be determined only as the job situation develops.

Fig. 9-64 shows a modified *tail mold*. Here the horse was set at the extreme end of the slipper, and the slipper was hollowed out to permit it to run over convex curves. This shows a typical modification that can be made to make a mold fit a special condition.

In Fig. 9-65 we have an example of a standard *panel mold* which has been modified to permit it to run on two pins set in the face of the slipper. This arrangement lets the mold run around any curved strip to form a circle, ellipse or other curved shape formed on a flat surface. Notice that the profile is set directly over the front pin.

One very popular mold which finds many uses is the *hanging mold* as shown in Fig. 9-66. The mold hangs on the cornice strip at the slipper side and bears up against the cornice

SLIPPER HOLLOWED OUT

TAIL HOLLOWED OUT

TAIL NIB

Fig. 9-64. Tail mold for use on curved surfaces.

427

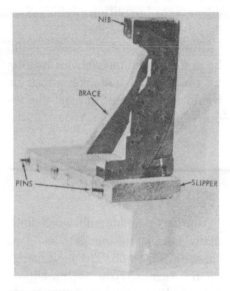

Fig. 9-65. Pin mold.

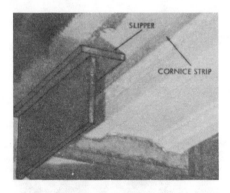

Fig. 9-66. Hanging soffit mold.

strip on the nib or other side of the beam. Part of the mold under the slipper must be cut to clear the cornice strip and the sticking material. This type of mold must be well braced to prevent chattering. A modification of this mold is used when

there is not enough clearance to hang the mold and both sides bear against the bottom of the strips. This type of mold is harder to run as it must be held up at all times—a condition that cannot always be avoided.

Fig. 9-67 shows a *boxed mold,* sometimes called a *double-slipper mold* in some areas. The purpose of the two slippers and the solid top board used to brace the mold is to provide the added rigidity needed in this type of tool where normal bracing cannot be used because it would extend so high that it would interfere with the molding. A common use for this type of mold is in making bench runs.

The *adjustable splay mold* is one that is seldom used, but when the need arises it can be a valuable asset to the plasterer. When a number of splays of varying angles must be run the adjustable mold will save a lot of time. The mold should be well constructed, using heavy sheet metal for the profile, which should be screwed to the horse to help stiffen it. Circular slots are necessary in this mold to allow the bolts to move in an arc, but these slots weaken the horse considerably. Good wood and solid bracing will help to overcome this problem.

Because the mold can be saved and used on many jobs, it is worthwhile to invest the time and materials needed to make it as strong as possible. By loosening the two bolts the

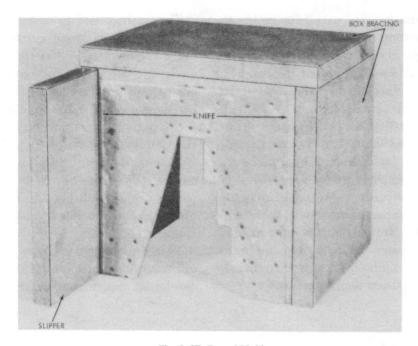

Fig. 9-67. Boxed Mold.

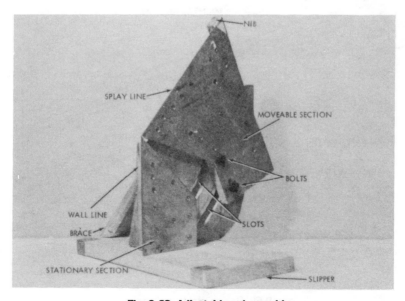

Fig. 9-68. Adjustable splay mold.

adjustable top part of the mold can be swung in an arc to any required angle. Sheet metal used for this mold should be 16 gauge or thicker. Fig. 9-68 shows this mold and most of the construction details.

Fig. 9-69 shows the layout drawing used to develop the two required *raking molds* from the normal mold. Raking molds are required whenever a cornice, run on an angle, intersects with the same cornice on a horizontal plane. Notice that the upper mold is completely different from the lower mold and both of them vary considerably from the normal mold.

The method used to develop raking molds follows. Draw out the profile of the regular mold at the angle or slope it will be run. Following this angle draw lines through each member, as shown in the drawing marked *A-B-C* etc. On large members additional lines are later used to help locate the contour. Next, at right angles to these lines or parallel to the normal mold's wall line, draw lines up from each member as shown, marked *1-2-3* etc.

Now, on a level line drawn above the area where the lower mold will be developed, measure out the distance

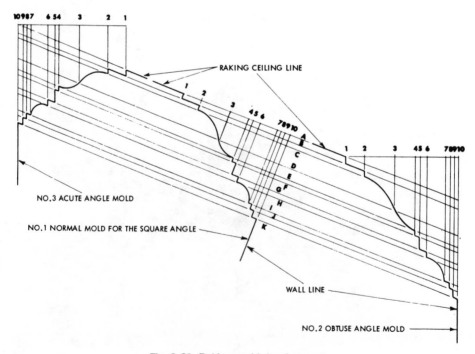

Fig. 9-69. Raking mold development.

between number *1* and *2* as found on the normal mold ceiling line. Continue transferring measurements from *1* through *10* as shown, and draw vertical lines down to the related letter lines. Line *1* runs from *A* to *B*, producing the first member of the raking mold. Line *2*, drawn between *B* and *C*, develops the second vertical member. Continue this procedure until all the members or points have been established. Then, using a French curve, draw the curved members through the points established.

Repeat this operation on the upper raking mold and produce its profile. The mold shown has quite simple members. If the mold is more complicated more points will have to be used to establish the profile. These profiles are now transferred to the sheet metal using the same techniques used to make a regular mold.

Fig. 9-70 shows this cornice and the miters completed using the three molds. This is a common setup for a cornice run on a stairway or similar sloping ceiling.

Fig. 9-71 shows the drawing devel-

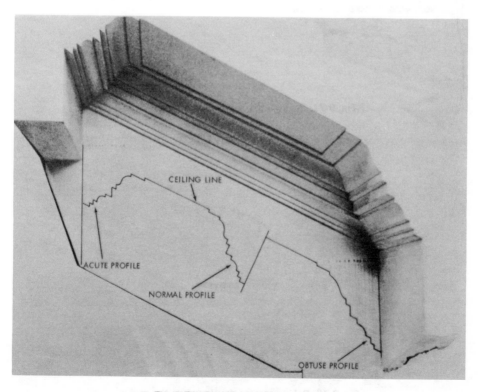

Fig. 9-70. Developed raking cornice.

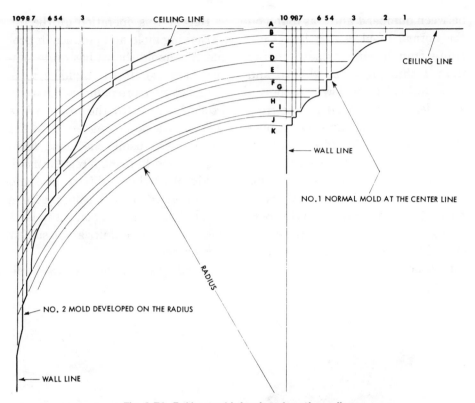

Fig. 9-71. Raking mold developed on the radius.

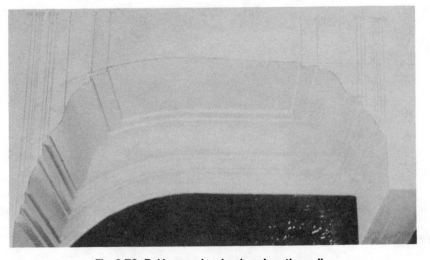

Fig. 9-72. Raking cornice developed on the radius.

Fig. 9-73. Muffled mold.

opment used for a raking mold on a radius. The procedure used to produce the raking profile is exactly the same as used in Fig. 9-69, except that the letter lines are drawn on the required radius. This condition is often found where a cornice is run around the end of a barrel or vaulted ceiling and then continues as an *impost molding* (one at the top of a column or wall supporting an arch or vault) along the springline of the vaulted ceiling.

Fig. 9-72 shows an example of this type of raking mold used in a stairway ceiling that curves down to the wall. Notice the extreme drawn-out length of raking mold required.

A special technique is used when running Portland cement cornices or extremely large plaster cornices. Because cement sets slowly, a cement cornice may take two or three days to complete. A *muffle* (a second profile fastened over the true profile and projecting between ¼″ to ½″ beyond it) is used to permit the cornice to be blocked out and left to set. Then the muffle is removed and the work can be completed without trouble. On large plaster cornices the muffle permits the cornice to be roughed out so that the normal swelling (occurring when large amounts of high-gauged plaster are used) can take place. With the muffle removed, the cornice can then be completed without interference. Fig. 9-73 illustrates a *muffled mold*.

Running Molds on Curves and Radiuses

At times the plasterer must run a mold on a curve or radius, such as a panel on a circular, elliptical or otherwise rounded ceiling or wall. If the area to be covered is large it is sometimes impractical to use a radius rod or *gig stick*, as the trade calls the arm that guides the mold. A simple solution to this problem is to describe the required curved line on the ceiling or wall. This line is set at a distance outside the actual molding line to allow for a wood strip and backing plaster or a metal pencil rod to be fastened.

If a wood strip is used it is usually a cornice strip that has been ripped in two lengthwise to make it flexible enough to be bent around the curve. The metal pencil rod is also very popular, as it can be bent around almost any curve. Care must be used in bending the rod not to kink it. As the kink cannot be completely removed, it will always show up in the finished run.

As shown in Fig. 9-74, lath nails

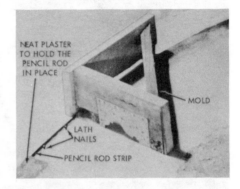

Fig. 9-74. Bent rod and mold setup for running an ellipse.

are driven into the screed on the line about 3″ to 4″ apart, and the rod is set to these nails and gradually bent around the curve. *Neat plaster* (all plaster, no lime) is then spread against the rod and over it to securely hold it in place. Trim away any excess plaster in front of the rod so the outside curve is smooth and flush with the metal.

The slipper on the mold is notched out so that only the nibs are in contact with the rod. Notice how the horse has also been cut so it will clear the rod and the plaster backing.

Fig. 9-75 shows the plasterer stuffing the mold with his rubber-gloved left hand while pushing the mold with the other hand. He must keep a steady pressure with the mold against the rod strip to keep the mold from running out of line.

Fig. 9-76 shows a finished elliptical panel. If there is not enough room on the outside of the run to set the running rod or strip, it can be set inside of the panel using the same technique but with the mold reversed.

Where the plasterer is called upon to form a simple elliptical panel for a ceiling or wall motif, he can run a double circle on the bench, cut the two circles into quarters, and then, using two pieces from each circle, form a pleasing ellipse. Fig. 9-77 shows how the gig stick is set up to run the two circles at once. Make up two profiles and file them together in the vise so they will be identical. Make up a radius stick long enough to swing the circles. Mount a slipper at one end and at the other end a notched strip of metal which will slip into a nail driven into the bench at the center point.

Fig. 9-75. Running mold around bent rod, forming an elliptical panel.

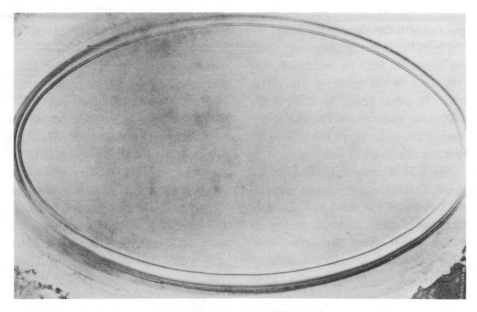

Fig. 9-76. Finished elliptical panel.

Fig. 9-77. Running two circles, using a double mold.

The profiles are marked on the stick at the required radius and the stick is then cut out to provide clearance behind the profile lines. Attach the metal profile and the gig-stick mold is ready to run the moldings. The diameters for the two circles are set by any of the many formulas for ellipses. This type of ellipse cannot have too great a difference between the major and minor axis (length and width) or the joining of the smaller circle to the larger circle will not be a smooth flowing line. The flatter ellipse requires more changes in radius and this method would not be practical.

For this method the diameter of the smaller circle should not be less than one half the diameter of the larger circle. The greater the difference in diameters the more elongated the ellipse will be.

Fig. 9-78 shows the two circles completed and cut into four quarters. Using two of the small quarters as the ends and two of the larger quarters as the top and bottom, the ellipse is formed. Fig. 9-79 shows the completed ellipse, which is now ready to be stuck in place and the joints pointed up.

When a true ellipse is required, the plasterer can best produce it using a trammel, either to run the mold or to lay out the strip lines. There are various other methods used by the trade when the degree of accuracy is not as important. Some of these are

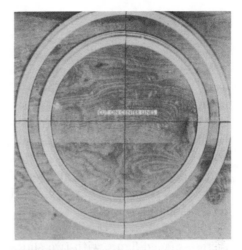

Fig. 9-78. Circles cut into quarters.

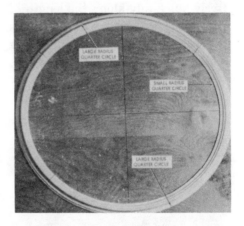

Fig. 9-79. Complete ellipse formed from four of the eight sections in two circles.

the *gardener's ellipse* developed by using a piece of string or wire fastened to two points on the major axis to guide a pencil. There are three, five and eight radius point approximate ellipses as well as numerous methods using circles, rectangles,

437

tangents, etc. all producing figures that closely approximate true ellipses.

Many plasterers call the ellipse an oval, but this is incorrect. The oval is an egg-shaped figure having one end larger than the other. Formulas for developing various ellipses and ovals can be found in many mathematics books. Appendix A at the end of the book gives some basic elliptical developments.

Fig. 9-80 shows a trammel. This is a cross-shaped figure with two slides that move in slots which hold them in place yet permit them to move freely but firmly along the length of the slots. The traveling arm connects the two slides and also holds the mold or pencil as the case may be.

The size of the ellipse that can be formed with a trammel is limited as follows: The size of the trammel itself will determine how small the ellipse can be, as the mold or pencil cannot move inside the cross arms. The outer limits are also based on how large the trammel is made and the stiffness of the traveling arm.

The long cross arm is called the major axis and the short arm is the minor axis. To set the pins of the slide to the traveling arm the following simple formula is used. Place the slide that moves in the minor axis at the exact center of the trammel, drill a hole in the traveling arm about 6″ from one end to fit the pin in the slide and slip this over the pin. Now mark on the other end of the arm the

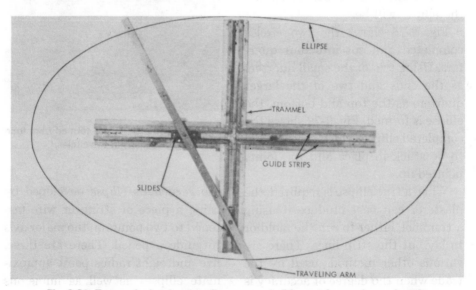

Fig. 9-80. Trammel, showing traveling arm and slides set for maximum length.

point at which the mold will be fastened or the hole will be drilled to hold the pencil to set the length of the ellipse. The distance from the center pin to the mark on the other end of the arm will be one half of the major axis of the ellipse.

Now move this slide down until the mark on the outer end of the traveling arm is in line with minor axis and on the mark that sets the width of the ellipse. Set the slide that moves in the major axis at the exact center and mark the pin location on the traveling arm. Drill the required hole and slip the arm on the pin. The

distance from this pin now at the center point to the mark at the end of the arm will be one half of the minor axis of the ellipse.

By studying the trammel and the traveling arm as shown in Fig. 9-80, the operation of this tool can be readily understood. If the trammel is to be used to run a mold, it should be made strong and smooth working. The trammel shown in Figs. 9-80 and 9-81 has flat iron top rails to provide positive alignment for the wood slides. Greasing these rails helps to let the slides move freely.

Fig. 9-81 shows the plasterers run-

Fig. 9-81. Running an elliptical panel, using a trammel.

ning an elliptical panel on the ceiling. Notice how the trammel has been tied through the ceiling using tie wire, also the edges of the arms have been stuck with high-gauge plaster. If the trammel cannot be tied in place it can be screwed to the wood joist, nailed to the brown coat and stuck at all the edges with high-gauged plaster. Support the trammel temporarily with props until it is centered and fastened securely. Various elliptical arches can also be run using the trammel.

One of the classics of the plastering craft is the running and mitering of intersecting arches. There are various methods employed to permit the mold to form these arches even though one run interferes with the running of the other. Fig. 9-82 shows one of the simplest and quickest methods used to permit the mold to run both arches and also develop the intersecting miter at the same time.

The center points have been located accurately on the spring line and a true center between the two arches to be run has been plumbed up on the screeds to be used later when cutting and planting the miter piece. Notice how the slipper line for both arches has been developed using the mold on its gig stick. The top and bottom intersection of these arcs will provide the true centerline between the arches.

The technique used in this method

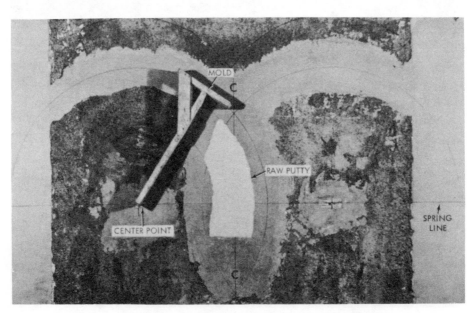

Fig. 9-82. Mold set on center point ready to run the first arch.

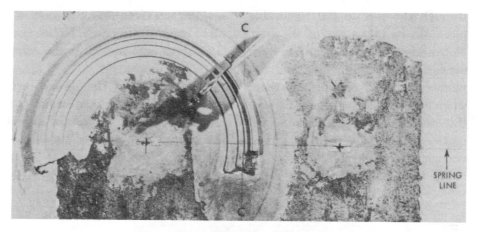

Fig. 9-83. Cutting the removable piece.

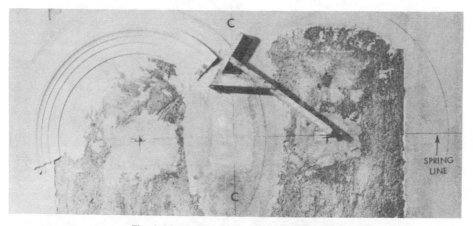

Fig. 9-84. Mold set to run the second arch.

of running the intersecting arches is to run one arch complete, then remove enough of the run molding to permit the other arch to be run and pass over where the section was removed. To permit the removal of the miter piece, raw putty is spread over the area that is to be removed. The slipper and molding lines show just where the raw putty is to be applied.

The center pins have to be set in small blocks of wood which were buried into the brown coat so as to provide solid immovable center points. Fig. 9-83 shows the first arch run and a saw is being used to cut

441

through the run so the removable piece can be taken out of the way. Notice also that this piece has been marked at the spring line and the plumb line so it can be cut there when it is laid on the bench. Be sure this run is made with high-gauged plaster so the piece will be strong enough to be removable.

Fig. 9-84 shows the section of the molding removed and the mold set up to run the other arch. Notice that the slipper will just clear the other arch.

Fig. 9-85 shows the second arch completed and the saw making a plumb cut which will be the actual miter line. The piece to the left of the saw will be removed and discarded. This is done by chopping it

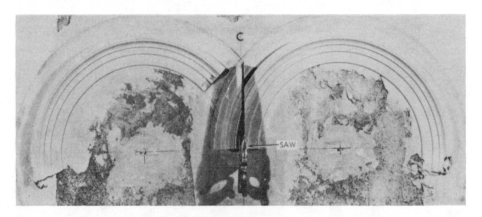

Fig. 9-85. Cutting the second arch at the miter line.

Fig. 9-86. Removable piece planted in place.

off with a sharp hatchet or hammer and chisel.

The first section removed is now fitted in place and checked to make sure it fits properly. Now plant this piece in place using high-gauged plaster as shown in Fig. 9-86. The arches are then cut to the spring line and the miter piece pointed up.

This technique can be used for a

Fig. 9-87. Large circular cornice run on the ceiling. Mold is turned on pin which fits into steel pipe held in temporary brace.

number of similar conditions where one mold must cross another or where the mold cannot run the full length due to an obstruction. In these cases the removable piece is run separately, removed and then planted where the mold could not run. Fig. 9-87 shows how a large cir-

cular cornice is run on the ceiling using a center point set in a temporary brace fastened across the opening. The center point is a ¾″ steel pipe with a hole drilled through it in which a removable pin is inserted. When the mold needs to be removed the pin is pulled out.

Running Niches

A niche is a recess or hollow in a wall formed to hold various decorative figures and at one time was popular as a telephone receptacle. The niche may be semi-circular, elliptical, or practically any shape desired. However, those that must be formed with a mold can present various problems depending upon their shapes.

Fig. 9-88 shows how a center-pin mold will form a semi-circular niche with a round dome top. The pin used here is a ¼″ iron rod fastened to the mold with strips of sheet metal which hold it firmly to the edge of the wood mold. The rod extends beyond the mold at the top and the bottom. The lower end of the rod rests on a wood block covered on top with a piece of flat iron to prevent the rod from cutting into the wood. This block, along with the strap holding the rod in place, will determine the floor height of the niche.

The upper end of the rod is held

by a removable slide bolt latch or similar retainer bearing. The mold is now free to turn on this rod, and

Fig. 9-88. Running a niche, using a center-pin mold.

the niche can be run just as though it were a cornice. As the niche is formed, the arris at the wall edge must be developed by trimming off the surplus material. The plasterer does this with a joint rod, which he slides on the wall screed and with an upward cutting motion, forming a sharp edge. This operation is repeated as the niche develops until both are completed.

In Fig. 9-88, the plasterer is shown blocking out the niche. A heavy wire loop is used to pull on the mold with one hand while the other hand pushes on the back of the mold. Note that by repeated trimming the arris

can be developed to a perfect edge.

Fig. 9-89 shows the elliptical niche almost finished. The plasterer has run the hinged mold up to the stop blocks and is now pushing the hinged section upward to form the top. The arris is developed in the same manner as in the center-pin method just described. Perfect gauging is the secret of success in these operations. If the material begins to set too soon the molds will chatter and ruin the job.

The elliptical niche with a circular top as shown in Fig. 9-90 requires two molds. The upper part of the niche is run from a center point set

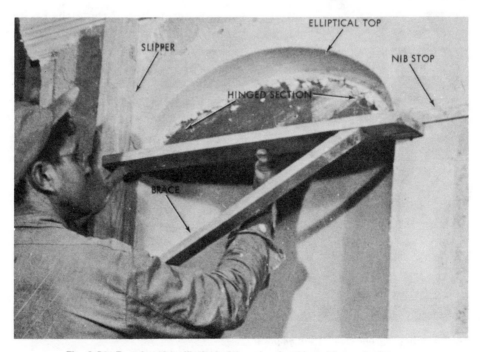

Fig. 9-89. Running the elliptical niche, showing hinged top swinging up.

Fig. 9-90. Running the circular top for an elliptical niche.

Fig. 9-91. Running one half of the elliptical niche.

at the spring line of the niche on its rear wall. The mold is elliptical in shape, but because it pivots on a center point produces a circular top with an elliptical back.

Fig. 9-91 shows the plasterer running one half of the elliptical shaped niche. This permits the same horse and profile to be used but a longer slipper is put on the mold to give it stability. The mold is run upside down to complete the other half of the niche. The joining where the nib ran on the screed must then be filled in and troweled smooth. A full mold could be made for this lower section and the upper profile filed to it so the two would match. The size of the niche would determine which method to use.

Arches

The plasterer is called upon to produce many types and shapes of arches. The variations are almost unlimited. A few of the basic methods used to run some of these arches will be covered. Each method shown can be used for many different arches. The shape of the arch or the type of ornamentation may vary but the method used to produce it will be the same.

For all arch work the first step is to form the screeds the mold will run on. Develop a plumb screed at each side of the arch, then form the screed around the arch working across the vertical screeds with the straightedge.

Fig. 9-92. Low crown Gothic arch, showing radius points and methods used to establish them.

Low Crown Gothic Arch

The low crown Gothic arch is formed by running the mold from two radius points set well within the span of the arch. The radius used and the location of the radius points will determine the curve and height of the arch. Formulas for setting the radii for the various arches to be shown can be found in various mathematics books. Appendix A at the end of the book gives detailed information on developing arches.

To permit the slipper and other members of the arch to run below the spring line so a clean cut can be made at this point, the cross arm with two uprights for radius points is set well below the spring line. Fig. 9-92 shows how this cross arm has two uprights located so the radius points can be centered on them. These uprights are firmly braced to the back wall and everything is stuck in place with high-gauged plaster.

The farther apart these radius

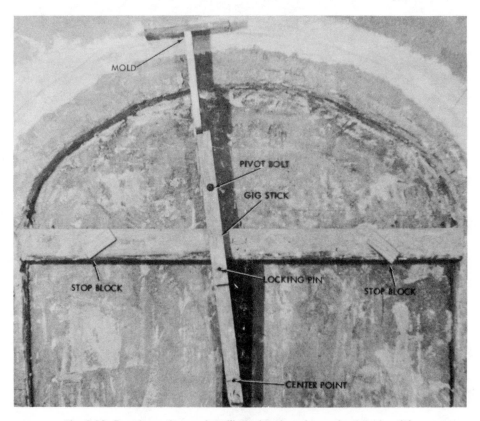

Fig. 9-93. Running a three-point elliptical arch, using a pivoting gig stick.

points are located from the center line the longer the radius arm must be, increasing the height and sharpness of the arch in relation to its width. For a simple mold the miter can be worked in by hand. If the mold has complicated members a section will have to be run first over raw putty, removed and cut to fit in the miter after the arch has been run.

Three Point Elliptical Arch Run Using A Pivoting Gig Stick

As shown in Fig. 9-93, the radius arm has a split section which permits the arm to pivot at the proper points and form the short radius as well as the long radius without changing the arms. This radius arm is now locked in position to run the long radius. When the arm reaches the left hand stop, the locking pin is pulled out and the upper part of the arm pivots on the pivot bolt to complete the short radius.

Fig. 9-94 shows the arm pivoting so the short radius can be run. The stop blocks are set so the arm will hit them when the pivot bolt crosses the spring line marked on the cross arm. The run is started at the right hand side with the long arm held solidly against the stop block, and

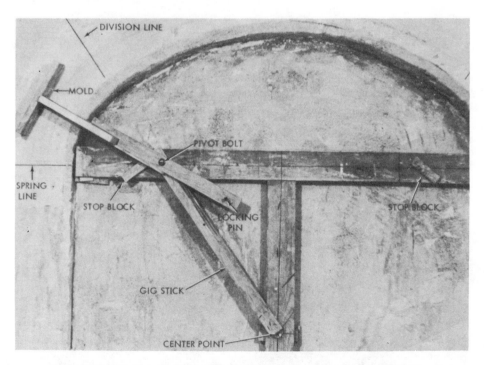

Fig. 9-94. Gig stick pivoting at the stop block to form the short radius run.

the locking pin removed so the mold can swing down to the spring line. With blocking material in place, run the mold up on the short radius until the two arms are over each other, slip the locking pin in place and continue the long run until the left hand stop block is reached. Pull the pin out and complete the short radius run. It will require some practice to make this a smooth working operation.

Elliptical Arch Run With a Trammel

The true elliptical arch can only be formed by using a trammel or by following a template or line established by the trammel. Use a regular trammel with one arm removed. The trammel is constructed to fit the condition and firmly stuck in place. Clearance is provided at the ends so the mold can run below the spring line. This is the smoothest operating setup that can be used to run this type of arch, also the quickest running because it works in a smoooth flowing motion with no stops or change of radius.

Arch Mold Run Against a Template

Fig. 9-95 shows a method that can be used to run arches, both face and soffit, when there is no room for a

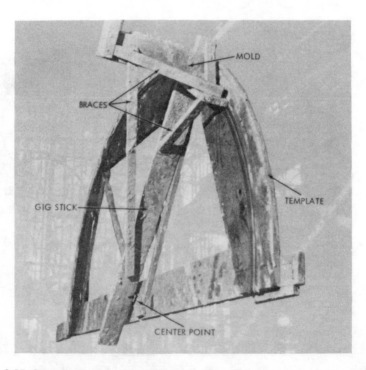

Fig. 9-95. Setup used to run the soffit of an arch, using a template as the screed.

screed or when running a series of arches of the same radius. The wood template provides a quick setup which can be placed in an arch opening and the complete arch run without screeding or setting radius points.

The template is set on wood blocks stuck to the arch opening just below the spring line and then plumbed at the correct wall line. The mold has a long slipper set on the center of the horse so the mold runs very smoothly. Because this is a complete unit setup, there is very little chance for error or poor runs due to bad screeding. The template setup is also very useful when a slipper run at the top of the mold would interfere with the arches already run or with structural framing.

Lunettes

A lunette is an opening in a vault or barreled ceiling, usually to provide for a window or a recess for decorative effect. The lunette as a rule is a circular penetration into the main vault with the resulting arris line producing an ellipse. However, there are various other shapes such as elliptical, raised crown, Gothic arch type, etc.

The basic problem for the plasterer in forming the lunette is to produce the arris line. If the rear portion is a blank wall, the half-circle lunette can be formed by swinging a large, square wood mold from a center point as shown in 9-96. The lower legs of the lunette are usually *stilted* or, in other words, the arris is vertical for about a foot before the curve starts.

As shown, the mold develops the circular shape of the lunette, forms the rear wall angle but does not finish the arris of the lunette. If this is to be a plain lunette without ornamentation the arris must be completed to a true line. There are a number of ways this can be done. One way is to form a true screed on the surface of the vaulted ceiling and then work up the arris by gradually finishing sections of the arris using a short flexible strip bent around the arris as a guide.

A second method, as shown in Fig. 9-97, can be used either to form a plain arris or to run a molding on the arris line. The round metal rods are set to marks established by measuring each way from the raw arris on the two surfaces. The rods are retained by nails driven in the screeds and backed up with high-gauged plaster. The mold is hinged at the center, and the hinge permits the

451

Fig. 9-96. Forming a lunette, using a square-arm mold run from a center point.

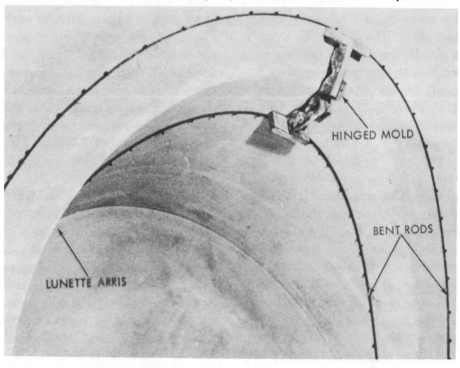

Fig. 9-97. Bent rod and hinged mold used to run lunette arris.

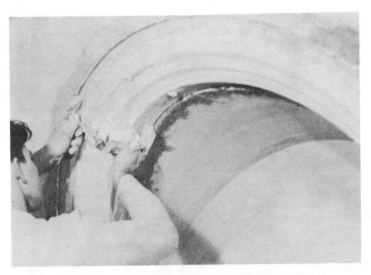

Fig. 9-98. Running a molding around the lunette arris.

mold to conform to the changing contour of the lunette.

Fig. 9-98 shows the plasterer running the mold on the lunette. This type of mold must be held very firmly and the pressure must be maintained outward against both strips at all times or the mold will leave the run. It is best to run each side up to the center, which means the left hand side of the run is made with the mold running backwards. The mold must therefore be cleaned out regularly.

Another method of forming the lunette arris line is to use a template constructed to the shape of the lunette and its arris line. Fig. 9-99 shows the first step in making this template. First make two templates equal to the radius of the main ceiling as shown in the picture. Next

make a semicircular solid top template or drum equal to the radius and shape of the lunette.

The two large templates are set up as shown, and the lunette template is set at right angles to the main templates. Make a marking stick with a pencil fastened at one end. This stick is guided over the two templates and the pencil marks a line on the lunette drum. This is the lunette arris line. The lunette template and the main ceiling templates must be set in line and fastened securely before the marking starts. This setup marks only one side of the lunette template or drum. The main ceiling templates must now be moved over to the other side, set up and the marking completed.

Fig. 9-100 shows the completed

453

Fig. 9-99. Marking the lunette arris line on the lunette drum.

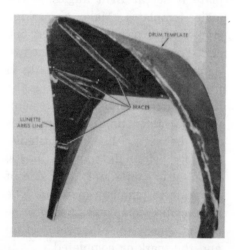

Fig. 9-100. Completed lunette template, showing shape and braces.

lunette template. The marked drum was carefully cut along the line marked on its surface and the lunette end braced underneath to support the projecting lip. The template should be shellacked or painted to protect it from the moisture of the plaster so it will not warp.

Fig. 9-101 shows a lunette template being set in place in the rough opening. The template is set to a line stretched across the opening at the bottom and upper section. The plasterers are shown tying the template to the rear wall of the lunette.

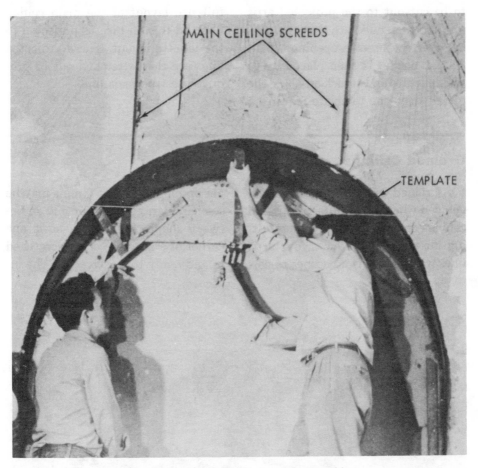

Fig. 9-101. Setting lunette template in position.

Extra-high-gauged plaster is then placed all around the arris line. It is forced into the space between the template and the rough opening to insure a good screed being formed on the lunette side as well as the main ceiling line. Always grease or oil the template before it is used so the plaster will not stick.

This method is a great time saver when there are a number of lunettes to form. It insures a perfect arris line with absolute accuracy for each lunette. If there are only a few lunettes to form, this method would be quite expensive and time-consuming because of the work required to make the template; however, it cannot be matched for accuracy.

There is still another method used to develop the arris line of a lunette. This calls for a sheet-metal template

455

to be cut out to the projected arris line and then nailed to the main ceiling over the lunette opening. High-gauged plaster is then placed up to the template and struck off along the metal edge. The template requires a full size layout of the main ceiling line and this, in turn, is projected to the lunette layout. These projected lines are then stretched out to produce the true arris line.

Groined Ceilings

A groined ceiling is the shape developed when two vaults or barreled ceilings cross each other. The resulting intersection produces four arris lines which cross from corner to corner on diagonal lines. Groins may be semicircular, elliptical or any in-between shape. Fig. 9-102 shows one corner of a groined ceiling scratched in and ready to be screeded. The in-

Fig. 9-102. Placing high-gauged plaster over template to form ceiling screed.

tersecting vaults are shown developing a sharp arris line diagonally across the area.

Templates are made to produce the vault screeds at each wall line or, if there are no walls such as crossing corridors, the screeds are formed on the vault where the vault wall lines cross. Fig. 9-102 also shows the template set on blocks leveled at each corner of the room. The plasterer is forcing high-gauged plaster over the template to form the screed. Repeat this operation on all four sides of the area.

In Fig. 9-103 the plasterers are shown rodding off the high-gauged plaster placed on one side of the groin. The rod must be held at right angles to the end screeds at all times. If the rod is permitted to slant it will distort the screed on the groin. When this one side is rodded off reverse the rod and run it over the opposite screeds to cut off any material that is projecting into this side. Repeat this operation at each side of the four groin arrises to form a perfect arris line. Go over the screeds with fresh material to smooth them to a true surface.

Another method used to produce the groin arris lines is to use a diagonal template as shown in Fig. 9-104. The template is set on the supporting blocks at the corner and tied up to the correct height at the center point or leveled at the bottom line

CEILING SCREED
AT WALL LINE

Fig. 9-103. Forming diagonal screeds, using straightedges.

Fig. 9-104. Forming the diagonal screeds, using a template.

of the template. Notice the brace nailed to the template and the ceiling at the mid point to steady the template.

Fig. 9-105 details the method for developing the diagonal template from the wall template. Lay out this template full size on a section of clean floor. Divide one half of the base line into any number of equal spaces. The greater the number of spaces the most accurate the transition will be. However, 6″ spaces will produce a good job on the average size template. Notice that the last space at the end of the template

has been re-divided into four equal spaces. This was done to provide more reference points at this sharp curve.

Now draw a series of lines up from the base line forming a square as shown by points *A, F, G, D*. Also a center line at *B* through *C* and at *C* parallel to the base line *A-F*, or in other words divide the square into quarters. Next draw diagonal lines from corner to corner through center *C*. Project the spacing marks *1* through *8* and subdivision of space *8*, up to the diagonal line.

Where these lines intersect with

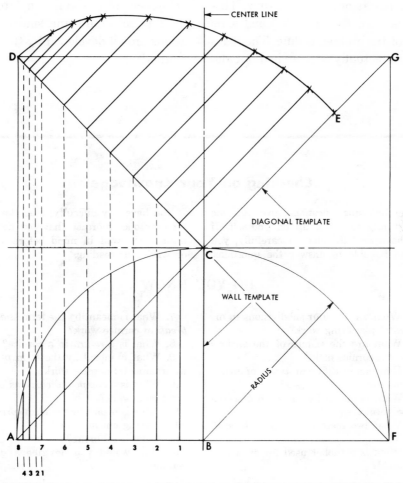

Fig. 9-105. Development of diagonal template for arris line from semi-circular wall arch, using ordinates.

the diagonal line draw lines running parallel to the opposite diagonal or at right angles to the first diagonal line and of lengths equal to each line drawn in the quarter circle. In other words, line *B-C* on the wall template is measured and marked as shown *C-E*. Then each succeeding line is measured and transferred to the diagonal line. This can be done quite easily with a steel tape and marking crayon or heavy pencil.

When all the lines have been developed, connect the marks, using a flexible strip like a cornice strip set on edge and bent to connect at least

459

three points or more at a time. This line is then the true diagonal arris line of the groined ceiling. The diagonal is actually a stretched-out development of the wall template.

Elliptical groins and similar variations are all developed by the same method.

Checking on Your Knowledge

The following questions give you the opportunity to check up on yourself. If you have read the chapter carefully, you should be able to answer the questions.

If you have any difficulty, read the chapter over once more so that you have the information well in mind before you go on with your reading.

DO YOU KNOW

1. What are the four subdivisions in ornamental plastering work?

2. What are the names of the various parts of a cornice mold?

3. How many different types of molds can you name?

4. What is meant by the term *blocking out* in cornice work?

5. What two methods are used to *stuff* a mold?

6. What is a rabbet used for on a cornice mold?

7. What is meant by the term *sticking a break* in cornice work?

8. What is a *return* in a cornice?

9. What is meant by the term *planting* in ornamental cornice work?

10. What is a *compound curve* as used in plastering work?

11. What is meant by the term *spring line* when laying out an arch?

12. What is a lunette?

13. How would you describe a groined ceiling?

Plastering Problems

The plastering trade, more than any other trade perhaps, must be aware of, and take precautions to protect itself against, conditions that can possibly ruin an otherwise fine job. In general, there are two main difficulties which, if not controlled, might make for a bad job of plastering. Then there is a third difficulty that affects the other two.

First, weather has a decided affect on newly applied plaster. If the weather is extremely wet or extremely dry, precautions must be taken to prevent injury to the plastering job. Newly applied plaster must cure properly. Very dry weather makes the plaster dry too quickly. Very wet weather causes the plaster to *rot*. (If plaster remains wet, after it sets for a prolonged time, it will cause the gypsum to break down and become weak.)

Freezing prevents the plaster from curing or setting properly.

Second, the proportioning and mixing of materials is another vitally important subject that affects the quality of the plastering job. The importance of proportioning and mixing of plaster mortar was shown in Chapter 4.

Each of these two main factors are linked with a third condition: the human element of personal error and misjudgment. A plasterer can misuse the materials he has been given to work with no matter how excellent their quality was to begin with. First, he can fail to proportion and to mix them properly. He can fail to apply the plaster mortar correctly. The plasterer can also fail to take the important precautions against weather changes.

Any of these conditions (weather,

461

materials and the human element of error) taken separately, or in combination, may spoil the work. Of the three elements or conditions, the personal human element is, perhaps, the most important, for in the final analysis it is the plasterer's personal knowledge and personal integrity that determine the outcome of the plastering job with respect to the quality of the work. It is the individual plasterer who can, because of his knowledge of the plastering craft, overcome the difficulties of uncertain weather conditions as well as proportioning and mixing problems.

Common Plaster Problems

Some of the problems that the plasterer may encounter are discussed here. The topics are dryouts, sweatouts, frozen plaster, oversanding, overly retarded plaster and color problems. Plaster cracking and its causes and, finally, repairs and alterations will be discussed later in the chapter.

Dryouts

Dryout is a condition found to occur on many jobs during hot weather. It also occurs on jobs done under unfavorable heating conditions in winter. Fig. 10-1 shows a section of plaster that has dried out.

During the summer when it is extremely hot and a dry wind is blowing, these spots will appear opposite open doors and windows and around these openings as well. A dry wind evaporates the water moisture quite rapidly from freshly mortared wall areas lying directly in its path.

When the water moisture in the plaster mortar is removed before the chemical reactions of setting have taken place, the plaster cannot set hard. Therefore it returns to its original condition before it was mixed; in other words, it is not mortar at all but calcined gypsum.

Hot air from a furnace, or steam radiators set too closely to a wall that has been plastered, will produce the same effect.

The method used to prevent this condition is to cover all openings so that a direct wind cannot strike the wet plaster. Muslin or any lightweight cloth will allow ventilation while preventing drafts. Furnace openings should be shielded so that the hot air is diffused. Steam radiators should be set out from the wall until the plastering is completed and dry.

If dryout spots do appear, the fault can be corrected in most cases by rewetting the area. Water sprayed on the dry spots will start the setting action again, and usually will harden these areas. Alum mixed with

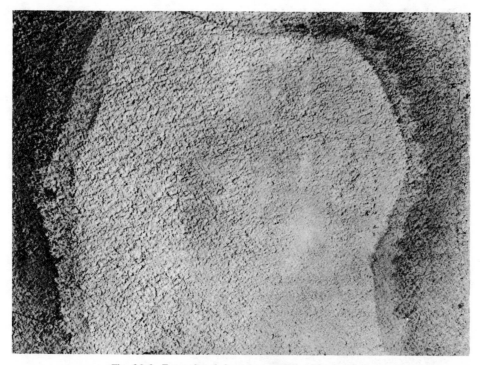

Fig. 10-1. Example of dryout caused by blasts of air.

water will assist in the action. If these procedures do not correct the condition, the spots will have to be scraped off and fresh material applied.

Veneer plastering, because the material is applied in one or two extremely thin coats, is very sensitive to dry winds and must be protected at all times from such conditions. If problems develop due to too fast drying, even with the windows and doors covered, the setting time of the material must be accelerated and smaller batches mixed so the material can be applied, finished and set before the dryout can occur.

Exposed aggregate applications, both exterior and interior, need protection from both wind and sun. These two can cause the matrix to stiffen too fast, thus preventing the proper bedding of the aggregate. Canvas covers should be hung around the scaffolds on exterior work to keep the hot sun from shining directly on the work and to prevent the wind from blowing over it. Working in the shade by following the sun around the building often helps.

Portland cement work should always be protected from strong drying winds and the hot sun. Drawing the moisture out of the cement mortar before it has set naturally will weaken the work considerably. If this condition should occur and the mortar can be dampened with a fine fog spray of water before it has dried out completely, the setting action will be reactivated and the cement will attain its proper set and strength.

Sweatouts

Sweatout is a condition that results from a situation just the opposite of that explained as the reason for dryout. Sweatout occurs where there is too much moisture. Here the plaster stays wet and will not set. This condition results from the lack of air circulation.

Normally, when plaster mortar is mixed, more water is used than the amount needed to set the plaster. It is necessary, however, to use the extra amount of water in order to produce a workable mortar. Proper circulation of air will remove the excess water gradually and allow the plaster to set properly.

When all of the doors and windows are shut tight, the excess water is

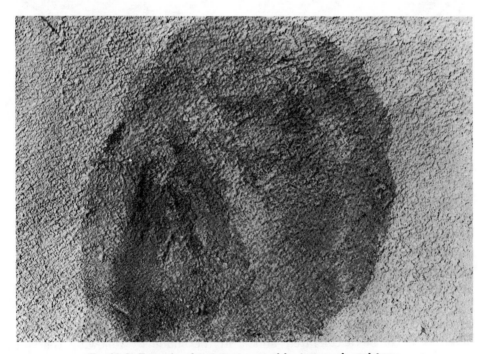

Fig. 10-2. Example of sweatout caused by too much moisture.

not permitted to escape, and the result is a wet area or room. This condition, if it is not corrected quickly, will cause the plaster to rot. Air must be allowed to enter and leave the building freely, even in the coldest of weather. Windows should be left open a little at the top and at the bottom until the moisture has been removed. In Fig. 10-2 the rotted condition resulting from the sweatout can be seen. The plaster is dark in color, moist, and without strength or ability to harden.

There is no cure for plaster that has rotted. The work must be removed and the area affected must be replastered.

Veneer plastering also, like all gypsum plastering, must be protected from extreme moisture conditions which would prevent the plaster from drying out within a reasonable time, usually overnight. Air must be allowed to move throughout the area without drafts so the excess moisture can be removed.

Non-gypsum materials such as Portland cement and plastics are not usually affected by extreme moisture conditions. The working conditions will be affected to some degree, but not the strength of the materials.

Frozen Plaster

When plaster freezes, it loses all of its strength. This happens, however, only if the freezing occurs while the plaster is still soft and has not

set. Usually, if the plaster is set, no damage will be done to the brown coat; the finish coat, though, will expand and scale off. Never apply putty coat over frozen brown mortar. Such work will peel off later.

Frozen plaster can usually be detected by the appearance of a white fuzz on the surface of the plaster area. Fig. 10-3 shows a section of plaster that has been frozen. Note the white spots. Proper heat and ventilation will prevent the occurrence of this condition. Never plaster in an unheated building when the temperature is near freezing.

Plaster that freezes *before* it has set will have to be removed. Plaster that freezes *after* it has set, usually can be thawed out by heating the building. Finish coats are more subject to damage by freezing than are brown mortar coats. Damage to the job results from expansion of the water retained by the plaster. The plaster bursts, then it peels and scales. This means that the job must be done again.

The new materials now on the market must all be protected from freezing. Even though some of them might seem to have resisted damage if applied during freezing weather without adequate heat, the effects of the freezing will show up the next year with the frozen areas crazing, blistering and peeling off.

In Portland cement work the greatest damage caused by light

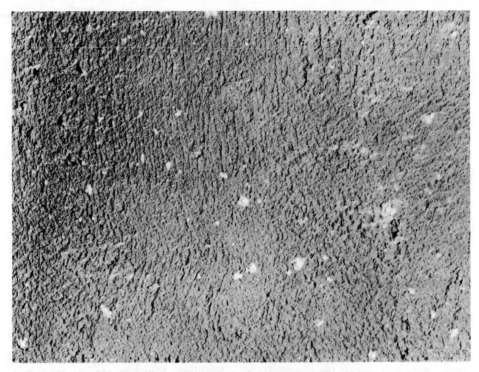
Fig. 10-3. Fuzz on face of plaster indicates frozen plaster.

freezing is in causing the various coats to separate later. Hard freezing will cause the mortar to crumble when warm weather occurs, because the frost prevented the water from chemically activating the setting action of the cement, thus leaving the cement and aggregate separated.

For all plastering materials, never use mixed materials that have frozen and then thawed out. This would include lime putty, sand finish, acoustical plasters, etc. The freezing action breaks down the cementing materials so that they no longer have any strength or holding power.

Oversanding

Oversanding is one of the evils common to the trade. Plasterers who seek to save on material cost sometimes use a few extra shovelfuls of sand per bag of plaster. In doing so, they weaken the mortar seriously. Cracking and loosening of the finish coat are the consequences of this practice. Little or no saving is effected because, using a poor mix, the mechanic drops more mortar than he would ordinarily if the proper mixture were used. Moreover, the material spreads with less ease if a poor mix is used.

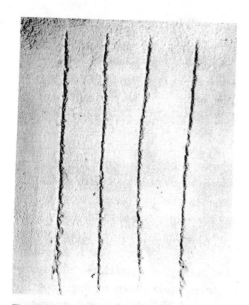

Fig. 10-4. Deep scratches indicate loose, oversanded mortar.

Fig. 10-5. Shallow, clean scratches indicate correct amount of sanding.

It is easy to test for this condition: just scratch the surface of the mortar after it has hardened. If the quantity of sand used was too great, the surface will scratch deeply and the mortar will crumble. See Fig. 10-4.

Good mortar will resist scratching; in fact, hardly a mark will show when it is rubbed hard with a nail. See Fig. 10-5. A sand-colored mortar is a good indication that the mix is of poor quality.

Most of the newer materials now used by the plasterer require very carefully measured quantities of the various ingredients. The chance of oversanding is reduced because of the small amount of sand needed and

because any oversanding will usually make the mix almost impossible to apply properly. This is particularly true in Portland cement work where oversanding would technically make the mortar stronger, but would be too poor to apply as a plaster.

Overly Retarded Plaster

Overly retarded plaster is the result of one of two possible conditions. Either the plaster contained too much retarder as it came from the mill or too much retarder was added to the mix by the mechanic at the job site. The result in either case is that the mortar does not set. If an overly retarded plaster is left on walls or ceilings, and no remedy for the condition is administered, it will dry out and become soft and powdery.

Ordinarily, plaster mortar should set within four hours on lath bases. It should set in less time when placed over masonry or scratched-in work. The rule to follow is that the mortar must be set before all of the water moisture has been drawn out of it by evaporation. As said before, water moisture is removed either by air currents or by suction. These factors cause wide variations in the time needed to accomplish the desired result. Therefore the plasterer must judge the time element for himself and retard or accelerate the mortar accordingly.

If over-retarding occurs, the remedy is this: spray the plaster either with a solution of alum and water or with zinc sulphate and water. Either of these solutions will accelerate the set of the plastering material. Lumps of retarder should not exist on the surface, particularly if the mortar has been well mixed. If lumps are found on the surface, they must be removed. If they are allowed to remain, they will burn through the finish material and the paint to be applied later.

Veneer plaster is extremely sensitive to over-retarding. If the material does not set within the required time (as soon as it has been troweled to a smooth, dense surface) the surface will become completely covered with small water blisters. This condition then requires much extra troweling after the material has set. It will never achieve the true smooth polished surface obtained when the set occurs at the right time.

Exposed aggregate matrixes of the epoxy, acrylic, and latex types are all subject to many problems if the materials are not proportioned correctly. The matrix, to be able to bed the aggregate properly, must not set or harden too fast; yet, if it stiffens too slowly, it may cause the aggregate to slide down or sag, creating spots of thickly massed aggregate with other areas having no aggregate. Equipment used to mix and apply these materials must be kept clean at all times. Keep mixed material covered as much as possible until it is applied.

Indiscriminate use of any retarding agent for all types of plastering materials is one of the most common causes of problems in applying and finishing them. The plasterer *must* follow the manufacturer's directions to the letter, and *must* make sure the plasterer tender follows these directions also. Gypsum-based plasters can stand more abuse than others in this respect and still produce a fair job. The newer materials cannot; their proportions must be measured accurately for each and every batch mixed because added ingredients produce an unsatisfactory mixture.

Color Problems

Any coloring added to plastering materials will have some weakening affect on the overall strength of the mix; therefore, the amount of color used must be very carefully measured and proportioned. For most Portland cement mixes the amount of mineral color added should not exceed 6 lbs. of color per bag of cement or 6 percent of the weight of the cement used.

To insure true color matching for the total area to be covered, measure all ingredients carefully. Even the water used must not vary between batches. Never add more water to the mortar while it is on the mortar board because this will wash out some of the color and change the final appearance. This same rule applies to water used to float or trowel the finished surface. Applying water at this point will also change the color, usually making the area cloudy or mottled in appearance.

Differences in the suction of the base in certain areas will also affect the final appearance of the color in those areas. High suction in a given area will hold or pull the color fast to that area, while a nearby area that is damp will permit the color to wash out or be floated and troweled out. To prevent this action the base or brown coat should be uniformly dampened the day before the finish coat is to be applied. This will provide a uniform suction over the whole surface for the finish coat application.

The newer exterior materials usually have carefully measured formulas for using color set by the manufacturers. These must be followed to the letter to insure a good job. Again, the sun and even the wind could affect the final color, so these conditions must be considered at all times.

Even the tools used by the individual plasterer may affect the final appearance. For example: identical float bases must be used by each plasterer or else a different texture and seemingly different color will show on a given area of work if the float base material (rubber, cork, wood, or plastic) varies.

Varying thickness in the finish coat may cause a difference in color, as the thin spots will usually appear duller and not true in color. Uniform thickness is important for uniform color appearance.

Not often encountered, but always a possibility, is shadow "photographing" on the finished surface so those portions dry slower. This is due to the scaffold being left in place too long after the work is finished. This creates uneven curing conditions by the shadows the scaffold forms on the plaster surface. Fog spraying or brushing of a factory-mixed color coating over the area with color variation will in many cases correct this problem.

Plaster Cracking and Its Causes

When cracking occurs in plaster, the fault may be due to any of four possible conditions and, in some cases, to a combination of all four. These conditions are: poor materials, structural strain, careless workmanship, and faulty construction.

Today, the most common cause of plaster cracking is the combination of faulty construction of the building and carelessness on the part of the plasterer. A coat of mortar so thin that it is jokingly referred to in the trade as "paperhanging" is clearly not ethical practice.

The average thickness of the brown coat on gypsum lath in most houses built today is ¼". The correct thickness is ½".

Poor Materials

The result of a poor mortar and a thin coat of regular compound (fibered gypsum plaster) is shown in Fig. 10-6. Here the careless workmanship was in evidence even before the building was finished. The mortar was oversanded to such an extent that it fell off the lath. Large cracks appeared at the lath joints because of the weakness of the mix and a thin spread.

Structural Strain

Structural strain produces crack-

Fig. 10-6. Very thin coat of weak mortar over gypsum lath.

ing in a large prominent crack extending across the surface and through the plaster, and in many cases through the backing as well. It is often found at the corner of a door or window: at the junction of two planes. It is caused by movement of the building structure and is not the fault of the plasterer.

Structural cracking is caused by direct transfer of stress from the structure to the plaster. These stresses usually result from movement of the structural members caused by warping, shrinkage, expansion, unequal settlement, or deflection.

It is often better to wait a reasonable amount of time before repairing this type of crack. For example, where this crack was caused by settlement of a building, it would be more practical to wait until the building has "found itself" and then patch and fill the crack.

Careless Workmanship

In finish coats, careless workmanship also furthers cracking of the plastered surface. Two common types of cracking found in the finish coat are chip cracking and map cracking.

Chip Cracking. Chip cracks are short cracks that take various directions and form patterns in the plastered surface. See Fig. 10-7. The

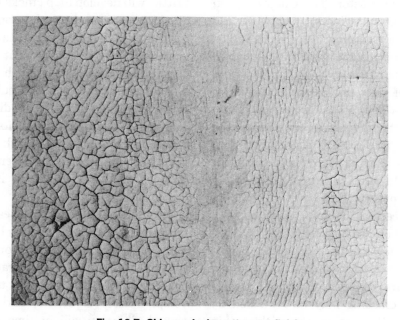

Fig. 10-7. Chip cracks in putty coat finish.

471

causes of chip cracking are numerous, but all of the causes are related to a single consideration, namely, that of suction. Chip cracking never occurs if the finishing plaster is applied over a damp base coat.

The factors most often responsible for chip cracking are: (1) over-use of retarder, which allows the water moisture to be drawn out of the plaster mortar by suction before it has time to set; (2) an excessively dry condition of the base coat, which causes too much suction and makes the finish shrink and crack; (3) hot, dry air directed upon the finish, which causes it to dry too quickly; (4) badly prepared putty that shrinks and cracks because of inadequate mixing; (5) insufficient use of gauging plaster (see Chapter 4 on mixing ingredients); and (6) not enough troweling.

With respect to this last condition, not enough troweling, if a poor job of troweling is done, the finish coat will be left in a spongy condition. Thorough troweling forces the putty into a hard, dense layer and, as the gauging plaster sets, the putty is held firm. Further shrinkage is then prevented.

Excessive chipping will cause a poor bond to be established between the finish and the base. Work that is chip cracked should be removed and the area refinished. Minor cases of chip cracking may be remedied by filling the fine cracks with a thin

consistency of finishing mortar. This work is done with the trowel. Be sure to wet the surface before filling the cracks. This is done in order to prevent the dry plaster from absorbing through suction, all of the moisture of the newly applied plaster.

Map Cracking. Map cracking, as the name signifies, is seen as a series of fine cracks that seemingly represent territorial boundaries of states or nations, because they are often long and irregular in shape. Depending upon the cause, the cracks may be fine, almost invisible, or quite open and conspicuous. Typical map cracking is shown in Fig. 10-8.

One common cause of this condition is weak brown mortar. The finish coat, being harder and more brittle, will develop map cracks if applied over a weak browning mortar. Such a mortar may be one that has been applied too thinly, or it may be one that contains too much sand and is, therefore, weak and subject to fault. Either of these conditions presents a weak foundation for the brittle and thinly applied finish coat. Temperature changes also tend to produce cracks of this type if the brown mortar is weak.

Another condition that causes map cracking is vibration. In many buildings situated along highways or near railways or factories, map cracking caused by vibration is likely to develop. Plaster is brittle, and continued vibration will in time cause

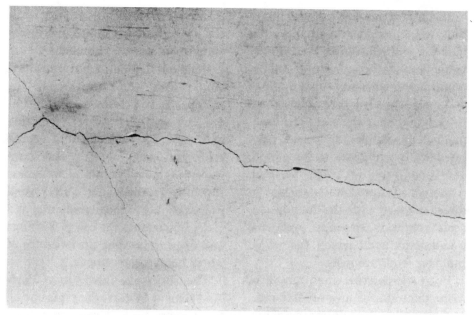

Fig. 10-8. Map cracks in putty coat finish.

the plaster to crack in large sections. The cracks may be so fine that they will be visible only when the area affected becomes soiled.

Suspended metal lath ceilings are susceptible to map cracking when exposed to great temperature changes. If the channel irons are wedged into the walls so tightly that there is little room for expansion, the ceiling may become sprung. The movement of the building will gradually cover plastered surfaces with a network of faintly visible cracks.

The use of vermiculite, perlite and similar lightweight aggregates may eliminate some of this type of cracking because of the greater flexibility of these materials. The best way to prevent map cracks is to insure adequate application of base-coat materials and to be sure that construction of the building itself is accomplished properly. Construction of the building in which plaster is to be placed is not the responsibility of the plasterer. A settling building will, of course, cause the walls to sag and dip.

Faulty Construction

The basic cause of all large cracks in a plastered surface is poor construction beneath lathing and plaster. If a building is put together with inferior materials and poor workmanship, the plasterer will not be able to produce a creditable job.

If construction is weak and materials used are poor in quality, the building will constantly develop defects. Foundations built of inadequate size or without proper footings will make the structure above them more liable to settlement. The basement walls will move and crack, plastered walls will cease to be plumb, and ceilings will no longer be level. Although plastered walls are not in direct contact with the foundation, their condition depends upon the foundations upon which the whole building structure rests.

Poor carpenter workmanship, more than almost any other factor, is responsible for plaster cracking. Improper bracing, poor corner construction, faulty framing of doors and windows, and weak joints that must carry partitions and other deflecting loads are other factors involved. Even the best job of plastering cannot stand up under structural fault or undue strain. The use of wet or green lumber can also cause cracking. A building should be permitted to dry and settle before the plaster is applied.

All the structural faults listed above for causing cracking in regular plastering work will, in most cases, cause cracking also in veneer plastering, so these faults must be corrected or overcome to insure a crack-free job. Chip cracking and map cracking are not as common in veneer plastering because the faster setting of the material prevents chip cracking while the stronger plaster used overcomes many map cracking problems.

Portland Cement. Exterior Portland cement work requires all the safeguards, to prevent cracking, that are followed for other plastering materials. These include: a strong structural background; proper reinforcement at all points of weakness (openings, corners, joints, etc.); good proportioning of the ingredients; and (very important) the use of flashings and drips to prevent the entrance of water behind the cement.

The shrinkage inherent in Portland cement mortar when it is curing and the cracks that can result from this condition can be minimized by: (1) the proper proportioning of the materials, (2) the use of good sharp sand properly graded as to size, (3) applying coats of mortar so the total thickness of this applied mortar is as monolithic as possible (machine application is very helpful in this respect), and (4) the proper curing of the cement during its hardening stages.

Using expansion joints (really contraction and expansion joints because they allow movement to take place in a controlled area) at regular intervals (some jobs have successfully used a 10 foot spacing) will prevent many of the large cracks that appear, after a period of time, on some jobs.

Proper thickness of the applied

mortar is an important factor in crack resistance. The total thickness should be one inch for cement work over metal reinforcement; masonry and monolithic concrete surfaces may require only ¾″ thickness. In all cases, it is important to keep the total coat thickness as uniform as possible. A varying thickness in certain places may create planes of weakness which will cause cracking at the thinner areas due to the difference in contraction at these points, or in bending movements across these points.

Another cause of cracking in Portland cement work is the use of different mixes for each of the coats used. Some plasterers and plasterer tenders think that the scratch coat should be made very rich so it will adhere better to the lath, then the brown coat can be made poor so it will rod and float better, and finally the finish coat should be made very rich so it will spread "like butter," as the expression goes.

Each of these differences in the mixes will set up a different rate of contraction during the hardening process. The end result will be a whole series of cracks because of the stresses that have developed. All coats should use the same ratio of cement to aggregate.

Good uniform scarifying of the surface between coats is also an important crack preventive. Scarifying must not cut too deep into the soft

mortar or it will weaken the coat, while too shallow a scarifying will prevent a good bond from developing, with the result that the coats of mortar may separate and crack later.

Fig. 10-9. Bad spacing of wood lath; mortar is unable to form good keys.

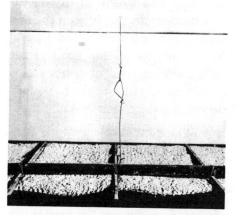

Fig. 10-10. Poor hanger construction and sagging laths.

Lathing. In Fig. 10-9 one of the forms of faulty lathing is shown. Note that the wood laths (not common today) are so close together that the mortar could not work between them to form a good mechanical key. Before the use of gypsum laths, wood laths were applied close together to save plaster mortar. The result was a plaster wall that soon cracked. Then plaster fell off.

Metal lathing can be made a poor plastering surface through faulty workmanship. Fig. 10-10 shows sagging laths and poor hanger construction. A condition of this sort allows the plastered ceiling to sag and to crack. The application of metal laths on wood joists or studs, using a size smaller than 5 penny lathing nails, is another common cause of falling ceilings and cracked wall plaster.

Other Problems

There are a number of other problems which can affect plastering, both interior and exterior. One of these is *nail pops,* where small areas of plaster directly over the nail head pop off exposing the nail head. This action can be caused by a number of reasons: nails not driven in far enough; wet lumber used during construction and which on drying exerts tremendous pressure on the nail shank, forcing it outward and pushing the plaster off the head of the nail; the wrong size nail head (too large); nails with shanks too short, preventing them from holding properly when normal drying of the lumber takes place.

Other pops or pitting can occur when foreign materials (that will expand when wet) are embedded in the plaster. Even though the plaster has dried, the foreign material continues to exert pressure until a small section is popped loose. Some of these foreign materials are: particles of burned or unsoaked lime, particles of iron, coal, seeds blown into the sand pile, plus many other impurities.

Stains and discoloration are caused by tar, creosote, soot, oils, and similar materials coming in contact with plaster either by getting mixed in the mortar or by smearing or soiling the base on which the plaster is applied. In most cases it will be impossible to correct this condition except by cutting out the stained plaster down to the lath or masonry base and removing the stain-causing material.

Efflorescence (a white powdery salt-like development on the surface of exterior or interior work) is caused by salts in either the mortar itself or in the base over which the plaster has been applied. Moisture getting behind the exterior plastering wall bring out the salts, if present, in masonry walls. When the salts are in the sand used to make the mortar, the moisture in the air will gradually bring the salt to the surface. There is no simple cure. Sometimes if a very small amount of salt is present, it can can be washed off until it has all been bleached out, or the area can be washed with a diluted (ten parts water to one part acid) muriatic acid solution. The acid must be flushed off with repeated washings of clean water or it will affect the color and strength of the finish material. Otherwise, the only cure is to cut out the area affected and replace it with salt-free material.

If the salt is in the masonry base, sometimes waterproofing the cement surface and caulking all places where water or moisture might get behind the mortar will eventually stop the efflorescence.

When plastering over foamed plastic lath, the lath manufacturer's directions must be followed to the letter. Some gypsum manufacturers disclaim this type of plastering base for their materials.

When applying putty coat finishes over lightweight aggregate basecoats, the material should include some lightweight fines or silica sand to help reduce the chance of map cracking.

When spraying decorative finishes directly over monolithic concrete ceilings, the concrete must be clean and free of all oils, grease, dirt and dust.

When spraying insulation, fireproofing or any material containing asbestos, the plasterer and all other persons in the work area must wear dust respirator masks to prevent the inhalation of any asbestos fibers.

Repairs and Alterations

The plastering industry devotes a great deal of its time to the problem of repairs and alterations. These are commonly referred to as *patching* in the trade. Patching jobs require as much workmanship as new plastering work. With care and experience, the workman can restore plastered surfaces to an excellent condition. Failures will be at a minimum if the suggestions made below are followed.

Three degrees of plaster failure, requiring patching, can occur in a plastered wall or ceiling: (1) the finish coat can become loose; (2) brown and finish coats can deteriorate; (3) lath, brown, and finish coats can become loose and all three coats can fail. These plaster failures are most often due to moisture problems.

Water Damage

In many cases when a roof leaks or water pipes break, the wetting of the plaster has little harmful effect if the plaster area affected can dry out rapidly. Repeated wetting, however, will rot the plaster and cause one or all three of the above-listed failures.

Sometimes wetting of the finish coat will cause large blisters to form. These result from the use of a particular type of lime. If dolomitic lime that has a high magnesium content is overburned, it will produce a putty with about 20 percent content of magnesium that remains unhydrated after the usual overnight soaking. If, years later, the putty coat used as a surface finish is wetted or if great humidity occurs, the magnesium in the lime will expand and cause blisters to form.

The best cure for this condition is to remove all of the finish coat and refinish with new putty coat. The use of autoclave lime as a refinish will prevent the reoccurrence of blisters because this type of lime contains almost completely hydrated magnesium.

If blistering occurs over a relatively small area and needs only to be patched, trim around the edges so as to reach the firmly bonded area. Trimming near a firmly bonded area provides a smooth running edge rather than a broken, irregular edge. To prevent the wet plaster used in making the repair from causing further blistering, shellac or paint the edges of the patch before applying any moisture to it.

Painting the edges of the patch with a liquid bonding agent is extremely helpful in preventing further blistering, and also will reduce the amount of shrinkage that may take place when the areas surrounding the patch absorb too much of the water from the freshly applied plaster. This will insure a good set and a secure bond at these edges.

Patching

Patching should be done with well-aged putty and enough gauging plaster to insure a hard, well-bonded patch. About 35 percent gauging plaster to 65 percent lime putty would be about right. All patches should be done in two steps, blocking in (filling in the area), and scraping it down straight to the existing work. When this base coat has set, finish it with some more of the same mortar material containing the same proportions as that used for the filling-in process. Then bring the work down to a smooth, level surface.

One-coat patching, however, usually shows up when completed as a slightly concave or dished surface. Later, when the wall is painted, patches can be easily detected. Irregular edges also produce a poor patch that is easily noticed after it is completed.

Separation of the final coat from

the brown coat, due to leakage, is a common reason for patching. In such cases, the loose finish coat should be removed and the area patched with a new finish coat.

Check the brown coat for firmness. Scratch it with the pointing trowel or any sharp tool. If the brown coat requires quite a bit of pressure to produce a mark, then the brown mortar is good. It may then be left as it is and the area may be patched with finish coat.

If the brown coat is also deteriorated, then all plaster must be removed down to the lathing. Then, of course, the various plaster coats must be replaced. Before replacing the brown coat, be sure that the lathing is in good condition. Gypsum laths are more subject to water damage than any other type of plaster base except perhaps lathing of the insulation board type.

The paper face and back of gypsum laths tend to separate from the plaster core if wetting occurs a number of times. To test whether insulation or gypsum boarding is damaged, feel and press the surface. Do not wet gypsum laths when patching. Wet only the surrounding plaster edges. Never patch any area without first wetting the old plaster edges except as mentioned earlier when patching putty coat finish that has blistered.

If the lathing is deteriorated, it must be removed and replaced. Such

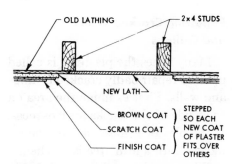

Fig. 10-11. New lath from stud to stud; new plaster stepped over old plaster.

areas must be removed back to a firm support, such as a joist or stud. Cut the laths so that they can be taken out, measuring from one support to the next one. Cut the lathing carefully in order not to crack the surrounding plaster area. Sawing or cutting may be accomplished by using a sharp hatchet or a wide-blade, sharp chisel.

Try to lap or step each operation. In other words, cut the brown coat out past the edges of the new lath to carry the mortar over and onto the old lath. Then step the finish coat in the same manner in order to lap the brown mortar joining with the finish coat. See Fig. 10-11.

This procedure will produce a finished patch that is strong and one that is invisible. If a patch is being made in a surface that is slightly wavy or bumpy, the contour of the patch should be made to conform to the irregularities of the existing work. A flat, straight patch in a wavy surface will show up readily.

479

Repairing Cracked Walls
and Ceilings

Many times the plasterer is called upon to repair badly cracked ceilings and walls. First examine the area on each side of the cracks for looseness. If the plaster is loose, it must be removed until a firm area is reached.

When cracks are large and due to the structural movement, it is wise to remove the plaster on each side of the crack to a width of about 6″ and down to the lathing or masonry. Place a strip of metal lath in the area cleared. The metal lath can be held by placing lightly driven lath nails into the sides of the existing plaster or through the lath. The placement of metal lath will reinforce the area and, in most cases, will prevent further cracking.

For the smaller cracks it is only necessary to open them up slightly by running a sharp pointed tool such as a knife or a chisel through them. Wet the crack and fill it with quick-setting finish coat. Never retard the set of patching material. Patching plaster should harden as quickly as possible in order to set before the old plaster has a chance to draw moisture out of it.

It is better to make a number of small batches of patching material and do a few cracks at a time than to attempt to fill in all cracks at once. Plaster that is retempered after it has partially dried or set will lose strength. A common mistake is the attempt to repair more cracks than can be finished at one time.

Proper wetting is most important. Wet the cracks three or four times before applying plaster patching mortar. Use a 3″ or 4″ painter's brush for this work. A painter's brush retains a lot of moisture, and water loss is at a minimum when it is used.

After the cracks or patches have been filled and have been finished, wash the area worked upon with clean water to remove the residue of lime and plaster left on the old work. When this has dried, rub it with a clean rag to remove the last trace of white dust remaining.

When patching over old wood lathing, clean out the old plaster completely, especially between the laths. Check old laths and renail them if necessary.

When the wet plaster touches the old dry wood laths, they usually twist and buckle. This occurs about four or five hours after the patch is finished. This buckling of the laths spoils the job. To overcome this trouble, wet the old laths four or five hours before applying the plaster or soak them thoroughly the night before.

As the plaster dries the wood laths dry too, and shrink at the same time. But drying and shrinking will not cause any trouble. Many plasterers nail metal laths over the wood laths to strengthen the patch. If the area is small, it can be filled in without wetting the laths, since so little water is present that no swelling of the wood will take place.

Metal laths should be checked for rusting. If metal is rusted, it is best to tie a new piece of metal lath over the old section before plastering.

Masonry surfaces require only cleaning and wetting before patching. However, if cracks show in the masonry work, a piece of metal lath should be placed over the crack to reinforce the patch.

Quite often the plasterer is called upon to fix a crack that runs through the center of a large ceiling area. Many times, after the patch has been made, cracks will develop again and the home owner is likely to feel that he received a poor repair job. Most often, however, he is mistaken. The fault often lies with the building structure rather than with the plasterer.

It will often be found, upon examination of the area above the cracked surface, that a partition or some other weight is bearing across the joist at the line of the crack. Deflection caused by roof loads upon these walls, or changes in moisture content in the wood joists, will cause movement and result in cracks.

To overcome the difficulty of cracks recurring because of load movement, two remedies are possible. Both of these courses require quite large operations. For the type of crack that shows only slightly, a wide section of the plaster is removed down to the laths. This should equal the width of a sheet of metal lath (or roughly 30") and should carry completely across the room.

Replaster by browning and finishing as in new work. This operation will remedy most of the minor types of cracks caused by seasonal movement.

For the large, wide crevices that usually occur during the dry, heated season of the year, the only cure is to completely relath and replaster the entire ceiling. To be successful, the old ceiling is first covered with furring strips nailed at right angles to the joist and spaced 16" on centers for gypsum laths or 12" on centers for metal laths. Furring strips, sizes 1" x 2" or 1" x 3", of white pine or fir are recommended. Nail the furring strips with ten-penny box nails, one to each joist and right through the old laths and plaster.

Over these strips nail gypsum laths or metal laths. Then plaster as you would for new work. This method will spread the movement of the joist over a larger area. Any excessive deflection is relieved at the wall angle where it will hardly show or may be hidden by a molding.

The above method of relathing and replastering is also the best solution for any badly cracked or damaged ceiling that is not fit to patch. Complete replastering over the old plaster saves time and creates much less dirt.

Complete replastering also insulates and gives the fire protection of a double ceiling.

On walls, it is possible to nail metal laths or gypsum laths directly over the old plaster. Use six-penny box nails and then replaster. Furring strips are not needed here. Door and window casings may be removed and replaced with a metal plaster stop. Metal plaster stops provide a neat modern trim around such openings. Baseboards are removed and the new lathing and plaster is carried right down to the floor line. Replastering the walls in this manner adds insulation and fire protection.

Veneer plastering can be repaired using the same techniques used for other plastering repairs, the only difference being the thickness of the material applied.

Exposed aggregate work is also repairable by removing any damaged material and replacing with new material of the same manufacturer. There is, however, the problem of slight changes in color due to aging and dirt; chemicals in the air and other agents tend over a period of time to affect the surface.

For any patching required on exposed aggregate surfaces it is best to first thoroughly wash and clean the whole area. This will make it as close to its original color as possible, then the new patch will blend in better. It is also possible, at times, to change the matrix color slightly for the patching material, so it will match the existing surface. Aggregate mix of stones can also be altered slightly to help this blending problem. Next, cut and pry out the damaged section as shown in Fig. 10-12, top left. Note that a rectangular cut is made. Then mask the adjoining area and apply matrix for seeded aggregate to same thickness as the matrix of the surrounding area (not including the aggregate), or for troweled-on aggregate to the same thickness as adjoining area, as shown in Fig. 10-12, top right.

Finally, strip tape and seed the area with the same type of aggregate. For troweled on aggregate, strip the tape when the new patch is setting, take a clean trowel and push into dry adjoining area to meet all irregularities of chipped edge and finish troweling as shown in Fig. 10-12, bottom.

Portland cement work can be patched just as well as the gypsum work; however, it will take much longer to do a good job. The area to be patched must be cleaned out thoroughly and all loose material removed. Then coat the edges and base, if masonry, with one of the bonding agents for cement work.

Mix enough material to fill in the area using one of the liquid latex admixes to help the patch adhere and resist shrinkage. (Make sure the mix used is not too rich as this will cause shrinkage problems later.) Fill in only as much as the original base

Fig. 10-12. Three steps for repairing exposed aggregate work. (Cement Enamel Development, Inc.)

coat thickness (over metal lath a scratch coat, over masonry bases a brown coat). Let this coat harden, and cure (7 days); then apply the succeeding coat or coats.

Plasterers who *spike* (add gypsum) cement mortar so they can complete a patch in one day are the cause of much of the loss of work in this field. Time is required in patching Portland cement work, each coat must be cured just like the original work. Unless this is done the patch will fail.

Repairing Untrue Surfaces

Sometimes the plasterer is called back to make alterations to his work so that another trade may fit his materials more evenly adjacent to the plastered surface. This is usually the case where cabinets are installed.

In the case where the surface of the plaster extends beyond a straight line to which something is to be installed, the plasterer should first determine the maximum area that should be chopped back. He will then mark this area and chop it out. After this he should patch it with fresh material, making sure the surface is straight so that the material to be installed will fit correctly.

If the surface of the plaster to which something is to be installed is shallow, the plasterer determines the minimum area necessary to be filled in. He will then add 4" to the length and 4" to the width and mark

with straight lines. He now carefully chips off the finish coat, using these lines as guides, from the outer marked edge toward the center, for approximately 4". He then paints the entire area to be filled in (and just a little beyond the outside edges) with a liquid bonding agent. He now fills it in with a high-gauge mix of the finish and straightens the area so that the material to be installed will fit correctly.

These two methods can also be used where a plasterer may have been careless in doing a straight, level and plumb job.

Altering Plaster Textures

Some home owners are confronted with the problems of removing or covering "antique finishes" popular a generation ago. Some of these finish coats are so rough in texture that the job of producing a smooth surface seems hopeless.

For the lighter or finer antique finishes the expert plasterer can apply two coats of spackling compound. Spackling compound is a plaster of Paris product containing additives that permit its adherence to any clean, painted surface.

Spackling compound is applied in two coats. The first step is to wash the old surface with a strong washing powder to remove any traces of dirt and grease. Then apply a coat of spackling compound thick enough to cover the irregularities. Smooth this down, using a clean strip of thin wood siding, plastic or metal. Allow the application to set over night. Apply a thin coat of material and trowel it down to a smooth surface on the following day.

If the antique finish is very rough, it must be chopped off and refinished. Using a sharp lather's hatchet, cut off the finish coat down to the brown mortar. By using a sharp hatchet and by chopping off little bits of the antique finish at a time, the finish coat can be removed without too much damage to the brown coat.

When this operation is complete, wet down the surface, using a spray gun. The garden tank type is ideal for this purpose. The surface must be wetted at least three times to achieve any penetration of water into the mortar.

When the area is well wetted, gauge up the finish coat and apply as for any new work. However, the finishing material must be gauged with more plaster than usual in order to insure a good bond.

If the brown coat is quite rough due to chopping, it is best to darby in the finish coat so as to level off these rough areas. Then immediately apply a thin coat of finishing material. Be sure to apply both coats of finishing material, using one batch of mortar. Do not mix separate batches of mortar for each coat. Such a practice, although easier, may lead to failure later because there is the danger that the two coats of finishing plaster may separate.

The new *liquid bonding agents* now make it possible to coat old rough antique finishes without chopping the old finish off. After washing the surface with a strong washing powder to remove dirt and grease, wash the surface again with clean water to remove the residue left by the washing powder. When the surface is dry, coat it with the bonding agent. Spraying it on will be best if the surface is very rough.

Now brown in the whole area using any of the mill-mixed gypsum-base coat plasters. When this coat is hard and dry, the finish coat is applied in the same manner as over any normal brown coat. The veneer two-coat technique and materials may also be used for this work if desired.

This method of using the new liquid bonding agents for the brown and finish coat is recommended over the *flanking in* (applying a thick layer of putty and plaster) with putty coat, because the brown coat is stronger and will provide a better base for a good finish coat application.

In most cases the coating can be done without removing the trim or baseboards. The new material only fills in the hollows and the final finish coat just covers the peaks of the antique finish, ending at the trim and base with very little extra thickness.

485

Checking On Your Knowledge

The following questions give you the opportunity to check up on yourself. If you have read the chapter carefully, you should be able to answer the questions. If you have any difficulty, read the chapter over once more so that you have the information well in mind before you go on with your reading.

DO YOU KNOW

1. What are the three main causes of plastering problems?

2. What is a *dryout* and how can it be corrected?

3. What is meant by the term *sweatout* and can it be corrected?

4. What is meant by *oversanding* and how is it detected?

5. Can over-retarding plaster damage the finished product?

6. What is the difference between chip and map cracking?

7. Is frozen plaster reusable?

8. How can structural cracking be overcome?

9. What problems listed above are usually not found in veneer plastering?

10. What problems listed above are not found in Portland cement plastering?

11. What new material can be used in altering rough textured plaster?

12. Why will flashings and drips help prevent many Portland cement problems?

Math for the Plasterer

This review can be used by plasterers and apprentices for quickly finding information to compute areas and volumes, develop geometric shapes, and make practical estimates of materials. For a more detailed study of mathematical problems and formulas, mathematics books should be consulted. Only basic information and items which have general use in the trade can be listed within the scope of this text, which includes:

Operations with Fractions and Decimals

Changing Fractions to Decimals and Decimals to Fractions

Metric-to-English and English-to-Metric Conversions

Computing Areas

Computing Volumes

Elementary Geometry

Development of Geometric Shapes

Estimating Areas, Volumes, and Materials

It is assumed that all readers understand the fundamental processes of adding, subtracting, multiplying and dividing whole numbers and are familiar with common units of measurement such as feet, yards, pounds, gallons, etc.

Operations With Fractions
Addition

To add fractions with the same *denominator* (lower part), such as $\frac{1}{8}$ + $\frac{2}{8}$, add the *numerators* (upper parts) and place this sum over the *common denominator* (bottom number). Thus, $\frac{1}{8}$ + $\frac{2}{8}$ = $\frac{3}{8}$. When the sum

487

of numerators is larger than the common denominator, such as $\frac{9}{8}$, divide this new numerator (top) by the denominator (bottom), giving a *mixed number* (whole number with fractional remainder). Thus, $\frac{9}{8} = 1\frac{1}{8}$.

To add fractions with different denominators, multiply both numerator and denominator of each fraction by a number that will make the denominators equal. Any multiplier may be used without changing the quantity of the fraction. For example, $\frac{1}{4} = \frac{2}{8} = \frac{4}{16} = \frac{8}{32}$, etc. Similarly, $\frac{1}{4} = \frac{3}{12} = \frac{9}{36}$, etc.

After all fractions to be added have been changed so as to have a common denominator, add the new numerators (top numbers) and place this sum over the common denominator (the bottom number). If the new fraction is larger than 1, it can be changed to a mixed number, as already explained. For example, if $\frac{1}{2}$, $\frac{3}{8}$ and $\frac{2}{3}$ were to be added, the common denominator (the bottom number) would be 24. Thus $\frac{1}{2}$ would become $\frac{12}{24}$, $\frac{3}{8}$ would become $\frac{9}{24}$ and $\frac{2}{3}$ would become $\frac{16}{24}$. To add: $\frac{12}{24} + \frac{9}{24} + \frac{16}{24} = \frac{37}{24} = 1$ and $\frac{13}{24}$. (Note that $\frac{37}{24}$ breaks down into $\frac{24}{24} + \frac{13}{24}$, thus since $\frac{24}{24} = 1$, we get 1 and $\frac{13}{24}$, or $1\frac{13}{24}$.

In some cases the various fractions to be added can be changed to have a common denominator by dividing instead of multiplying both numerators and denominators by the same number. Again, the quantities would

not be changed. For example, $\frac{8}{32} = \frac{4}{16} = \frac{2}{8} = \frac{1}{4}$.

Although both numerator and denominator of *each individual fraction* must be multiplied or divided by the same number, it is not necessary that the same multiplier or divisor be used for *all fractions* to be added. Thus, to add $\frac{1}{5} + \frac{2}{3}$, the first fraction could be multiplied by $\frac{3}{3}$ giving $\frac{1}{5} = \frac{3}{15}$. The second fraction could be multiplied by $\frac{5}{5}$, giving $\frac{2}{3} = \frac{10}{15}$. The addition would then be $\frac{3}{15} + \frac{10}{15} = \frac{13}{15}$.

Subtraction

Change the fractions to have a common denominator as in the case of addition. Subtract the smaller of the new numerators from the larger and place this subtracted number over the common denominator. (It is assumed that no negative fractions will be used, such as $-\frac{3}{8}$ or "minus three-eighths," in plastering applications).

Multiplication

To multiply fractions there is no need to change them first so as to have a common denominator. Simply multiply all the numerators to obtain a new numerator and multiply all denominators to obtain a new denominator. Thus, $\frac{2}{3} \times \frac{1}{8} \times \frac{3}{5} = \frac{6}{120}$. (For the numerator: $2 \times 1 \times 3 = 6$; for the denominator: $3 \times 8 \times 5 = 120$.) This can be simplified to $\frac{1}{20}$ by dividing both numerator and denominator by 6.

Another method to simplify calculations is called *cancellation*. In multiplying a series of fractions with no intervening subtractions or additions, it often happens that the same digit appears in a numerator and a denominator. In such cases both numerator and denominator can be cancelled. Thus, in the above case, the digit 3 appears as the denominator in $\frac{2}{3}$ and as the numerator in $\frac{3}{5}$. These 3's cancel each other and disappear in the multiplication. Thus, $\frac{2}{3} \times \frac{1}{8} \times \frac{3}{5} = \frac{2}{40}$. This is readily simplified by dividing both numerator and denominator by 2, giving $\frac{1}{20}$ as the answer, exactly as before.

Division

Division of fractions is very simple, but is best understood by example. Consider the problem $\frac{1}{2} \div \frac{2}{3} = ?$ Here the *dividend* (number to be divided) is $\frac{1}{2}$ and the *divisor* (number that divides it) is $\frac{2}{3}$. To do this operation, re-write the dividend without change, change the division sign $\div$ to a multiplication sign $\times$, invert numerator and denominator of the divisor $\frac{2}{3}$ to read $\frac{3}{2}$, and proceed to multiply as the sign directs. Thus $\frac{1}{2} \div \frac{2}{3} = \frac{1}{2} \times \frac{3}{2} = \frac{3}{4}$.

It comes as a surprise to some people not used to calculations with fractions that generally the quantities decrease when they are multiplied and increase when they are divided. In common speech, when a person speaks of "half an apple" he is multiplying $\frac{1}{2} \times 1$ apple, and thus decreasing the quantity.

Operations With Decimals

Decimal numbers, unlike ordinary fractions, have only numerators. Denominators are implied by the place of the last digit to the right of the decimal point. Thus, $0.1 = \frac{1}{10}$, $0.01 = \frac{1}{100}$, $0.001 = \frac{1}{1000}$, etc.

Any number of zeros may follow the significant digits in the decimal number without increasing the quantity. Thus, $0.68 = 0.680 = 0.680000000000$. Zeros to the left of the significant digits and immediately following the decimal point are another matter; the quantity decreases by a factor of 10 for each zero preceding these digits. Thus, $0.75 = \frac{75}{100}$, $0.075 = \frac{75}{1000}$, and $0.0075 = \frac{75}{10000}$.

(Note: (It is generally considered good practice to place a zero before the decimal point where no whole number is included with the decimal remainder.)

Operations with decimals differ in no way from those used with whole numbers except in placement of the important decimal point. Because United States money is based on the decimal system, these operations are already familiar to nearly everyone.

Addition

Line up decimal numbers to be added so that the decimal points are directly under each other. Add as with whole numbers and place the decimal point in the sum in exactly the same location as in the numbers added. This is a case where neatness counts.

For example, add 3.236, 0.75, and 107.205. Arrange as:

$$\begin{array}{r} 3.236 \\ 0.750 \\ 107.205 \\ \hline \end{array}$$ — Adding zero to right

Adding gives 111.191

This is read as "one hundred eleven and one hundred ninety-one thousandths."

Subtraction

As with addition, arrange the decimal numbers with the decimal points directly underneath each other. Subtract as with whole numbers, and place the decimal point directly under those of the listed numbers.

For example, subtract 0.9 from 2.356. Arrange as:

$$\begin{array}{r} 2.356 \\ -0.900 \\ \hline 1.456 \end{array}$$

Subtracting gives — Add 2 zeros to right

Multiplication

Multiplication of decimal numbers differs from that of whole numbers only in placement of the decimal point in the product. Multiply the numbers, then place the decimal point as many places to the left of the significant digits as the sum of such places in the numbers multiplied.

For example, multiply $0.9 \times 0.3 \times 0.5$. The digits in the product will be 135. Since there are three places to the right of the decimal point in the numbers multiplied, the answer is pointed off as 0.135. Here the zero to the left of the decimal point has no meaning other than to assure that there is no whole number. The answer is read as "one hundred thirty-five thousandths."

Division

Division of decimal numbers differs from that of whole numbers only in placement of the decimal point in the *quotient* (the number resulting from the division). Consider the problem $7.835 \div 0.5 = ?$ Here 7.835 is the dividend and has three places of decimals to the right of the decimal point. The divisor, 0.5, has one place to the right of the decimal point. In this particular case the dividend has *two more places* of decimals than the divisor. This difference will determine the placement of the decimal point in the quotient. Dividing gives the digits 1567, the quotient, which is pointed off *two places* as

15.67, read as "fifteen and sixty-seven hundredths."

Changing Fractions to Decimals

A common fraction, such as ⅛, is an instruction to perform an operation. It says "divide 1 by 8." Doing this operation, as shown, converts it to a decimal number.

Set down and divide:

$$
\begin{array}{r}
0.125 \\
8\,)\,\overline{1.000} \\
8 \\
\hline
20 \\
16 \\
\hline
40 \\
40 \\
\hline
\end{array}
$$

The quotient, 0.125 (read as "one hundred twenty-five thousandths), is

the conversion. Similarly, ⅜ is an instruction to divide 3 by 8:

$$
\begin{array}{r}
0.375 \\
8\,)\,\overline{3.000} \\
2\,4 \\
\hline
60 \\
56 \\
\hline
40 \\
40 \\
\hline
\end{array}
$$

Thus, ⅜ = 0.375 (three hundred seventy-five thousandths).

Conversions of this type are required so frequently that tables such as the one shown in Table 1 have been compiled for quick reference.

TABLE 1 DECIMAL EQUIVALENTS OF COMMON FRACTIONS OF AN INCH

Common Fractions	Decimal Equivalents	Common Fractions	Decimal Equivalents	Common Fractions	Decimal Equivalents	Common Fractions	Decimal Equivalents
1/64	.015625	17/64	.265625	33/64	.515625	49/64	.765625
1/32	.03125	9/32	.28125	17/32	.53125	25/32	.78125
3/64	.046875	19/64	.296875	35/64	.546875	51/64	.796875
1/16	.0625	5/16	.3125	9/16	.5625	13/16	.8125
5/64	.078125	21/64	.328125	37/64	.578125	53/64	·828125
3/32	.09375	11/32	.34375	19/32	.59375	27/32	.84375
7/64	.109375	23/64	.359375	39/64	.609375	55/64	.859375
1/8	.125	3/8	.375	5/8	.625	7/8	.875
9/64	.140625	25/64	.390625	41/64	.640625	57/64	.890625
5/32	.15625	13/32	.40625	21/32	.65625	29/32	.90625
11/64	.171875	27/64	.421875	43/64	.671875	59/64	.921875
3/16	.1875	7/16	.4375	11/16	.6875	15/16	.9375
13/64	.203125	29/64	.453125	45/64	.703125	61/64	.953125
7/32	.21875	15/32	.46875	23/32	.71875	31/32	.96875
15/64	.234375	31/64	.484375	47/64	.734375	63/64	.984375
1/4	.25	1/2	.50	3/4	.75		

Not all conversions are as neat as the ones illustrated. For instance, $\frac{1}{12} = 0.083333333333333333\ldots$ (approx.) As many significant figures are retained in the conversion as practically requires. For most applications 0.083 would be acceptable.

Changing Decimals to Fractions

Translate the decimal number to fractional form, then reduce this fraction to its simplest form by dividing both numerator and denominator by the same number. For example, $0.84 = \frac{84}{100}$. This can be reduced to $\frac{42}{50}$ (dividing by 2), and again to $\frac{21}{25}$. No number can evenly divide both 21 and 25, so $\frac{21}{25}$ is the simplest form and the fractional conversion.

Metric-to-English and English-to-Metric Conversions

Rapid expansion of trade and industry on an international basis in the past two decades has increased the need for understanding of both the *metric* or CGS (Centimeter-Gram-Second) system used by nearly all countries of the world and the *English* or FPS (Foot-Pound-Second) system used by the United States and some other English-speaking countries.

If the co-existence of two systems seems inconvenient, as it is, remember that in respect to worldwide agreement we are the exception. In view of the increasing need for a universal system to measure lengths, areas, volumes, weights, temperatures, etc., it now seems likely that the CGS system will ultimately replace the FPS system despite immense costs and problems that will be involved in making the changeover.

Table 2 lists factors for converting units from metric to English, while Table 3 lists factors for converting from English to metric units.

To convert a quantity from *metric* to *English* units:

1. Multiply by the factor shown in Table 2.
2. Use the resulting quantity "rounded off" to the number of decimal digits needed for practical application.
3. Wherever practical in semi-precision measurements, convert the decimal part of the number to the nearest common fraction.

TABLE 2 CONVERSION OF METRIC TO ENGLISH UNITS

LENGTHS:		WEIGHTS:	
1 MILLIMETER (mm)	= 0.03937 IN.	1 GRAM (g)	= 0.03527 OZ (AVDP)
1 CENTIMETER (cm)	= 0.3937 IN.	1 KILOGRAM (kg)	= 2.205 LBS
1 METER (m)	= 3.281 FT OR 1.0937 YDS	1 METRIC TON	= 2205 LBS
1 KILOMETER (km)	= 0.6214 MILES	**LIQUID MEASUREMENTS:**	
AREAS:		1 CU CENTIMETER (cc)	= 0.06102 CU IN.
1 SQ MILLIMETER	= 0.00155 SQ IN.	1 LITER (= 1000 cc)	= 1.057 QUARTS OR 2.113 PINTS OR 61.02 CU INS.
1 SQ CENTIMETER	= 0.155 SQ IN.	**POWER MEASUREMENTS:**	
1 SQ METER	= 10.76 SQ FT OR 1.196 SQ YDS	1 KILOWATT (kw)	= 1.341 HORSEPOWER
VOLUMES:		**TEMPERATURE MEASUREMENTS:**	
1 CU CENTIMETER	= 0.06102 CU IN.	TO CONVERT DEGREES CENTIGRADE TO DEGREES FAHRENHEIT, USE THE FOLLOWING FORMULA: DEG F = (DEG C X 9/5) + 32	
1 CU METER	= 35.31 CU FT OR 1.308 CU YDS		

SOME IMPORTANT FEATURES OF THE CGS SYSTEM ARE:

1 cc OF PURE WATER = 1 GRAM. PURE WATER FREEZES AT 0 DEGREES C AND BOILS AT 100 DEGREES C.

TABLE 3 CONVERSION OF ENGLISH TO METRIC UNITS

LENGTHS:		WEIGHTS:	
1 INCH	= 2.540 CENTIMETERS	1 OUNCE (AVDP)	= 28.35 GRAMS
1 FOOT	= 30.48 CENTIMETERS	1 POUND	= 453.6 GRAMS OR 0.4536 KILOGRAM
1 YARD	= 91.44 CENTIMETERS OR 0.9144 METERS	1 (SHORT) TON	= 907.2 KILOGRAMS
1 MILE	= 1.609 KILOMETERS	**LIQUID MEASUREMENTS:**	
AREAS:		1 (FLUID) OUNCE	= 0.02957 LITER OR 28.35 GRAMS
1 SQ IN.	= 6.452 SQ CENTIMETERS	1 PINT	= 473.2 CU CENTIMETERS
1 SQ FT	= 929.0 SQ CENTIMETERS OR 0.0929 SQ METER	1 QUART	= 0.9463 LITER
1 SQ YD	= 0.8361 SQ METER	1 (US) GALLON	= 3785 CU CENTIMETERS OR 3.785 LITERS
VOLUMES:		**POWER MEASUREMENTS:**	
1 CU IN.	= 16.39 CU CENTIMETERS	1 HORSEPOWER	= 0.7457 KILOWATT
1 CU FT	= 0.02832 CU METER	**TEMPERATURE MEASUREMENTS:**	
1 CU YD	= 0.7646 CU METER	TO CONVERT DEGREES FAHRENHEIT TO DEGREES CENTIGRADE USE THE FOLLOWING FORMULA: DEG C = 5/9 (DEG F - 32)	

To convert a quantity from *English* to *metric* units:

1. If the English measurement is expressed in fractional form, change this to an equivalent decimal form.

493

2. Multiply this quantity by the factor shown in Table 3.
3. Round off the result to the precision required.

Relatively small measurements, such as 17.3 cm, are generally expressed in equivalent millimeter form. In this example the measurement would be read as 173 mm.

Computing Areas

Computation of areas is of primary importance in the plastering trade because the greater part of all materials used is applied to *plane* (flat) surfaces, and the areas of these surfaces must be known before the work can proceed. Planes may be vertical, as interior and exterior walls, or horizontal, as flat ceilings. They may also be angled. In all cases, however, they are flat or two-dimensional, having height and length or width and length. Areas, in terms of square feet, square yards, or other square units, are computed by multiplying two lineal measurements to obtain the number of square units of like kind. For instance, consider a simple wall in the form of a *rectangle* 8 ft. high and 20 ft. long. The area is 8 ft. × 20 ft. = 160 sq. ft.

A *square* is a special rectangle in

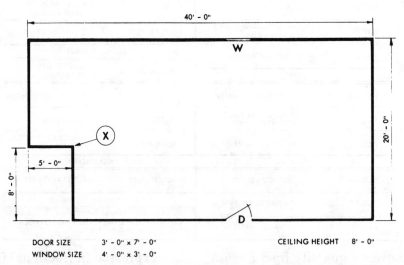

DOOR SIZE   3' - 0" x 7' - 0"   CEILING HEIGHT   8' - 0"
WINDOW SIZE   4' - 0" x 3' - 0"

Fig. 1. An area requiring plastering of walls and ceiling. An outside (exterior) corner is shown at X.

which the height and length, or width and length, are equal. If a square had a side of 8′ then *all* sides would be 8′. The square feet would be found by multiplying the 8′ measurement of one side by itself. Thus 8 ft. × 8 ft. = 8 ft.^{2} (eight ft. *squared*) = 64 sq. ft.

The number of possible shapes is, of course, unlimited, and computation of their areas can be quite involved. More detailed information is contained in the sections *Elementary Geometry* and *Development of Geometric Shapes* which follow.

The most important areas to remember, in addition to those of the rectangle and square, are:

Area of a circle = 0.785 × square of diameter.

Area of a triangle = base × ½ perpendicular height.

In actual practice, many areas of unusual shape can be divided into rather simple sections. The areas of these sections can be separately computed and then added together to obtain the total area required.

A common practice in figuring the wall area of a room is to first compute the total rectangular area of the four walls disregarding any openings, then compute the areas of the openings, and finally subtract the areas of the openings. See Fig. 1. Here it is interesting to note that the exterior corner marked "X" does not affect the calculations and can be omitted; the wall area of this irregularity is the same as the missing corner it replaces.

Full length: 40′ + 40′ + 20′ + 20′ = 120′

Height: 8′

Full rectangular area: 120′ × 8′ = 960 sq. ft.

Door area: 21 sq. ft.
Window area: *12 sq. ft.*
Total openings: 33 sq. ft.

Net area: 960 sq. ft. − 33 sq. ft. = 927 sq. ft. or *103 sq. yds.* (Divide by 9 to convert sq. ft. to sq. yds. There are 9 sq. ft. in one sq. yd.)

To facilitate calculations of rectangular room areas, Tables 4, 5, 6, 7 and 8 have been included. These tables assume that side walls and ceilings are to be plastered solid (disregarding openings) and convert the lineal dimensions of length in feet to total areas in square yards, for rooms with 7½, 8, 8½, 9, and 10 ft. ceilings, respectively.

TABLES 4, 5, 6, 7, AND 8 TO FOLLOW

TABLE 4

NUMBER OF SQUARE YARDS IN ROOMS WITH 7½ FOOT CEILINGS

LENGTH IN FEET → · ← WIDTH IN FEET

Width ＼ Length	3	4	5	6	7	8	9	10	11	12	13	14	15	16	17	18	19	20	21	22	23	24	25	26	27	28	29	30
3	11.0	13.0	15.0	17.0	19.0	21.0	23.0	25.0	27.0	29.0	31.0	33.0	35.0	37.0	39.0	41.0	43.0	45.0	47.0	49.0	51.0	53.0	55.0	57.0	59.0	61.0	63.0	65.0
4	13.0	15.1	17.2	19.3	21.4	23.5	25.6	27.7	29.8	32.0	34.1	36.2	38.3	40.4	42.5	44.6	46.7	48.8	51.0	53.1	55.2	57.3	59.4	61.5	63.6	65.7	67.8	70.0
5	15.0	17.2	19.4	21.6	23.8	26.1	28.3	30.5	32.7	35.0	37.2	39.4	41.6	43.8	46.1	48.3	50.5	52.7	55.0	57.2	59.4	61.6	63.8	66.1	68.3	70.5	72.7	75.0
6	17.0	19.3	21.6	24.0	26.3	28.6	31.0	33.3	35.6	38.0	40.3	42.6	45.0	47.3	49.6	52.0	54.3	56.6	59.0	61.3	63.6	66.0	68.3	70.6	73.0	75.3	77.6	80.0
7	19.0	21.4	23.8	26.3	28.7	31.2	33.6	36.1	38.5	41.0	43.4	45.8	48.3	50.7	53.2	55.6	58.1	60.5	63.0	65.4	67.8	70.3	72.7	75.2	77.6	80.1	82.5	85.0
8	21.0	23.5	26.1	28.6	31.2	33.7	36.3	38.8	41.4	44.0	46.5	49.1	51.6	54.2	56.7	59.3	61.8	64.4	67.0	69.5	72.1	74.6	77.2	79.7	82.3	84.8	87.4	90.0
9	23.0	25.6	28.3	31.0	33.6	36.3	39.0	41.6	44.3	47.0	49.6	52.3	55.0	57.6	60.3	63.0	65.6	68.3	71.0	73.6	76.3	79.0	81.6	84.3	87.0	89.6	92.3	95.0
10	25.0	27.7	30.5	33.3	36.1	38.8	41.6	44.4	47.2	50.0	52.7	55.5	58.3	61.1	63.8	66.6	69.4	72.2	75.0	77.7	80.5	83.3	86.1	88.8	91.6	94.4	97.2	100.0
11	27.0	29.8	32.7	35.6	38.5	41.4	44.3	47.2	50.1	53.0	55.8	58.7	61.6	64.5	67.4	70.3	73.2	76.1	79.0	81.8	84.7	87.6	90.5	93.4	96.3	99.2	102.1	105.0
12	29.0	32.0	35.0	38.0	41.0	44.0	47.0	50.0	53.0	56.0	59.0	62.0	65.0	68.0	71.0	74.0	77.0	80.0	83.0	86.0	89.0	92.0	95.0	98.0	101.0	104.0	107.0	110.0
13	31.0	34.1	37.2	40.3	43.4	46.5	49.6	52.7	55.8	59.0	62.1	65.2	68.3	71.4	74.5	77.6	80.7	83.8	87.0	90.1	93.2	96.3	99.4	102.5	105.6	108.7	111.8	115.0
14	33.0	36.2	39.4	42.6	45.8	49.1	52.3	55.5	58.7	62.0	65.2	68.4	71.6	74.8	78.1	81.3	84.5	87.7	91.0	94.2	97.4	100.6	103.8	107.1	110.3	113.5	116.7	120.0
15	35.0	38.3	41.6	45.0	48.3	51.6	55.0	58.3	61.6	65.0	68.3	71.6	75.0	78.3	81.6	85.0	88.3	91.6	95.0	98.3	101.6	105.0	108.3	111.6	115.0	118.3	121.6	125.0
16	37.0	40.4	43.8	47.3	50.7	54.2	57.6	61.1	64.5	68.0	71.4	74.8	78.3	81.7	85.2	88.6	92.1	95.5	99.0	102.4	105.8	109.3	112.7	116.2	119.6	123.1	126.5	130.0
17	39.0	42.5	46.1	49.6	53.2	56.7	60.3	63.8	67.4	71.0	74.5	78.1	81.6	85.2	88.7	92.3	95.8	99.4	103.0	106.5	110.1	113.6	117.2	120.7	124.3	127.8	131.4	135.0
18	41.0	44.6	48.3	52.0	55.6	59.3	63.0	66.6	70.3	74.0	77.6	81.3	85.0	88.6	92.3	96.0	99.6	103.3	107.0	110.6	114.3	118.0	121.6	125.3	129.0	132.6	136.3	140.0
19	43.0	46.7	50.5	54.3	58.1	61.8	65.6	69.4	73.2	77.0	80.7	84.5	88.3	92.1	95.8	99.6	103.4	107.2	111.0	114.7	118.5	122.3	126.1	129.8	133.6	137.4	141.2	145.0
20	45.0	48.8	52.7	56.6	60.5	64.4	68.3	72.2	76.1	80.0	83.8	87.7	91.6	95.5	99.4	103.3	107.2	111.1	115.0	118.8	122.7	126.6	130.5	134.4	138.3	142.2	146.1	150.0
21	47.0	51.0	55.0	59.0	63.0	67.0	71.0	75.0	79.0	83.0	87.0	91.0	95.0	99.0	103.0	107.0	111.0	115.0	119.0	123.0	127.0	131.0	135.0	139.0	143.0	147.0	151.0	155.0
22	49.0	53.1	57.2	61.3	65.4	69.5	73.6	77.7	81.8	86.0	90.1	94.2	98.3	102.4	106.5	110.6	114.7	118.8	123.0	127.1	131.2	135.3	139.4	143.5	147.6	151.7	155.8	160.0
23	51.0	55.2	59.4	63.6	67.8	72.1	76.3	80.5	84.7	89.0	93.2	97.4	101.6	105.8	110.1	114.3	118.5	122.7	127.0	131.2	135.4	139.6	143.8	148.1	152.3	156.5	160.7	165.0
24	53.0	57.3	61.6	66.0	70.3	74.6	79.0	83.3	87.6	92.0	96.3	100.6	105.0	109.3	113.6	118.0	122.3	126.6	131.0	135.3	139.6	144.0	148.3	152.6	157.0	161.3	165.6	170.0

The amount indicated includes side walls and ceilings.

THOMAS B. TEAGAN COMPANY

TABLE 5

NUMBER OF SQUARE YARDS IN ROOMS WITH

8 FOOT CEILINGS

LENGTH IN FEET

WIDTH	3	4	5	6	7	8	9	10	11	12	13	14	15	16	17	18	19	20	21	22	23	24	25	26	27	28	29	30
3	11.6	13.7	15.8	18.0	20.1	22.2	24.3	26.4	28.5	30.6	32.7	34.8	37.0	39.1	41.2	43.3	45.4	47.5	49.6	51.7	53.9	56.0	58.1	60.2	62.3	64.4	66.5	68.7
4	13.7	16.0	18.2	20.4	22.6	24.8	27.1	29.3	31.5	33.7	36.0	38.2	40.4	42.6	44.8	47.1	49.3	51.5	53.7	56.0	58.2	60.4	62.7	64.9	67.1	69.3	71.5	73.8
5	15.8	18.2	20.5	22.8	25.2	27.5	29.8	32.2	34.5	36.8	39.2	41.5	43.8	46.2	48.5	50.8	53.2	55.5	57.8	60.2	62.6	64.9	67.2	69.5	71.8	74.2	76.5	78.9
6	18.0	20.4	22.8	25.3	27.7	30.2	32.6	35.1	37.5	40.0	42.4	44.8	47.3	49.7	52.2	54.6	57.1	59.5	62.0	64.4	66.8	69.3	71.7	74.2	76.6	79.1	81.5	84.0
7	20.1	22.6	25.2	27.7	30.3	32.8	35.4	38.0	40.5	43.1	45.6	48.2	50.7	53.3	55.8	58.4	61.0	63.5	66.1	68.6	71.2	73.7	76.3	78.8	81.4	84.0	86.5	89.1
8	22.2	24.8	27.5	30.2	32.8	35.5	38.2	40.8	43.5	46.2	48.8	51.5	54.2	56.8	59.5	62.2	64.8	67.5	70.2	72.8	75.5	78.2	80.8	83.5	86.2	88.8	91.5	94.2
9	24.3	27.1	29.8	32.6	35.4	38.2	41.0	43.7	46.5	49.3	52.1	54.8	57.6	60.4	63.2	66.0	68.7	71.5	74.3	77.1	79.9	82.6	85.4	88.2	91.0	93.8	96.5	99.3
10	26.4	29.3	32.2	35.1	38.0	40.8	43.7	46.6	49.5	52.4	55.3	58.2	61.1	64.0	66.8	69.7	72.6	75.5	78.4	81.3	84.2	87.1	90.0	92.8	95.8	98.7	101.5	104.4
11	28.5	31.5	34.5	37.5	40.5	43.5	46.5	49.5	52.5	55.5	58.5	61.5	64.5	67.5	70.5	73.5	76.5	79.5	82.5	85.5	88.6	91.5	94.5	97.5	100.6	103.6	106.5	109.5
12	30.6	33.7	36.8	40.0	43.1	46.2	49.3	52.4	55.5	58.6	61.7	64.8	68.0	71.1	74.2	77.3	80.4	83.5	86.6	89.7	92.9	96.0	99.1	102.2	105.3	108.4	111.5	114.7
13	32.7	36.0	39.2	42.4	45.6	48.8	52.1	55.3	58.5	61.7	65.0	68.2	71.4	74.6	77.8	81.1	84.3	87.5	90.7	94.0	97.2	100.4	103.7	106.9	110.1	113.3	116.5	119.8
14	34.8	38.2	41.5	44.8	48.2	51.5	54.8	58.2	61.5	64.8	68.2	71.5	74.8	78.2	81.5	84.8	88.2	91.5	94.8	98.2	101.5	104.8	108.2	111.5	114.8	118.2	121.5	124.9
15	37.0	40.4	43.8	47.3	50.7	54.2	57.6	61.1	64.5	68.0	71.4	74.8	78.3	81.7	85.2	88.6	92.1	95.5	99.0	102.4	105.8	109.3	112.8	116.2	119.7	123.1	126.5	130.0
16	39.1	42.6	46.2	49.7	53.3	56.8	60.4	64.0	67.5	71.1	74.6	78.2	81.7	85.3	88.8	92.4	96.0	99.5	103.1	106.6	110.2	113.7	117.3	120.9	124.4	128.0	131.5	135.1
17	41.2	44.8	48.5	52.2	55.8	59.5	63.2	66.8	70.5	74.2	77.8	81.5	85.2	88.8	92.5	96.2	99.8	103.5	107.2	110.8	114.5	118.2	121.9	125.5	129.2	132.9	136.5	140.2
18	43.3	47.1	50.8	54.6	58.4	62.2	66.0	69.7	73.5	77.3	81.1	84.8	88.6	92.4	96.2	100.0	103.7	107.5	111.3	115.1	118.8	122.6	126.4	130.2	134.0	137.8	141.5	145.3
19	45.4	49.3	53.2	57.1	61.0	64.8	68.7	72.6	76.5	80.4	84.3	88.2	92.1	96.0	99.8	103.7	107.6	111.5	115.4	119.3	123.2	127.1	131.0	134.9	138.8	142.7	146.5	150.4
20	47.5	51.5	55.5	59.5	63.5	67.5	71.5	75.5	79.5	83.5	87.5	91.5	95.5	99.5	103.5	107.5	111.5	115.5	119.5	123.5	127.5	131.5	135.5	139.6	143.6	147.6	151.5	155.6
21	49.6	53.7	57.8	62.0	66.1	70.2	74.3	78.4	82.5	86.6	90.7	94.8	99.0	103.1	107.2	111.3	115.4	119.5	123.6	127.7	131.8	136.0	140.1	144.2	148.3	152.4	156.5	160.7
22	51.7	56.0	60.2	64.4	68.6	72.8	77.1	81.3	85.5	89.7	94.0	98.2	102.4	106.6	110.8	115.1	119.3	123.5	127.7	132.0	136.2	140.4	144.7	148.9	153.1	157.3	161.5	165.8
23	53.8	58.2	62.5	66.8	71.2	75.5	79.8	84.2	88.5	92.8	97.2	101.5	105.8	110.2	114.5	118.8	123.2	127.5	131.8	136.2	140.6	144.9	149.2	153.6	157.9	162.2	166.5	170.9
24	56.0	60.4	64.8	69.3	73.7	78.2	82.6	87.1	91.5	96.0	100.4	104.8	109.3	113.7	118.2	122.6	127.1	131.5	136.0	140.4	144.9	149.3	153.8	158.2	162.7	167.1	171.5	176.0

WIDTH IN FEET

The amount indicated includes side walls and ceilings.

THOMAS B. TEAGAN COMPANY

TABLE 6

8½ FOOT CEILINGS

NUMBER OF SQUARE YARDS IN ROOMS WITH

LENGTH IN FEET

WIDTH	3	4	5	6	7	8	9	10	11	12	13	14	15	16	17	18	19	20	21	22	23	24	25	26	27	28	29	30
3	12.3	14.5	16.7	19.0	21.2	23.4	25.6	27.8	30.1	32.2	34.5	36.7	39.0	41.2	43.4	45.6	47.8	50.1	52.3	54.5	56.8	59.0	61.2	63.4	65.7	67.9	70.1	72.3
4	14.5	16.8	19.2	21.5	23.8	26.2	28.5	30.8	33.2	35.5	37.8	40.2	42.5	44.8	47.2	49.5	51.8	54.2	56.5	58.8	61.2	63.6	65.9	68.2	70.6	72.9	75.2	77.6
5	16.7	19.2	21.6	24.1	26.5	29.0	31.4	33.8	36.3	38.7	41.2	43.6	46.1	48.5	51.0	53.4	55.8	58.3	60.7	63.2	65.7	68.1	70.6	73.0	75.4	77.9	80.3	82.8
6	19.0	21.5	24.1	26.6	29.2	31.7	34.3	36.8	39.4	42.0	44.5	47.1	49.6	52.2	54.7	57.3	59.8	62.4	65.0	67.5	70.1	72.7	75.2	77.8	80.3	82.9	85.4	88.0
7	21.2	23.8	26.5	29.2	31.8	34.5	37.2	39.8	42.5	45.2	47.8	50.5	53.2	55.8	58.5	61.2	63.8	66.5	69.2	71.8	74.6	77.2	79.9	82.6	85.2	87.9	90.6	93.2
8	23.4	26.2	29.0	31.7	34.5	37.3	40.1	42.8	45.6	48.4	51.2	54.0	56.7	59.5	62.3	65.1	67.8	70.6	73.4	76.2	79.0	81.8	84.6	87.3	90.1	92.9	95.7	98.4
9	25.6	28.5	31.4	34.3	37.2	40.1	43.0	45.8	48.7	51.6	54.5	57.4	60.3	63.2	66.1	69.0	71.8	74.7	77.6	80.5	83.4	86.3	89.2	92.1	95.0	97.9	100.8	103.7
10	27.8	30.8	33.8	36.8	39.8	42.8	45.8	48.8	51.8	54.8	57.8	60.8	63.8	66.8	69.8	72.8	75.8	78.8	81.8	84.8	87.8	90.9	93.9	96.9	99.9	102.9	105.9	108.9
11	30.1	33.2	36.3	39.4	42.5	45.6	48.7	51.8	55.0	58.1	61.2	64.3	67.4	70.5	73.6	76.7	79.8	83.0	86.1	89.2	92.3	95.4	98.6	101.7	104.8	107.9	111.0	114.1
12	32.2	35.5	38.7	42.0	45.2	48.4	51.6	54.8	58.1	61.3	64.5	67.7	71.0	74.2	77.4	80.6	83.8	87.1	90.3	93.5	96.8	100.0	103.2	106.4	109.7	112.9	116.1	119.3
13	34.5	37.8	41.2	44.5	47.8	51.2	54.5	57.8	61.2	64.5	67.8	71.2	74.5	77.8	81.2	84.5	87.8	91.2	94.5	97.8	101.2	104.6	107.9	111.2	114.6	117.9	121.2	124.6
14	36.7	40.2	43.6	47.1	50.5	54.0	57.4	60.8	64.3	67.7	71.2	74.6	78.1	81.5	85.0	88.4	91.8	95.3	98.7	102.2	105.7	109.1	112.6	116.0	119.4	122.9	126.3	129.8
15	39.0	42.5	46.1	49.6	53.2	56.7	60.3	63.8	67.4	71.0	74.5	78.1	81.6	85.2	88.7	92.3	95.8	99.4	103.0	106.5	110.1	113.7	117.2	120.8	124.3	127.9	131.4	135.0
16	41.2	44.8	48.5	52.2	55.8	59.5	63.2	66.8	70.5	74.2	77.8	81.5	85.2	88.8	92.5	96.2	99.8	103.5	107.2	110.8	114.6	118.2	121.9	125.6	129.2	132.9	136.6	140.2
17	43.4	47.2	51.0	54.7	58.5	62.3	66.1	69.8	73.6	77.4	81.2	85.0	88.7	92.5	96.3	100.1	103.8	107.6	111.4	115.2	119.0	122.8	126.6	130.3	134.1	137.9	141.7	145.4
18	45.6	49.5	53.4	57.3	61.2	65.1	69.0	72.8	76.7	80.6	84.5	88.4	92.3	96.2	100.1	104.0	107.8	111.7	115.6	119.5	123.4	127.3	131.2	135.1	139.0	142.9	146.8	150.7
19	47.8	51.8	55.8	59.8	63.8	67.8	71.8	75.8	79.8	83.8	87.8	91.8	95.8	99.8	103.8	107.8	111.8	115.8	119.8	123.8	127.9	131.9	135.9	139.9	143.9	147.9	151.9	155.9
20	50.1	54.2	58.3	62.4	66.5	70.6	74.7	78.8	83.0	87.1	91.2	95.3	99.4	103.5	107.6	111.7	115.8	120.0	124.1	128.2	132.3	136.4	140.6	144.7	148.8	152.9	157.0	161.1
21	52.3	56.5	60.7	65.0	69.2	73.4	77.6	81.8	86.1	90.3	94.5	98.7	103.0	107.2	111.4	115.6	119.8	124.1	128.3	132.5	136.8	141.0	145.2	149.4	153.7	157.9	162.1	166.3
22	54.5	58.8	63.2	67.5	71.8	76.2	80.5	84.8	89.2	93.5	97.8	102.2	106.5	110.8	115.2	119.5	123.8	128.2	132.5	136.8	141.2	145.6	149.9	154.2	158.6	162.9	167.2	171.6
23	56.7	61.2	65.6	70.1	74.5	79.0	83.4	87.8	92.3	96.7	101.2	105.6	110.1	114.5	119.1	123.4	127.8	132.3	136.7	141.2	145.7	150.1	154.6	159.0	163.4	167.9	172.3	176.8
24	59.0	63.5	68.1	72.6	77.2	81.7	86.3	90.8	95.4	100.0	104.5	109.1	113.6	118.2	122.7	127.3	131.8	136.4	141.0	145.5	150.1	154.6	159.2	163.8	168.3	172.9	177.4	182.0

The amount indicated includes side walls and ceilings.

THOMAS B. TEAGAN COMPANY

TABLE 7

NUMBER OF SQUARE YARDS IN ROOMS WITH

9 FOOT CEILINGS

LENGTH IN FEET → WIDTH IN FEET ↓

W \ L	3	4	5	6	7	8	9	10	11	12	13	14	15	16	17	18	19	20	21	22	23	24	25	26	27	28	29	30
3	13.0	15.3	17.6	20.0	22.3	24.6	27.0	29.3	31.6	34.0	36.3	38.6	41.0	43.3	45.6	48.0	50.3	52.6	55.0	57.3	59.6	62.0	64.3	66.6	69.0	71.3	73.6	76.0
4	15.3	17.7	20.2	22.6	25.1	27.5	30.0	32.4	34.8	37.3	39.7	42.2	44.6	47.1	49.5	52.0	54.4	56.8	59.3	61.7	64.2	66.6	69.1	71.5	74.0	76.4	78.8	81.3
5	17.6	20.2	22.7	25.3	27.8	30.4	33.0	35.5	38.1	40.6	43.2	45.7	48.3	50.8	53.4	56.0	58.5	61.1	63.6	66.2	68.7	71.3	73.8	76.4	79.0	81.5	84.1	86.6
6	20.0	22.6	25.3	28.0	30.6	33.3	36.0	38.6	41.3	44.0	46.6	49.3	52.0	54.6	57.3	60.0	62.6	65.3	68.0	70.6	73.3	76.0	78.6	81.3	84.0	86.6	89.3	92.0
7	22.3	25.1	27.8	30.6	33.4	36.2	39.0	41.7	44.5	47.3	50.1	52.8	55.6	58.4	61.2	64.0	66.7	69.5	72.3	75.1	77.8	80.6	83.4	86.2	89.0	91.7	94.5	97.3
8	24.6	27.5	30.4	33.3	36.2	39.1	42.0	44.8	47.7	50.6	53.5	56.4	59.3	62.2	65.1	68.0	70.8	73.7	76.6	79.5	82.4	85.3	88.2	91.1	94.0	96.8	99.7	102.6
9	27.0	30.0	33.0	36.0	39.0	42.0	45.0	48.0	51.0	54.0	57.0	60.0	63.0	66.0	69.0	72.0	75.0	78.0	81.0	84.0	87.0	90.0	93.0	96.0	99.0	102.0	105.0	108.0
10	29.3	32.4	35.5	38.6	41.7	44.8	48.0	51.1	54.2	57.3	60.4	63.5	66.6	69.7	72.8	76.0	79.1	82.2	85.3	88.4	91.5	94.6	97.7	100.8	104.0	107.1	110.2	113.3
11	31.6	34.8	38.1	41.3	44.5	47.7	51.0	54.2	57.4	60.6	63.8	67.1	70.3	73.5	76.7	80.0	83.2	86.4	89.6	92.8	96.1	99.3	102.5	105.7	109.0	112.2	115.4	118.6
12	34.0	37.3	40.6	44.0	47.3	50.6	54.0	57.3	60.6	64.0	67.3	70.6	74.0	77.3	80.6	84.0	87.3	90.6	94.0	97.3	100.6	104.0	107.3	110.6	114.0	117.3	120.6	124.0
13	36.3	39.7	43.2	46.6	50.1	53.5	57.0	60.4	63.8	67.3	70.7	74.2	77.6	81.1	84.5	88.0	91.4	94.8	98.3	101.7	105.2	108.6	112.1	115.5	119.0	122.4	125.8	129.3
14	38.6	42.2	45.7	49.3	52.8	56.4	60.0	63.5	67.1	70.6	74.2	77.7	81.3	84.8	88.4	92.0	95.5	99.1	102.6	106.2	109.7	113.3	116.8	120.4	124.0	127.5	131.1	134.6
15	41.0	44.6	48.3	52.0	55.6	59.3	63.0	66.6	70.3	74.0	77.6	81.3	85.0	88.6	92.3	96.0	99.6	103.3	107.0	110.6	114.3	118.0	121.6	125.3	129.0	132.6	136.3	140.0
16	43.3	47.1	50.8	54.6	58.4	62.2	66.0	69.7	73.5	77.3	81.1	84.8	88.6	92.4	96.2	100.0	103.7	107.5	111.3	115.1	118.8	122.6	126.4	130.2	134.0	137.7	141.5	145.3
17	45.6	49.5	53.4	57.3	61.2	65.1	69.0	72.8	76.7	80.6	84.5	88.4	92.3	96.2	100.1	104.0	107.8	111.7	115.6	119.5	123.4	127.3	131.2	135.1	139.0	142.8	146.7	150.6
18	48.0	52.0	56.0	60.0	64.0	68.0	72.0	76.0	80.0	84.0	88.0	92.0	96.0	100.0	104.0	108.0	112.0	116.0	120.0	124.0	128.0	132.0	136.0	140.0	144.0	148.0	152.0	156.0
19	50.3	54.4	58.5	62.6	66.7	70.8	75.0	79.1	83.2	87.3	91.4	95.5	99.6	103.7	107.8	112.0	116.1	120.2	124.3	128.4	132.5	136.6	140.7	144.8	149.0	153.1	157.2	161.3
20	52.6	56.8	61.1	65.3	69.5	73.7	78.0	82.2	86.4	90.6	94.8	99.1	103.3	107.5	111.7	116.0	120.2	124.4	128.6	132.8	137.1	141.3	145.5	149.7	154.0	158.2	162.4	166.6
21	55.0	59.3	63.6	68.0	72.3	76.6	81.0	85.3	89.6	94.0	98.3	102.6	107.0	111.3	115.6	120.0	124.3	128.6	133.0	137.3	141.6	146.0	150.3	154.6	159.0	163.3	167.6	172.0
22	57.3	61.7	66.2	70.6	75.1	79.5	84.0	88.4	92.8	97.3	101.7	106.2	110.6	115.1	119.5	124.0	128.4	132.8	137.3	141.7	146.2	150.6	155.1	159.5	164.0	168.4	172.8	177.3
23	59.6	64.2	68.7	73.3	77.8	82.4	87.0	91.5	96.1	100.6	105.2	109.7	114.3	118.8	123.4	128.0	132.5	137.1	141.6	146.2	150.7	155.3	159.8	164.4	169.0	173.5	178.1	182.6
24	62.0	66.6	71.3	76.0	80.6	85.3	90.0	94.6	99.3	104.0	108.6	113.3	118.0	122.6	127.3	132.0	136.6	141.3	146.0	150.6	155.3	160.0	164.6	169.3	174.0	178.6	183.3	188.0

The amount indicated includes side walls and ceilings.

THOMAS B. TEAGAN COMPANY

TABLE 8

NUMBER OF SQUARE YARDS IN ROOMS WITH

10 FOOT CEILINGS

WIDTH IN FEET / **LENGTH IN FEET**

Width \ Length	3	4	5	6	7	8	9	10	11	12	13	14	15	16	17	18	19	20	21	22	23	24	25	26	27	28	29	30
3	14.3	16.8	19.4	22.0	24.5	27.1	29.6	32.2	34.7	37.3	39.8	42.4	45.0	47.5	50.1	52.6	55.2	57.7	60.3	62.8	65.4	68.0	70.5	73.1	75.7	78.2	80.8	83.3
4	16.8	19.5	22.2	24.8	27.5	30.2	32.8	35.5	38.2	40.8	43.5	46.2	48.8	51.5	54.2	56.8	59.5	62.2	64.8	67.5	70.2	72.9	75.5	78.5	80.9	83.5	86.2	88.9
5	19.4	22.2	25.0	27.7	30.5	33.3	36.1	38.8	41.6	44.4	47.2	50.0	52.7	55.5	58.3	61.1	63.8	66.6	69.4	72.2	75.0	77.8	80.5	83.3	86.1	88.9	91.7	94.4
6	22.0	24.8	27.7	30.6	33.5	36.4	39.3	42.2	45.1	48.0	50.8	53.7	56.6	59.5	62.4	65.3	68.2	71.1	74.0	76.8	79.8	82.7	85.5	88.4	91.3	94.2	97.1	100.0
7	24.5	27.5	30.5	33.5	36.5	39.5	42.5	45.5	48.5	51.5	54.5	57.5	60.5	63.5	66.5	69.5	72.5	75.5	78.5	81.5	84.5	87.5	90.5	93.5	96.5	99.5	102.5	105.5
8	27.1	30.2	33.3	36.4	39.5	42.6	45.7	48.8	52.0	55.1	58.2	61.3	64.4	67.5	70.6	73.7	76.8	80.0	83.1	86.2	89.3	92.4	95.5	98.7	101.8	104.9	108.0	111.1
9	29.6	32.8	36.1	39.3	42.5	45.7	49.0	52.2	55.4	58.6	61.8	65.1	68.3	71.5	74.7	78.0	81.2	84.4	87.6	90.8	94.1	97.3	100.5	103.8	107.0	110.2	113.4	116.7
10	32.2	35.5	38.8	42.2	45.5	48.8	52.2	55.5	58.8	62.2	65.5	68.8	72.2	75.5	78.8	82.2	85.5	88.8	92.2	95.5	98.9	102.2	105.5	108.9	112.2	115.5	118.9	122.2
11	34.7	38.2	41.6	45.1	48.5	52.0	55.4	58.8	62.3	65.7	69.2	72.6	76.1	79.5	83.0	86.4	89.8	93.3	96.7	100.2	103.7	107.1	110.5	114.0	117.4	120.9	124.3	127.8
12	37.3	40.8	44.4	48.0	51.5	55.1	58.6	62.2	65.7	69.3	72.8	76.4	80.0	83.5	87.1	90.6	94.2	97.7	101.3	104.8	108.4	112.0	115.5	119.1	122.7	126.2	129.8	133.3
13	39.8	43.5	47.2	50.8	54.5	58.2	61.8	65.5	69.2	72.8	76.5	80.2	83.8	87.5	91.2	94.8	98.5	102.2	105.8	109.5	113.2	116.9	120.5	124.2	127.9	131.5	135.2	138.9
14	42.4	46.2	50.0	53.7	57.5	61.3	65.1	68.8	72.6	76.4	80.2	84.0	87.7	91.5	95.3	99.1	102.8	106.6	110.4	114.2	118.0	121.8	125.5	129.3	133.1	136.9	140.7	144.4
15	45.0	48.8	52.7	56.6	60.5	64.4	68.3	72.2	76.1	80.0	83.8	87.7	91.6	95.5	99.4	103.3	107.2	111.1	115.0	118.8	122.8	126.7	130.5	134.4	138.3	142.2	146.1	150.0
16	47.5	51.5	55.5	59.5	63.5	67.5	71.5	75.5	79.5	83.5	87.5	91.5	95.5	99.5	103.5	107.5	111.5	115.5	119.5	123.5	127.5	131.5	135.5	139.5	143.5	147.5	151.5	155.5
17	50.1	54.2	58.3	62.4	66.5	70.6	74.7	78.8	83.0	87.1	91.2	95.3	99.4	103.5	107.6	111.7	115.8	120.0	124.1	128.2	132.3	136.4	140.5	144.7	148.8	152.9	157.0	161.1
18	52.6	56.8	61.1	65.3	69.5	73.7	78.0	82.2	86.4	90.6	94.8	99.1	103.3	107.5	111.7	116.0	120.2	124.4	128.6	132.8	137.1	141.3	145.5	149.8	154.0	158.2	162.4	166.7
19	55.2	59.5	63.8	68.2	72.5	76.8	81.2	85.5	89.8	94.2	98.5	102.8	107.2	111.5	115.8	120.2	124.5	128.8	133.2	137.5	141.9	146.2	150.5	154.9	159.2	163.5	167.9	172.2
20	57.7	62.2	66.6	71.1	75.5	80.0	84.4	88.8	93.3	97.7	102.2	106.6	111.1	115.5	120.0	124.4	128.8	133.3	137.7	142.2	146.7	151.1	155.5	160.0	164.4	168.9	173.3	177.8
21	60.3	64.8	69.4	74.0	78.5	83.1	87.6	92.2	96.7	101.3	105.8	110.4	115.0	119.5	124.1	128.6	133.2	137.7	142.3	146.8	151.4	156.0	160.5	165.1	169.7	174.2	178.8	183.3
22	62.8	67.5	72.2	76.8	81.5	86.2	90.8	95.5	100.2	104.8	109.5	114.2	118.8	123.5	128.2	132.8	137.5	142.2	146.8	151.5	156.2	160.9	165.5	170.2	174.9	179.5	184.2	188.9
23	65.4	70.2	75.0	79.7	84.5	89.3	94.1	98.8	103.6	108.4	113.2	118.0	122.7	127.5	132.3	137.1	141.8	146.6	151.4	156.2	161.0	165.8	170.5	175.3	180.1	184.9	189.7	194.4
24	68.0	72.8	77.8	82.6	87.5	92.4	97.3	102.2	107.1	112.0	116.8	121.8	126.6	131.5	136.4	141.3	146.2	151.1	156.0	160.8	165.8	170.7	175.5	180.4	185.3	190.2	195.1	200.0

The amount indicated includes side walls and ceilings.

THOMAS B. TEAGAN COMPANY

Computing Volumes

Computation of volumes can be quite complicated, and reference to a mathematics book is necessary in difficult cases. Fortunately, however, most volumes required in plastering practice can be obtained easily by multiplying the area by the desired depth. This gives the 3-dimensional content in terms of cu. ft. or cu. yds.

To simplify computations of volume, the plastering trade makes wide use of *slices* as basic units of calculation. The slice is conceived of as a flat piece ⅛" thick, generally 1 sq. yd. in area (that is, 9 sq. ft.). This thickness or depth is convenient because it is the thinnest coat of material used in plastering, and all common plastering thicknesses are multiples of the ⅛" slice. See Fig.

2. More will be said about the use of slices in the section on *Estimating*, later in this appendix.

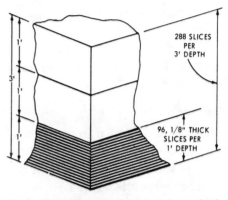

Fig. 2. The ⅛" thick *slice* is used as a basic unit of calculation because it is the thinnest coat of material used in plastering. One cubic yard of material yields 288 slices that will normally cover an area of 288 square yards.

Elementary Geometry

Much of the study of geometry from an academic standpoint has to do with theorems and proofs. From a trade standpoint, however, the properties of lines, figures, etc. are of greater concern.

Lines

The *horizontal line* is a level line. It is the opposite of a vertical line. (Fig. 3.)

A *perpendicular line* is a line at right angles to another line. (Fig. 3.)

A *level line* and a *plumb line* produce a square and are perpendicular to each other. (Fig. 3.)

The *diagonal line* is one joining two opposite angles. (Fig. 3.)

Parallel lines are those having the same direction and are an equal distance from each other at all points. (Fig. 3.)

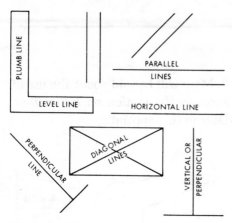

Fig. 3. Straight lines that differ because of their placement.

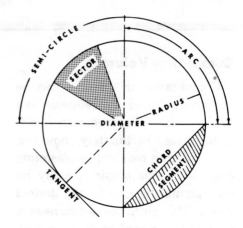

Fig. 4. A circle and its parts.

The Circle

The *circle* is drawn from the center and is a continuous curved line, being of an equal distance from the center at all points. (Fig. 4.)

An *arc* of a circle is any part of its circumference. (Fig. 4.)

The *chord* is a straight line joining two points of the circumference. (Fig. 4.)

The *semi-circle* is one-half of the complete circle. (Fig. 4.)

The *circumference* is the entire distance around the circle.

The *diameter* is the distance across the circle thru the center. (Fig. 4.)

The *radius* is half the diameter or the distance from the center to any point of the circumference. (Fig. 4.)

A *sector* is a portion of a circle between two radii and the circumference. (Fig. 4.)

A *segment* is a portion of a circle contained by a straight line and the circumference which it cuts off. (Fig. 4.)

The *tangent* is a straight line which touches a circle or curve but does not cut it and is at right angle to a straight line from the center. (Fig. 4.)

Circle Measurements

Circumference of a circle equal diameter × 3.1416. (3.1416 = Pi or the Greek letter π.)

Area of a circle equals diameter squared (dia.^{2}) × .7854.

Length of arc equals degrees in arc × radius × .01745. *Example:* 45° × 4' radius = 180 × .01745 = 3.141' length of arc.

Degree of arc equals length/radius × .01745. *Example:* 4' radius × .01745 = .0698, 3.141' length ÷ .0698 = 45°.

Radius of arc equals length/degrees × .01745. *Example:* 45° ×

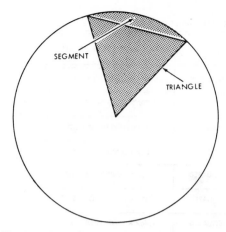

Fig. 5. Area of segment found by subtracting area of triangle from area of sector.

.01745 = .78525, 3.141' length ÷ .78525 = 4' radius.

To find the *area of a sector* of a circle: 3.1416 × radius squared × degrees of the sector ÷ 360.

To find the *area of a segment* of a circle: Find area of sector and subtract area of included triangle. Fig. 5 illustrates both sector and segment of the circle.

Spherical Measurements

Surface area equals diameter squared × 3.1416. *Example:* dia. 4' sq. = 16' × 3.1416 = 50.26 sq. ft. surface area.

Volume equals diameter cubed × .5236. *Example:* dia. 4' cubed = 64 cu. ft. × .5236 = 33.5104 cu. ft.

Elliptical Measurements

Elliptical *surface area* equals width × height × .7854. *Example:*

width 36" × height 24" = 864 sq. in. × .7854 = 678.59 sq. in. area.

Triangular Measurements

When two *angles* of a triangle are known, the third can be found by subtracting the sum of the two known angles from 180°. (The sum of the angles of a triangle equals 180°.)

The square of the *hypotenuse* (longest side) of a right triangle is equal to the sum of the squares of the other two sides.

Area of a triangle equals ½ height × base.

Pyramid Measurements

Area equals, ½ perimeter of base × slant height + area of base.

Volume equals, area of base × ⅓ height or altitude.

Trapezoid Measurements

Area equals height × ½ the sum of its parallel sides.

Rectangular Measurements

Volume = width × length × height. For volume in gallons divide cubic content *in inches* by 231; for cubic content *in feet* divide by 7.48.

Measurements of Regular Polygons

Regular polygons, also called equilateral polygons, are those having

TABLE 9 REGULAR POLYGONS: NAME, AREA AND RADIUS OF INCLOSING CIRCLE

NAME OF FIGURE	NUMBER OF SIDES	AREA EQUALS:	RADIUS OF CIRCLE EQUALS:
EQUILATERAL TRIANGLE	3	0.433 X 1 SIDE SQUARED	0.577 X LENGTH OF 1 SIDE
SQUARE	4	1.000 X 1 SIDE SQUARED	0.707 X LENGTH OF 1 SIDE
PENTAGON	5	1.720 X 1 SIDE SQUARED	0.851 X LENGTH OF 1 SIDE
HEXAGON	6	2.598 X 1 SIDE SQUARED	1.000 X LENGTH OF 1 SIDE
HEPTAGON	7	3.634 X 1 SIDE SQUARED	1.152 X LENGTH OF 1 SIDE
OCTAGON	8	4.828 X 1 SIDE SQUARED	1.307 X LENGTH OF 1 SIDE
NONAGON	9	6.182 X 1 SIDE SQUARED	1.462 X LENGTH OF 1 SIDE
DECAGON	10	7.694 X 1 SIDE SQUARED	1.618 X LENGTH OF 1 SIDE
UNDECAGON	11	9.365 X 1 SIDE SQUARED	1.775 X LENGTH OF 1 SIDE
DODECAGON	12	11.196 X 1 SIDE SQUARED	1.932 X LENGTH OF 1 SIDE

equal sides. All can be inscribed in circles so that all vertices (corners) exactly touch the circle's circumference. It follows that all angles of a regular polygon are equal, as well as its sides.

Table 9 lists the more common regular polygons and formulas for calculating their areas.

Domes

Area of perfect dome in square yards = square of the diameter (in feet) × .175. *Example:* Perfect dome having 40 ft. diameter, then 40′ × 40′ = 1600 sq. ft.; 1600 × .175 = 280 sq. yds.

Area of an oval dome ceiling in square yards:

1. Deduct width from length.
2. Multiply the remainder by the width.
3. To this add the square of the width.
4. Multiply the total of steps 2 and 3 by .175.

Example: Oval dome ceiling 60 ft. long and 45 ft. wide:

1. 60′ − 45′ = 15′.

2. 15′ × 45′ = 675 sq. ft.

3. 45′ × 45′ = 2,025 sq. ft.
   Total 2,700 sq. ft.

4. 2,700 × .175 = 472.5 sq. yds.

Development of Geometric Shapes

This section outlines methods and procedures for developing geometric shapes such as arches, domes and ellipses that are of particular importance to the plastering trade.

To Find Radius Point of an Arc by T-square Method. (Fig. 6.) A simple method can be used to establish the center point of any circular arch or arc. This method is particularly useful for finding the center point of an arch or arc already formed in place.

The curved line represents the arch or arc. Use a gage made of light lumber in the form of a T-square with the long leg set so that its edge is on the center line of the short leg, as shown at *C*. Place this gage anywhere in the arch or arc, letting

points *A* and *B* touch the soffit or members of the arch. A line drawn along the long leg will cross the center point of the arch. Repeat this operation at least once more to establish the true center as shown at *D*.

To Find Radius Point of an Arc by Method of Bisected Chords. (Fig. 7.) This is a good method for locating the center point of an arc when the span and rise are known. Lay off span *A-B*, and at its center establish the rise at point *C*. Draw diagonals from *A* to *C* and *C* to *B*. Bisect these diagonals, using points *C* and *A* as centers and any convenient length as a radius. Draw a line through points *D* and *E* until it crosses center line at *F*. Repeat this operation for the other diagonal. This will check

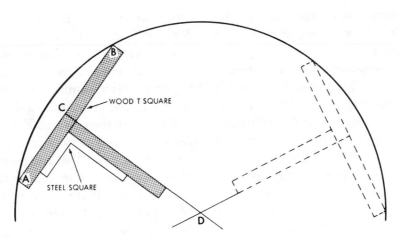

Fig. 6. Finding radius point of an arc by T-square method.

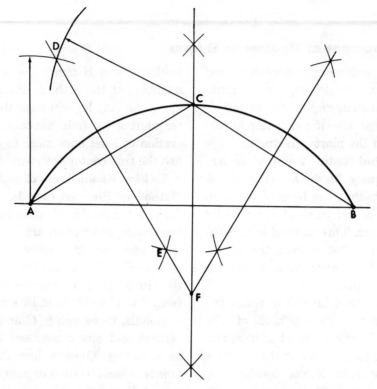

Fig. 7. Finding radius point of an arc using bisected chords method.

and prove the center point of the arc. Now draw arc *A-C-B* using *F* as the center.

Segmental Arch Layout Obtained by Quarter Rise Method. (Fig. 8.) Used where layout space is limited, this is a good method for laying out a long-span, low-rise arch. Lay out span *A-B*, bisect and draw perpendicular *C-D*, equal to the rise of arch. Divide rise into 4 equal parts, draw line *A-C*, bisect this line and, on a perpendicular at this point, measure off *G-E* equal to one of the spaces *E*

found on line *C-D*. Divide *G-F* into 4 equal parts and draw lines from *A* to *G* and *G* to *C*. Bisect each of these lines and at these points measure up a distance equal to one of the spaces *H* found on line *G-F*. This will establish rises *I-J* and *K-L*. Now draw a curved line from points *A* to *I* to *G* to *K* to *C*. This completes ½ of the arc. Repeat these operations for the other half. A cornice strip can be used on edge to form the curve, bending the strip to fit the points established.

506

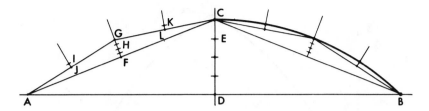

Fig. 8. Segmental arch layout obtained by quarter rise method.

The larger the arc the more points should be established to develop the arc. Each division can be redivided to establish additional points.

Segmental Arch Layout Obtained by Ordinates. (Fig. 9.) A good method to use when space is limited. Lay out span *A-B*, then bisect it and draw a perpendicular up to point *C*, which is the rise. Draw diagonal line A-C, and at 90° from the diagonal draw line *A-F*. Now draw line *D-E* parallel to *A-B* at rise *C*. Divide distance *F-C* into any number of parts, in this case six. Next divide half the span into the same number of parts. Draw line *A-G* perpendicular to *A-B*

and divide this also into the same number of parts. Now connect the points on line *F-C* to points of the same number on line *A-B*. Then draw lines from point *C* to each of the divisions on line *A-G*. The intersection of these lines establishes the curve. Draw the curve through these points, such as: *1-1, 2-2, 3-3*, etc. Repeat this operation for the other half of the arc.

The more divisions used on the layout lines the more accurate the curve will be. A cornice strip on edge can be bent to connect the points and then used as a guide to draw the curve.

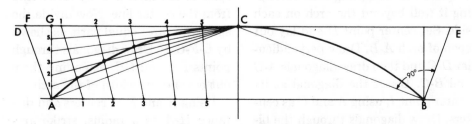

Fig. 9. Segmental arch layout obtained by ordinates.

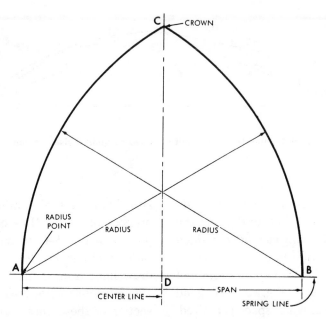

Fig. 10. Layout for equilateral arch.

Layout For An Equilateral Arch. (Fig. 10.) Establish the spring line of the arch *A-B* and the center line *C-D*. Using a radius equal to the span of the arch and radius point A, strike arc *B-C*; then with radius point *B* strike arc *A-C*, completing the arch.

Layout for an Acute Gothic Arch. (Fig. 11.) Draw spring line, extending it well beyond the arch on each side. Set center point *D* and lay out span of arch *A-B*. Draw perpendicular *D-C* and then draw diagonals *A-C* and *B-C*. Bisect the diagonal at its center point *I*, using *A* and *C* as centers. Draw diagonals through the bisecting arcs down to the spring line

at *F* to establish the radius point. Now draw arc *A-C* from radius point *F* to form one side of the arch. Repeat the operation on the other side to establish radius point *E* and complete the arch.

Developing a Tudor Arch. (Fig. 12.) Lay out spring line and draw the center line *C-D*. Establish the span of the arch *A-B* by measuring from the center line. Now divide the span into three equal parts as shown by *E-F-G-H*. Draw diagonals through points *FH* and *GE*, extending them out beyond the width of the arch.

Using *H* and *E* as centers and distance *H-A* as a radius, strike arcs *A-I* and *B-J*. Then, using *G* and *F*

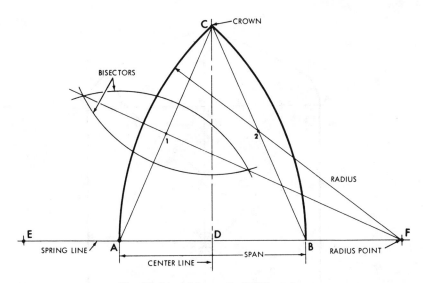

Fig. 11. Layout for acute Gothic arch.

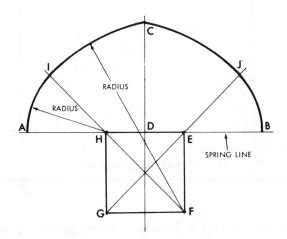

Fig. 12. Developing a Tudor arch.

as centers and distance *F-I* as a radius, strike arcs *I-C* and *J-C*, completing the arch.

Developing a Flat Tudor Arch. (Fig. 13.) Lay out spring line of the arch *A-B* and the center line *C-D* extending down well below the spring line. Divide the span of the arch *A-B* into six equal parts, working out from the center line each way and

509

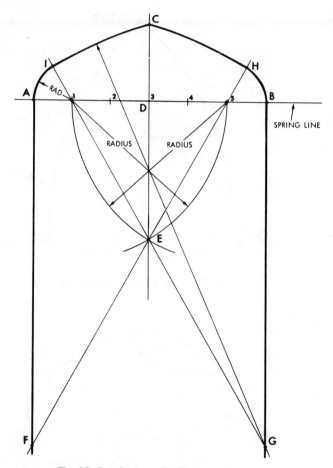

Fig. 13. Developing a flat Tudor arch.

mark points, *1, 2, 3, 4* and *5.* Now using points *1* and *5* as centers and the distance from *1* to *5* as a radius, draw arcs *5* to *E* and *1* to *E.* Draw diagonals through *1* and *E* and *5* through *E,* extending these diagonals until they cross lines dropped down parallel to the center line from points *A* and *B,* establishing centers *F* and *G.*

Using point *1* as a center and the distance *1* to *A* as a radius, strike the arc *A-I.* Next, with point *5* as a center, strike arc *B-H.* Now, using *G* as a center and the distance *G-I* as a radius, strike the arc *I-C,* and with *F* as a center strike the arc *H-C,* completing the arch.

Layout for Three Point Elliptical Arch. (Fig. 14.) Establish the

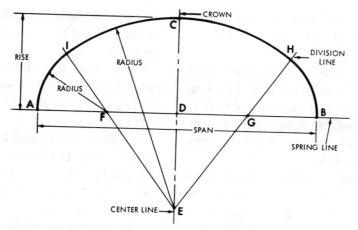

Fig. 14. Layout of three point elliptical arch.

spring line of the arch, then find the center point *D* and lay out the span of the arch *A-B*, measuring each way from the center. *A-B* equals the map or axis of the ellipse in which *C-D* equals ½ of the minor axis or rise of the arch. *E-D* equals *A-B* minus twice *C-D*.

E-D and *G-D* each equal ¾ of *E-D*. Point *E* is the radius center for arc *H-C-I*. Point *G* is used to strike arc *B-H*, and point *F* is the radius center for arc *I-A* completing the elliptical arch.

The basic formula for any size ellipse of this size is: major axis minus minor axis equals radius point for the long arcs laid out on the minor axis from the center point. Three quarters of the long radius equals the short arc radius measured from the center point on the major axis.

Developing an Approximate Ellipse from Two Circles. (Fig. 15.) Draw center lines *A-B* and *C-D*. On center line *A-B* draw two circles whose combined diameters equal the major axis or length of the ellipse. *E-G* and *E-F* are each equal to ¼ of the major axis measured out each way from the center point *E*. Now

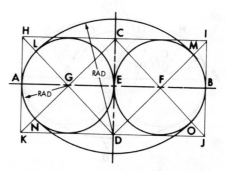

Fig. 15. Approximate ellipse developed from two circles.

develop a rectangle inclosing the two circles as shown with points *K*, *J*, *I*, *H*.

Next, draw the diagonals *K-C*, *H-D*, *J-C* and *I-D*. Using *D* as a radius point and a radius equal to *D-L*, strike arc *L-M*; then using *C* as a radius point, strike arc *N-O* to complete the ellipse.

Developing an Ellipse from Three Circles. (Fig. 16.) Draw center lines *A-B* and *C-D*. Divide the major axis into four equal parts, working from the center line at *E* and producing radius points *G*, *E*, *F*. Draw three circles, each with the radius equal to *G-A* as shown. Now draw division lines from *G* through *N* and extending up until it intersects with the vertical center line at point *C* and down to point *K*; also extend a line from *C* through *O-F* and down to *J*. Then *F* through *M* to intersection *D* and extended up through *I*. Next *D* through *L-G* and *H*. Points *C* and *D*

are now established as radius points for the arcs *H-I* and *K-J*, completing the ellipse.

Notice that this three-circle ellipse produces a flatter shape than the two circle one but it is still not a true ellipse.

An Ellipse Developed From Two Circles. (Fig. 17.) Draw center lines *A-B* and *FG*. The center point is now used to draw two circles. The smaller circle's diameter equals the minor axis of the ellipse, while the larger circle equals the major axis. Divide the outer circle into as many parts as is desired. The more divisions used, the more points will be established along the perimeter of the ellipse. Draw lines from center point *E* to divisions points as shown at *E-J* and *E-K*. Now draw vertical lines down from points *J* and *K*. Next, draw horizontal lines out from points *I* and *H*.

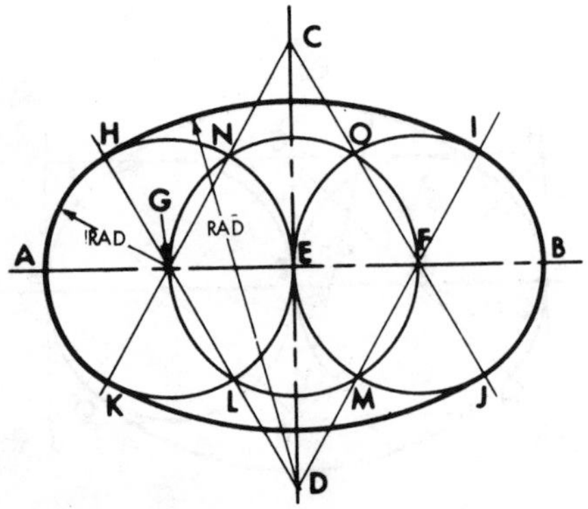

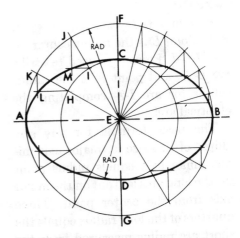

Fig. 16. Approximate ellipse developed from three circles.

Fig. 17. Second method for developing an ellipse from two circles.

The intersections of these lines develop the outline of the ellipse. Notice how in the quarter circle *B-F* the sectors were redivided so as to establish more points, making it easier to develop a good curve for the ellipse.

Ellipse Developed by Tangents. (Fig. 18.) Lay out a rectangle equal to the length and width of the desired ellipse indicated by letters *H*, *F*, *G*, *I*. Draw center lines *A-E-B* and *C-E-D*. Divide *A-F* into any number of parts, dividing *C-F* into the same number of parts. Now connect points *1-1*, *2-2*, *3-3*, and so on, with lines. These lines or tangents will form the ellipse in a series of short, straight sections. The greater number of divisions used the better the curve will be developed. Repeat the operation on all four quarters or make a template from the first quarter and use this template to draw the other three quarters.

Developing an Ellipse from a Quarter Circle, Using Ordinates.

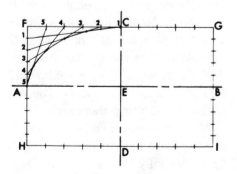

Fig. 18. Ellipse developed by tangents.

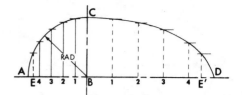

Fig. 19. Ellipse developed from a quarter circle, using ordinates.

(Fig. 19.) Lay out center lines *A-D* and *B-C*. *A-B* equals ½ of the minor axis, and *B-D* equals ½ of the major axis. Strike arc *A-C* using radius point *B* as the center. Divide *A-B* into as many points as desired. To insure accurate development of the ellipse at the ends, divide the last space into two or more divisions, as shown at *E*. Divide *B-D* into the same number of parts and also subdivide last part as shown at *E'*.

Draw perpendiculars at each division in the quarter circle as shown by lines *1*, *2*, *3*, *4* and *E*. Now transfer each line, using its exact length, to section *B-D* as shown. The ends of these lines will now be used to layout the elliptical curve. The more divisions used, the better the curve will be defined.

Ellipse Developed by Using a String. (Fig. 20.) Establish center lines *A-B* and *C-D*. Using a compass or steel tape, depending upon the size of the ellipse, start at point *C* as a center and with a radius equal to *A-E* mark the major axis at points *F* and *G*. Now loop a string or fine wire

513

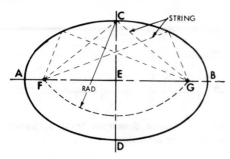

Fig. 20. Ellipse developed by using a string.

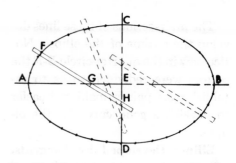

Fig. 21. Ellipse developed with a graduated stick.

around a nail set at point *F*, then bring it taut to point *C* and down to another nail set at point *G*. Using a pencil held against the string or wire and keeping it tight at all times, work it back and forth to form the ellipse.

This is a practical and quick method to form a true ellipse. This method, along with the trammel and similar methods, is based on the fact that the ellipse is a curve whose path is described by a point moving in such a way that the sum of its distances from two fixed points remains the same. These two fixed points are called *foci* points.

Developing an Ellipse Using a Graduated Stick. (Fig. 21.) Lay out center lines *A-B* and *C-D*. Then prepare a straight stick as shown, setting mark *F* at point *A* and at point *E* make mark *H*. Now move the stick around so mark *F* is at point *C* and make mark *G* at point *E*.

Keeping mark *G* on the horizontal or major axis line and mark *H* on the vertical or minor axis line move the stick gradually, starting at point *A*. Make marks at close intervals as the marks *G* and *H* are carefully kept on their respective lines.

The stick is moved completely around, establishing the outline of the ellipse as shown. This produces the same true ellipse as the trammel and string methods.

Developing an Oval Around a Circle. (Fig. 22.) Lay out center lines *A-B* and *C-D*. Using their intersection *E* as a center with a radius equal to the required lower end of the oval, describe a circle. Draw divisional lines from *A* through *F* and *B* through *F*. Now using *A* as a center and *A-B* as a radius draw arc *B-G* then with *B* as the center draw arc *A-H*. Next using *F* as a center and *F-G* as the radius describe the arc *G-C-H* completing the oval.

Many tradesmen and other people use the word oval to describe an ellipse. Actually, the oval is an egg-shaped figure, and to use the word

514

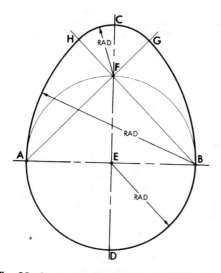

Fig. 22. Oval developed around a circle.

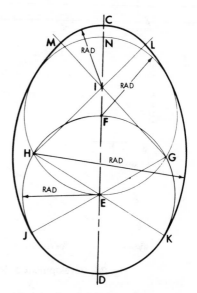

Fig. 23. Developing an oval around two circles.

oval when speaking of ellipses shows the speaker is not well acquainted with either of these figures.

Developing an Oval Around Two Circles. (Fig. 23.) Draw center line C-D. Using E and F as centers and a radius equal to the desired lower end of the oval, draw two circles overlapping each other. Through E, the center and the intersection of the two circles, draw division lines J-E-G and K-E-H. Using points H and G as centers and a radius to H-K, draw the two arcs K-L and J-M.

Now divide the distance F-N into two equal parts and establish center point I. Draw division lines H-I-L and G-I-M. Then, with I as the center and a radius equal to I-M draw arc M-L, completing the oval. This oval is more elongated than the

one developed around one circle and truer to the actual egg shape.

Delevopment of the Three Templates for an Elliptical Groin Ceiling. (Fig. 24.) Three templates are required to develop an elliptical groin ceiling. Rectangle A-D-I-E represents the wall lines of the room. Draw diagonals to each corner and center lines each way. The given or regular ellipse is shown as A-C-D. Using one half of its span, or A-B, divide this line into any number of parts as shown. Notice that the last part has been redivided into four parts to provide additional reference points for the sharp rising curve. Now extend lines up from these points to the diagonal line D-E.

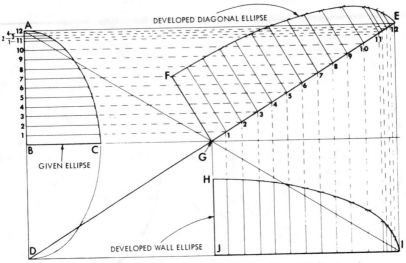

Fig. 24. Development of the three templates for an elliptical groin.

Starting at midpoint *G*, and along the diagonal line *G-F*, draw perpendiculars at each point up from the diagonal line. Line *G-F* equals line *B-C* in length. Each line on the given ellipse *1, 2, 3*, and so forth, is measured and transferred to the diagonal line. The curved line *F-E* is then developed through these points, producing the elliptical groin intersection or diagonal template.

Next, from these same points, *1, 2, 3, etc.* on the diagonal line, drop lines from the long wall ellipse to base line *J-I* and rise *J-H*. Make each division identical in length to the corresponding lines used for the other two templates. As can be seen, one set of ordinates produces the three templates, which are identical except for their lengths. This vari-ance is governed by the slant of the diagonal line. Center heights are the same for each template.

This same technique can be used in developing any template that must vary in width but otherwise conform to another template.

Elliptical Dome, Showing Development of the Required Templates. (Fig. 25.) This is a type of dome that cannot be run or formed by the radius rod methods. Only the use of a series of templates, each made to the contour of the ellipse at a certain point in the dome, will produce it.

The elliptical outline of the dome *A-A'-B-B'* is drawn full size, using the center lines as the major and minor axis. One quarter of the ellipse is now divided into any number of parts as shown at *A, C, D, E, F, G,*

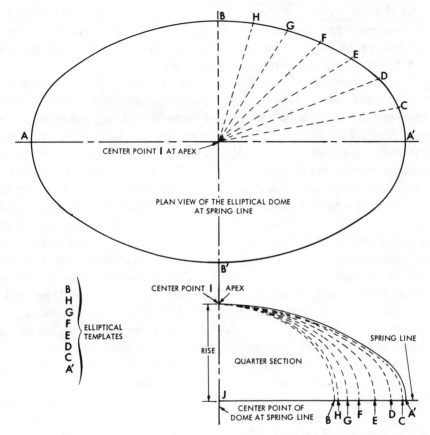

Fig. 25. Development of the required templates for an elliptical dome.

H, B. Use enough divisions to insure a good development of the dome.

An ellipse is now drawn for each line, measuring from the apex point *I* to the point on the elliptical line. When the required number of templates are made they are used in series in each quarter of the dome. When all the screeds have been formed, the area between the screeds is filled in using a short flexible strip to shape the dome to its curve.

Estimating Areas, Volumes, and Materials

What is an estimate? Basically an estimate represents a careful study of the plans and specifications for a building or unit, giving a detailed account of every item of labor, material, business operating cost and contractor's profit required to complete the work. The estimator who prepares the estimate must be skilled in mathematics, plan reading, and material and labor costs, as well as knowing the different types of construction and methods used to do work under varying conditions.

It will be helpful to the plasterer to have at least a basic understanding of how the material estimate for lathing and plastering is made. The authors recommend that the apprentice obtain one or more books on estimating in order to understand the various steps used in preparing the estimate.

The plasterer is mainly concerned with the amount of material required to do a given job. He works with many materials, and each of these requires a specific formula to estimate the area it will cover per bag or per unit of volume or weight.

To help the plasterer figure the material he needs for any given job, the trade has developed numerous units of area and volume. There are two methods in common use in the trade for computing the material needs for a job. One is the unit of weight, or what a ton (2,000 lbs.) of a given material will cover. Another is the amount of material required to cover 100 square yards of work. For small jobs or patch work it is common to figure how many bags of

material will be required for the job, based on the formulas developed to tell how much a bag of each material will cover.

Units Used. Gypsum plaster is generally packaged in 100 lb. bags; lime in 50 lb. bags; and Portland cement in 100 lb. (net 94 lbs.) bags. Sand is sold by the cubic yard or by the ton. A cubic yard of sand has an average weight of 2,700 lbs. or 100 lbs. per cubic foot. Vermiculite, Perlite and similar lightweight aggregates are usually packaged in bags holding from three to four cubic feet.

The plasterer's common measure of area is the square yard, which equals 9 square feet. See Fig. 26. All plain surfaces, such as walls and ceilings, are figured in square yards. Beams, pilasters, window and door revels, cornice work and various miscellaneous items are measured in

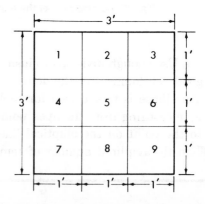

Fig. 26. The plasterer's common measure of area for large, flat surfaces such as walls and ceilings is the square yard, containing 9 square feet.

square feet. Arrises, corner beads, channel iron, and similar units are measured in linear feet.

When measuring beams, pilasters and cornices, it is customary to measure the *girth* (circumference or distance around an object). For a beam, add the number of lineal feet in the width of the two sides and the bottom (or soffit) and multiply this sum by the length to compute the area in square feet. Cornice work is usually measured in six inch increments, so that any size up to six inch *girth* (distance from the ceiling across the cornice members to the wall) is classified as a six inch cornice. Next, any cornice up to twelve inch girth is measured as a twelve inch cornice, and so forth. The required girth is then multiplied by the length to determine the area.

In figuring material needs, two methods are used in the trade to allow for openings in walls or ceilings, such as windows, doors or skylights. The most common method is to disregard these openings and figure all walls and ceilings as solid. The material not used in the openings makes up generally for the droppings and other waste that occurs. This method also allows for the extra time required for such additional tasks as cleaning grounds or corner beads in working around doors and windows.

The second method, used mostly in large commercial jobs, is to deduct every opening and then to figure separately any revels or jambs, by the square foot measurement, to cover the extra work involved.

Estimating Mortars

In jobs requiring sand as the aggregate, the number of cubic yards of sand is estimated first, and most of the other material requirements are worked out from that figure.

Sand and plaster mixed in proportions of 3 sand to 1 plaster will retain the same volume as the sand alone. That is, 1 cubic yard of sand and 9 bags of plaster when combined will still contain only 1 cubic yard; the plaster fills the voids between the sand particles. When this ratio is increased to 2 to 1, such as for mortar for use on gypsum or metal lath, the added plaster bulks the mixture so that a greater volume is produced.

As stated before, plastering sand has an average weight of 2,700 lbs. per cubic yard. Thus, for a 2 to 1 mix the amount of plaster required would be 2,700 lbs. $\div$ 2 = 1,350 lbs., or 13½ bags containing 100 lbs. each.

Now, to find out how great an area a cubic yard of material will cover, find the number of $\frac{1}{8}''$ *slices* in the height of the cubic yard. The $\frac{1}{8}''$ slice is used as a basic unit of calculation because it is the thinnest coat of material used in plastering; moreover, all common plastering thicknesses are multiples of the $\frac{1}{8}''$ slice. *The cubic yard of material can be cut into 8 × 12 × 3 = 288 slices, each ⅛ of an inch thick.* **Refer back to Fig. 2.** For the average masonry wall, where the mortar is ¾″ thick, this formula would be divided by six because there are six-eighths in three quarters inch thickness. This, then, figures 288 ÷ 6 = 48, or *48 slices of mortar ¾″ thick can be made from 1 cubic yard of sand and 9 bags of plaster.*

When richer (more plaster) mortar is required, such as 2 to 1 mix, more volume is produced, adding about 1 ft. *additional depth* of material to the standard cubic yard. See Fig. 27. When multiplying the slices

for this "expanded cube" the formula would be 8 × 12 × 4 = 384 slices ⅛″ thick. To bring this into the average plaster thickness for gypsum lath, divide the figure by three, as most two-coat work on gypsum lath is ⅜″ thick. Example: 384 ÷ 3 = 128 square yards of plaster ⅜″ thick per cubic yard of sand and 13½ bags of plaster. (See Fig. 27, showing a cubic yard and an expanded cube.)

These two formulas will take care of all the mortar requirements for sand aggregate plaster. Once the units are worked out they are then used to compute material needs. To reduce this unit to a still more suitable basis, it may be broken down to pounds of sand and plaster required per square yard of mortar.

To find the *sand* required to plaster 1 square yard ⅛″ thick, divide 288 sq. yds. into 2,700 lbs. of sand (1 cubic yard). The result will be 9.72 lbs. or for practical purposes, 10 lbs. of sand will be required for each square yard.

Plaster needed to cover 1 square yard ⅛″ thick, can be found by dividing the sand by three. In this case 9.72 ÷ 3 = 3.24 lbs. of plaster per square yard. For the 2 to 1 mix, divide the sand requirements by two, or 9.72 ÷ 2 = 4.86 lbs. Using these two units and multiplying them by the thickness wanted and, again, multiplying by the square yardage will produce the material needs for any job.

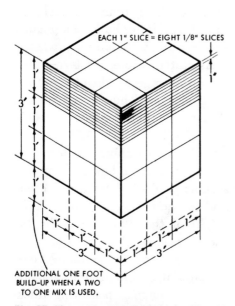

EACH 1″ SLICE = EIGHT 1/8″ SLICES

ADDITIONAL ONE FOOT BUILD-UP WHEN A TWO TO ONE MIX IS USED.

Fig. 27. The cubic yard divided into 27 cubic feet, and the "extended cubic yard" used as a convenient unit of measurement by the plastering trade for certain mortar mixtures.

Lightweight Aggregates. The lightweight aggregates come packaged in paper bags containing either three or four cubic feet. The manufacturer's specifications as to the correct amount of plaster to add for each operation are usually printed on the bag. Most lightweight aggregates call for two 100 lb. bags of plaster to one 4-cubic-foot bag of aggregate.

When plastering over high-suction masonry bases (soft brick, gypsum partitions, tile, etc.) with lightweight aggregates, the aggregate must be increased to better than three to one to prevent shrinkage cracking.

Portland Cement. *Portland cement* plastering materials are based upon the same basic formulas as used for estimating gypsum plastering materials.

Finish Materials. *Finish materials* are much harder to estimate by mathematical means owing to the differences in spreading capacity of the lime putty. This is caused by chemical variations in the lime, and the weather or job conditions. Finish lime will cover from 12 to 17 square yards per bag, depending upon the condition of the brown coat. Commercial work using 3 to 1 mortar will produce a coarse surface that requires more finishing material to cover it. Residential work using a 2 to 1 mortar mix for the brown coat will produce a dense, smooth surface requiring much less lime.

Most large commercial contractors

figure 500 square yards coverage per ton of lime. In the residential field 700 square yards per ton of lime is the average. Thickness of the finish coat needed to cover the brown coat is the most difficult to estimate because chemical differences, degree of dryness of the base coat and weather variation all affect the quantity of material needed for proper coverage.

Gauging plaster is figured at 25 percent of the lime putty for plain work and 50 percent for cornice work. One 50 lb. bag of lime will make approximately 1.14 cubic feet of putty. This, when mixed with 25 percent of gauging plaster and 4 quarts of water to soak the plaster, will make 1.39 cubic feet of finishing material. When put in a measure 12″ square, this material will fill the measure to a height of 15 inches. This quantity, when sliced into ⅛″ slices 1 foot square, will produce $8 \times 15 = 120$ slices or 120 square feet of putty coat. When divided by 9, this equals 13.3 square yards per bag of lime.

The finishing material requirements are based on the amount of lime needed. When the volume of the lime is found, the plaster is computed from it. Thus, two materials constitute the base from which all other materials are worked out: *the sand for the base coats and the lime for the finish coats.* Special materials should be checked out on the manufacturer's specifications and these should be followed exactly.

521

Estimating Lathing Materials

Gypsum lath is usually packed 6 sheets to a bundle and equals 32 square feet. (Sheets = 16″ × 48″ = 768 sq. in. × 6 sheets = 4608 sq. in. ÷ 144 sq. in. per ft. = 32 sq. ft. per bundle.) Some areas pack 5 sheets to the bundle equaling 3 square yards. This lath is ordered either by the square foot or by the bundle. Thicknesses available are ⅜″ and ½″.

Due to a slight gain made at the joinings of the boards, and disregarding all openings, it is customary to deduct about 7 percent from the total. For ceiling work only, the full total must be used. Good mechanics will have practically no waste, as small pieces are used up in the closets and other narrow areas.

Metal lath is ordered by the square yard or bundle. Bundles vary in content depending upon lath width. The two common widths and lengths are 24″ × 96″ and 27″ × 96″. The 27″ × 96″ sheet equals 2 square yards and is packed 10 sheets to the bundle, totaling 20 square yards. Sheets 24″ × 96″ are packed 9 to the bundle, totaling 16 square yards. Weight of the lath runs 2.5, 3.0, 3.4, and 4.0 pounds per square yard for standard diamond mesh. Other types vary in weight according to their type and purpose. To cover end and side laps, about 7 percent is added to the figured yardage.

Channel irons are measured and ordered in lineal feet. The channels are made in various lengths. These are 13, 16, 18 and 20 feet long. Three-fourths inch channels are packed 20 pieces to the bundle; 1″, 1½″ and 2″ channels are packed 10 pieces to the bundle. Channels are made in both hot and cold rolled types. A certain amount of waste must be allowed for lapping and cutting. On small jobs 10 percent is needed for this purpose; on larger jobs, openings and other gains usually balance this factor. For special work such as groins, domes, and light troughs, add 50 percent.

To estimate the lathing material requirements for a standard suspended metal lath and channel iron ceiling, the items used are as follows:

No. 8 galvanized wire hangers, 4 feet on center.

1½″ carrier channels, 4 feet on center.

¾″ furring channels, 12 inches on center.

3.4 lb. metal lath, diamond mesh. 18 gauge tie wire.

To figure the *hanger-wire requirements,* figure the length of the hanger from the supports to the ceiling, then add at least 6 inches to each end for tying to the channels and "pigtailing" (coiling about two turns around a 1½″ pipe) in the concrete slab or wrapping around the steel trusses. Example: A ceiling suspended 3 feet below a concrete floor would require a hanger 4′ long. When spaced 4′ on

TABLE 10 NAILS COMMONLY USED IN THE PLASTERING TRADE

DESCRIPTION OF NAILS	LENGTH OR SIZE	GAUGE OR DESCRIPTION	NUMBER PER POUND
LARGE-HEADED NAILS, 7/16" HEAD	1"	11	275
LARGE-HEADED NAILS, 7/16" HEAD	1 1/2"	11	180
PLASTER BOARD, BLUED, 3/8" FLAT HEAD	1 1/8"	13	380
PLASTER BOARD, BLUED, EGG–CASE HEAD	1 1/8"	15	738
WOOD LATH NAILS, BLUED, SMALL HEAD	1 1/8" 3d	FINE	770
COMMON NAILS	4d	12 1/2	300
COMMON NAILS	6d	11 1/2	180

centers, each hanger gives 16 square feet of supported ceiling; the number of hangers required may therefore be found by dividing 16 into the square footage of the ceiling area. Also, if the square *yardage* is known, multiply that figure by 0.57 to obtain the number of hangers needed. This figure is then multiplied by the length of the hangers needed to find the total number of lineal feet needed.

For 1½" *channels*, divide the square foot ceiling area by 4. (Example: room 15' × 20' = 300 sq. ft. ÷ 4 = 75 lineal feet of 1½" channels required.)

For ¾" *channels* the number of lineal feet needed is the same as the area of the ceiling in square feet. Metal lath requirements will be the square foot area plus 7 percent for lapping. Tiewire used to tie the channel iron and metal lath together will equal approximately 2 lbs. of 18 gauge wire per square yard of area.

For nailing metal lath to wood studs or joists spaced 16" on center and nails spaced 6" on center, figure 18 nails per square yard. For nailing gypsum lath to wood framing, each lath should be secured with four nails to each framing member, requiring 16 nails per board.

After estimating the approximate number of nails of each type and size required for a given job, it is a simple matter to convert these quantities into pounds by referring to the tabulation in Table 10.

Finally, there are many miscellaneous, varied and specialized items used in both lathing and plastering operations so infrequently it would be impossible to list them all. Manufacturers' catalogs can be consulted regarding their uses and specifications. In doing this make sure the catalogs are up-to-date, as they are constantly being revised.

Appendix B

The field of plastering has become so vast that no book or set of books can encompass all of the material. New products are reaching the market constantly and new techniques are being developed each year. Printed information covering most of these products and techniques, however, is available in one form or another. The person interested in following plastering as an occupation should be alert to all of the material which is published.

New textbooks, reference books, and books in related fields are reaching the market constantly. Some have a general broad purpose, others are written in depth in a limited field. A list of some of the more useful reference books is given in this appendix.

Progressive manufacturers are generally very willing to supply information about their products and how they should be used correctly. They feel that it is to their advantage to have the plasterer know about them and how they should be applied. Manufacturers often join forces by forming associations or institutes.

A subject directory of some of the manufacturers is also given in this appendix.

Reference Books

Dalton, Byron William. *Practical Plastering, Cement Finishing and Related Subjects*. Byron William Dalton, Publisher. (330 S. Mansfield Ave., Chicago, Ill.), 1949.

Diehl, John R., A.I.A. *Manual of Lathing and Plastering*. M.A.C. Publishers Association, produced by Stahley Thompson Associates. (New York, N.Y.), 1960.

Meyer, Franz Sales. *Handbook of Ornament*. The Architectural Book Publishing Co. (New York, N.Y.), no date.

Meilach, Donna Z. *Creating With Plaster*. Reilly & Lee Co. (114 W. Illinois, Chicago, Ill. 60610), 1966.

Miles, John W. *Techniques of Casting*. Van Nostrand-Reinhold Books. (450 W. 33rd St., New York, N.Y.), 1967.

Miller, William. *Plastering, Plain and Decorative*. Darien Press. (Edinburgh, Great Britain), 1927.

Taylor, J. B. *Plastering*. George Godwin, Ltd., Publisher. (4 Catherine St., London, England, WC 2), 1970.

Verrall, W. *Solid and Fibrous Plastering*. Chemical Publishing Co., Inc. (221 Park Ave. S., New York, N.Y. 10003)

Tools and Equipment

Plasterers Hand Tools, Trowels, Pointers, etc.

Goldblatt Tool Co.
511 Osage
Kansas City, Kansas 66110

Marshalltown Trowel Co.
Box 738
Marshalltown, Iowa 50158

Rods, Featheredges, Slickers & Darbies

Goldblatt Tool Co.
511 Osage
Kansas City, Kansas 66110

Wha-Lite Products
6014 N. Keystone Ave.
Chicago, Illinois 60646

Scaffold

Gadsden Scaffold Co., Inc.
139 Ewing Ave., Box 571
Gadsden, Alabama 35902

Jackson Scaffolding Ltd.
1870 West 1st Ave.
Vancouver 9, British Columbia

Patent Scaffolding Co.
(Div. of Harsco Corp.)
34th Ave. & 12th St.
Long Island City, N.Y. 11106

Safway Steel Products Div.
A-T-O, Inc.
6228 West State St.
Milwaukee, Wisconsin 53213

Waco Scaffold and Shoring Co.
P.O. Box 66415
O'Hare International Airport,
Ill. 60666

Mixing Machines

Anchor Manufacturing Co.
2922 West 26th St.
Chicago, Ill. 60623

Essick Div. A-T-O Inc.
1950 Santa Fe Ave.
Los Angeles, California 90021

Gilson Brothers Co.
P.O. Box 345
Oostburg, Wisconsin 53070

Goldblatt Tool Co.
511 Osage
Kansas City, Kansas 66110

Muller Machinery Co., Inc.
248 Whitman Avenue
Metuchan, New Jersey 08840

Norton Construction Products Div.
Clipper Manufacturing Co., Inc.
P.O. Box 9604
Kansas City, Missouri 64134

Plastering Machines

Essick Div. A-T-O Inc.
1950 Santa Fe Ave.
Los Angeles, California 90021

Glover Manufacturing Co.
15226 Stagg Street
Van Nuys, Calif. 91405

Goldblatt Tool Co.
511 Osage
Kansas City, Kansas 66110

Muller Machinery Co., Inc.
Whitman Ave.
Metuchen, New Jersey 08840

Norton Construction Products Div.
Clipper Manufacturing Co., Inc.
P.O. Box 9604
Kansas City, Missouri 64134

Quickspray, Inc.
Rt. No. 2 West
Port Clinton, Ohio 43452

Smith & Kanzler Co.
1414 East Linden Ave.
Linden, New Jersey

Thomsen Division (Royal Industries)
130 West Victoria St.
Gardena, Calif. 90247

Universal Insulating Machine Co.
P.O. Box 1371
Winter Haven, Florida 13881

Texture Machines

Glover Manufacturing Co.
15226 Stagg St.
Van Nuys, Calif. 91405

Goldblatt Tool Co.
511 Osage
Kansas City, Kansas 66110

Muller Machinery Co., Inc.
248 Whitman Ave.
Metuchen, New Jersey 08840

Aggregate Guns

Cement Enamel Development, Inc.
26765 Fullerton Ave.
Detroit, Michigan 48239

Hansen Rock Gun Co.
11800 Valley Blvd.
El Monte, Calif. 91732

Quickspray, Inc.
Box 327
Port Clinton, Ohio 43452

Power Tools

Bostitch Div. of Textron, Inc.
815 Briggs Drive
East Greenwich, Rhode Island 02818

Goldblatt Tool Co.
511 Osage
Kansas City, Kansas 66110

Milwaukee Electric Tool Corp.
13167 West Lisbon Rd.
Brookfield, Wisconsin 53005

Gypsum Products

Blue Diamond Gypsum Concrete
Materials Div., Flintkote Co.
1650 South Alameda Street
Los Angeles, California 90021

Canadian Gypsum Co., Ltd.
790 Bay Street
Toronto 1, Ontario, Canada

Celotex Corporation
1500 N. Dale Mabry Hwy.
Tampa, Florida 33607

Domtar Construction Materials, Ltd.
2210 Place Ville Marie
Montreal, Canada

Fibreboard Corp.
53 Francisco St.
San Francisco, Calif. 94133

Grand Rapids Gypsum Co.
1007 North Division Ave.
Grand Rapids, Michigan 49501

Kaiser Cement & Gypsum Corp.
Kaiser Industries Corp.
300 Lakeside Drive
Oakland, California 94604

National Gypsum Company
325 Delaware Avenue
Buffalo, New York 14202

United States Gypsum Co.
101 South Wacker Drive
Chicago, Illinois 60606

Western Gypsum Limited
2650 Lakeshore Highway
Clarkson, Ontario, Canada

Lime and Lime Products

Domtar Construction Material, Ltd.
2210 Place Ville Marie
Montreal, Canada

Chas. Pfizer & Co., Inc.
235 E. 42nd St.
New York, N.Y. 10016

Grand Rapids Gypsum Co.
1007 North Division Ave.
Grand Rapids, Michigan 49501

Highland Stucco and Lime Prod., Inc.
15148 Oxnard Street
Van Nuys, California 91401

Kaiser Gypsum Co.
Kaiser Industries Corp.
300 Lakeside Drive
Oakland, California 94604

National Gypsum Co.
325 Delaware Ave.
Buffalo, New York 14202

Ohio Lime Company
Woodville, Ohio 43469

United States Gypsum Co.
101 South Wacker Drive
Chicago, Illinois 60606

U.S. Lime Div. Flintkote Co.
2244 Beverly Boulevard
Los Angeles, California 90057

Western Gypsum Limited
2650 Lakeshore Highway
Clarkson, Ontario, Canada

Portland Cement

Florida Portland Cement
Suite 907 DuPont Plaza Center
Miami, Florida 33131

General Portland Cement Co.
P.O. Box 324
Dallas, Texas 75221

Ideal Cement Company
821 17th St.
Denver, Colo. 80202

Lone Star Cement Corp.
1 Greenwich Plaza
Greenwich, Conn. 06830

Lehigh Portland Cement Company
Allentown, Pennsylvania 18105

Louisville Cement Co.
501 South 2nd
Louisville, Kentucky 40202

Medusa Portland Cement Co.
P.O. Box 5668
Cleveland, Ohio 44101

Veneer Plastering

California Products Corp.
167 Waverly Street
Cambridge, Massachusetts 02139

Flintkote Company
480 Central Ave.
East Rutherford, N.J. 07083

Grand Rapids Gypsum Co.
1007 North Division Ave.
Grand Rapids, Michigan 49501

Highland Stucco & Lime Products
15148 Oxnard St.
Van Nuys, California 91401

Kaiser Gypsum Company
Kaiser Industries Corp.
300 Lakeside Drive
Oakland, California 94604

Merlex Materials & Mfg., Inc.
2911 Orange-Olive Road
Orange, California 92665

National Gypsum Co.
325 Delaware Ave.
Buffalo, New York 14202

United States Gypsum Co.
101 South Wacker Drive
Chicago, Illinois 60606

Western Gypsum Limited
2650 Lakeshore Highway
Clarkson, Ontario, Canada

Swimming Pool Finishes

Cota Industries, Inc.
5512 S.E. 14th St.
Des Moines, Iowa 50320

L & M Surco Mfg., Inc.
P.O. Box 35472
Dallas, Texas 75235

Pre-Mix Concrete, Inc.
Kennewick, Wash. 99336

Materials

Cota Industries, Inc.
5512 S.E. 14th Street
Des Moines, Iowa 50320

Domtar Construction Mat., Ltd.
2210 Place Ville Marie,
Montreal, Canada

Grefco, Inc., Dicalite Div.
(General Refractories Co.)
630 Shatto Place
Los Angeles, Calif. 90005

Kaiser Gypsum Co., Inc.
Kaiser Industries Corp.
300 Lakeside Drive
Oakland, California 94604

Lahabra Products, Inc.
1631 West Lincoln Ave.
P.O. Box 3700
Anaheim, California 92803

National Gypsum Co.
325 Delaware Avenue
Buffalo, New York 14202

Ohio Lime Co.
General Refractories Co.
Woodville, Ohio 43469

Chas. Pfizer & Co., Inc.
235 East 42nd St.
New York, N.Y. 10016

United States Gypsum Co.
101 South Wacker Drive
Chicago, Illinois 60606

United States Mineral Products Co.
Stanhope, New Jersey 07874

Additives

Atlas Chemical Co.
4801 N.W. 77th Ave.
Miami, Florida 33144

Cement Enamel Development, Inc.
26765 Fullerton Avenue
Detroit, Michigan 48239

Construction Aggregates Corp.
120 S. LaSalle St.
Chicago, Illinois 60602

Dow Chemical Co.
(Construction Materials Sales)
Midland, Michigan 48640

National Chemsearch Corp.
Mohawk Labs
P.O. Box 10087
Dallas, Texas 75061

Bonding Agents

Atlas Chemical Co.
4801 N.W. 77th Ave.
Miami, Florida 33144

Cota Industries, Inc.
5512 S.E. 14th St.
Des Moines, Iowa 50320

Larsen Products Corporation
5420 Randolph Road
Rockville, Maryland 20852

National Chemsearch Corp.
Mohawk Labs, P.O. Box 10087
Dallas, Texas 75061

Preferred Chemical Products, Inc.
200 E. Walton
Chicago, Illinois 60611

Standard Dry Wall Products, Inc.
Dept. TR-71, 7800 N.W. 38th St.
Miami, Fla. 33166

Acoustical Suspension Systems and Tile Adhesives

Celotex Corporation
1500 N. Dale Mabry Hwy.
Tampa, Florida 33607

Down Products
700 Basset Road
Westlake, Ohio 44091

Johns Manville
22 East 40th Street
New York, N.Y. 10016

Kaiser Gypsum Co.
Kaiser Industries Corp.
300 Lakeside Drive
Oakland, California 94604

Miracle Adhesives Corp.
Devices, Inc.
250 Pettit Avenue
Bellemore, Long Island, N.Y. 11710

National Gypsum Company
325 Delaware Avenue
Buffalo, New York 14202

United States Gypsum Co.
101 South Wacker Drive
Chicago, Illinois 60606

Architectural Aggregates

Colonna & Co. of Colorado, Inc.
P.O. Box 860
Cannon City, Colorado 81212

Manhattan Terrazzo Brass Strip Co.
General Stone and Materials Corp.
1401 Franklin Road, S.W.
Roanoke, Virginia 24016

Terrazzo & Stone Supply Company
645 N.W. 42nd
Seattle, Wash. 98107

Willingham-Little Stone Div.
The Georgia Marble Co.
11 Pryor St., S.W.
Atlanta, Ga. 30303

Lightweight Aggregates

American Vermiculite Corp.
527 Madison Ave.
New York, N.Y. 10022

Atlantic Perlite Company
1919 Kenilworth Ave. N.E.
Beaver Heights, Maryland
Washington, D.C. 20027

Domtar Construction Materials, Ltd.
2210 Place Ville Marie
Montreal, Canada

Filter-Media Company
P.O. Box 19156-TR
Houston, Texas 77024

W. R. Grace & Co.
(Construction Materials Div.)
62 Whittemore Ave.
Cambridge, Massachusetts 02140

Grefco, Inc., Dicalite Div.
(General Refractories Co.)
630 Shatto Place
Los Angeles, California 90005

Lahabra Products, Inc.
1631 West Lincoln Ave.
P.O. Box 3700
Anaheim, California 92803

Mica-Pellets, Inc.
1008 Oak Street
DeKalb, Illinois 60115

Perlite Products Co., Inc.
Boro & Secane Avenues
Primos, Pa. 19018

Silbrico Corporation
6300 River Road
Hodgkins, Illinois 60525

Exposed Aggregate Surfacing

Cement Enamel Development, Inc.
26765 Fullerton Ave.
Detroit, Michigan 48239

Ceram-Traz (Concrete Chemical
  Products Corp.)
6500 Oxford
Minneapolis, Minnesota 55426

Construction Aggregates Corp.
120 S. LaSalle St.
Chicago, Illinois 60636

Cota Industries, Inc.
5512 S.E. 14th Street
Des Moines, Iowa 50320

Finestone Corporation
11846 E. McNichols Road
Detroit, Michigan 48205

Hallmark Chemical Corp. (Div. of
  National Chemsearch Corp.)
2730 Carl Road
Irving, Texas 75060

MD Corporation
1829 S. Stewart Ave.
Springfield, Missouri 65804

Fireproofing

Albi Mfg. Dept., Cities Service Co.
100 E. Main St.
Rockville, Conn. 06066

Asbestospray Corporation
605-T Broad Street
Newark, New Jersey 07102

Baldwin-Ehret-Hill, Inc.
Keene Corp.
Route No. 1
Princeton, New Jersey 08540

Domtar Construction Materials, Ltd.
2210 Place Ville Marie
Montreal, Canada

Keene Corporation
  (Sound Control Division)
U.S. Route No. 1
Princeton, New Jersey 08540

National Gypsum Company
325 Delaware Ave.
Buffalo, New York 14202

Smith and Kanzler Company
1414 East Linden Avenue
Linden, New Jersey 07036

United States Gypsum Co.
101 South Wacker Drive
Chicago, Illinois 60606

United States Mineral Products Co.
Stanhope, New Jersey 07874

W. R. Grace and Co.
  (Construction Materials Div.)
62 Whittemore Ave.
Cambridge, Massachusetts 02140

Lath "Metal"

Alabama Metal Industries Corp.
3239 Fayette Avenue
Birmingham, Alabama 35208

Bostwick Steel Lath Company
Niles, Ohio 44446

Ceco Corp.
5601 West 26th Street
Cicero, Illinois 60650

K-Lath Corporation
204 West Pomona Ave.
Monrovia, California 91016

National Gypsum Co.
325 Delaware Ave.
Buffalo, New York 14202

Penn Metal Corp. of Pennsylvania
Alan Wood Steel Co.
Wood Road,
Conshohocken, Pa. 19428

United States Gypsum Co.
101 South Wacker Drive
Chicago, Illinois 60606

Western Metal Lath Co.
15220 Canary Ave.
La Mirada, California 90638

Index

Numerals in **bold type** refer to illustrations.

F

Faulty construction, 473-474, **475**
Faulty lathing, **475**, 476
Featheredge, 70, **70, 71**
   using, 252, **253**
Felt finishing brush, 44-46, **44**
Fiber admixtures, 108-109
Fibered gypsum, 92
Fibered lime, 96
Fiberglass mesh covering, 271, **271**
File, 63, **64**
Finish coat, 285-309, **286-293, 296, 297, 300**
   machine applied, 232,
   mixing, 132-141, **134-141**
Finishing board, 25
Finishing brush, felt, 44-46, **44**
Finishing tool, trio, 66, **66**
Fire insulation, direct-applied, 356-359, **357, 358**
Fireproofing, 8, **9**, 356-359, **357, 358**
   machine, 80-83, **81**
   materials, contact, 112-113
   materials, direct-applied, 156
   structural steel, 184-185, **184**
Fire rating with lightweight aggregate plaster, **273**
Fire shield, 8, **9**
Fire shield plaster, **112**
Flashings, 201-206, **202-205**
Flexible corner tool, 65-66, **66**
Floating angles, 287, **287**

Floating, finish, 291-292, **291**, 296-297, **296**
Floating rough, 290-291, **291**
Floats, 46, **46**
Floor, swimming pool, 355-356, **355**
Foam plastics, plastering, 265
Foaming agents, 111
Folding rules, 48, **48**
Foldstir mixer, 84-85, **85**
Foreman, 14
Formulas, base coat materials, 123-132, **124, 126, 128**
Fractions, changing to decimals, 491-492, **491**
Fractions, operations with, 487-489
Frozen plaster, 465-466, **466**
Furring channels, 194-197, **196**
Furring nails, **200**, 201
Furring, wall and ceiling, 197-199, **198**

G

Gages, wall and ceiling, 228-229, **229**
Gauging plaster, 92
   estimating, 521
Gelatin, 108
eGneral Superintendent, 14
Geometrical panel construction, 425-427, **425, 426**
Geometric shapes, development of, 505-517
Geometry, elementary, 501-504
Gig stick, 449-450, **449**
Glitter gun, 77-79, **78**

Gloves, rubber, 27, 61-62
Goggles, safety, 24
Groined ceilings, 456-460
Groins, 456-460
Grid, exposed, 369-371, **369, 370**
Grounds, 192, 221
Gun, aggregate seeder, 83-84, **84**
Gun, catalyst additive, 80-81, **82**
Gun, glitter, 77-79, **78**
"Gun Lath", 190-191, **190**
Gypsum, 11, 89-91
   backer boards, 167, **171**
   board lathing, 261-262, **261, 262**
   brown coat on various bases, 266-268, **266-268**
   calcined, 1-2
   lath, insulating, 166
   lath, perforated, 166, **170**
   lath, radiant heat, 167
   lath, radiant heat, 167
   lath, veneer plaster, 166-167
   mortar, bond coat, 127
   mortar, fibered, 125-127, **126**
   mortar, unfibered, 125-127, **126**
   mortar with perlite aggregate, 127
   processing, 89-91
   products, 91-93
   rock, mining, 89, **89**
   scratch coat on various bases, 259-265, **260-263**
   sound-deadening boards, 168
   studs, 168
   wallboards, 167

H

Hair admixtures, 108-109